Numerical Calculations in Clifford Algebra

Numerical Calculations in Clifford Algebra

A Practical Guide for Engineers and Scientists

Andrew Seagar
PO Box 192
Griffith University Post Office
QLD 4222, Australia

Registered Offices
John Wiley & Sons, Inc., 111 River Street, Hoboken, NJ 07030, USA
John Wiley & Sons Ltd, The Atrium, Southern Gate, Chichester, West Sussex, PO19 8SQ, UK

For details of our global editorial offices, customer services, and more information about Wiley products visit us at www.wiley.com.

Wiley also publishes its books in a variety of electronic formats and by print-on-demand. Some content that appears in standard print versions of this book may not be available in other formats.

Library of Congress Cataloging-in-Publication Data:

Names: Seagar, Andrew, author. | John Wiley & Sons, publisher.
Title: Numerical calculations in Clifford algebra : a practical guide for
    engineers and scientists / Andrew Seagar.
Description: Hoboken, NJ : Wiley, 2023. | Includes index.
Identifiers: LCCN 2022053609 (print) | LCCN 2022053610 (ebook) | ISBN
    9781394173242 (hardback) | ISBN 9781394173266 (adobe pdf) | ISBN
    9781394173259 (epub)
Subjects: LCSH: Clifford algebras. | Algebra–Data processing.
Classification: LCC QA199 .S43 2023 (print) | LCC QA199 (ebook) | DDC
    512/.57–dc23/eng20230203
LC record available at https://lccn.loc.gov/2022053609
LC ebook record available at https://lccn.loc.gov/2022053610

Cover design: Wiley
Cover image: © J Studios/Getty Images

Set in 9.5/12.5pt STIXTwoText by Straive, Chennai, India
Printed and bound by CPI Group (UK) Ltd, Croydon, CR0 4YY

C9781394173242_170523

To Cathy, Laura, Jon and Axil

Contents

List of Figures

List of Tables

Preface

Clifford algebra is a tool, and like all tools should only be used for tasks to which it is well suited. One sign that a tool is well suited to a task is that by using the tool the task is simpler than by using a different tool. However, like most tools, with Clifford algebra there is the need to invest some time to develop properly the skill to use it.

Clifford developed his algebra with a particular task in mind, namely the solution of Maxwell's equations of classical electromagnetism. At the time two other tools were in use, Cartesian components and quaternions. Vector calculus came later, out of quaternions.

Clifford called his algebra a 'geometric algebra'. He constructed it from quaternions, pinning their behaviour at a fundamental level on the algebra of the orthogonal elementary vector units e_p introduced earlier by Grassmann. As a consequence, in addition to the multiplication of quaternions, Clifford's geometric algebra inherits the many and varied multiplications and other operations supported by Grassmann's extensive[1] algebra. The list includes (as named by Grassmann at the time) inner and outer multiplication, progressive and regressive multiplication, central multiplication, and the unary supplement operation.

Today, other names are also used. Grassmann himself changed outer multiplication to progressive multiplication when attempting to combine different types of multiplication within a single framework[2]. That name was not adopted generally, with exterior multiplication now as another alternative.

Regressive multiplication may be found in the form of the 'meet' operation within the context of computer graphics, where it is used to formulate the intersection between geometric descriptions of objects. Central (or middle) multiplication is currently called Clifford multiplication somewhat losing sight of Clifford's algebra as supporting many kinds of multiplication, not just one. Grassmann's supplement operation is now called the Hodge star (or Hodge's dual) operation.

As have the names changed over time, so have the notations. Grassmann himself used different notations at different times for the same operations as his works matured. Take as example inner multiplication, defined by Grassmann as the outer multiplication of one number by the supplement of another. At the time, Grassmann used (square) brackets [ ] to indicate the outer multiplication of the numbers in between, and a vertical bar | to

---

1 The meaning of 'extensive algebra' is in the sense that the algebra supports the extension from any particular dimensional space to the next higher-dimensional space.
2 For now, Clifford algebra does that for us.

represent the supplement of the number following. Thus, the notation [$a \mid b$] denotes inner multiplication as the outer multiplication of number a by the supplement of number b.

Today, the inner multiplication of two numbers is written in at least two notations. The first notation is (a , b), where Grassmann's brackets have been replaced by parentheses and the vertical bar has withered to a comma. With this notation, sight of the underlying outer multiplication and supplement is lost.

The second notation is $a \wedge * \; b$, where the brackets have been replaced by the infix wedge operator[3] $\wedge$ to indicate the outer multiplication of the numbers on either side, and the vertical bar has been replaced by the Hodge star operator. This notation is more in keeping with Grassmann's original.

Some works on geometric algebra focus more on the algebra than on the geometry and more on the theory than the practice. It is a mistake to concentrate wholly on the algebra because that loses sight of the geometrical application. It is equally a mistake to consider only vectors because they are too simple, by themselves being insufficient to represent the full geometry of two or higher dimensional spaces. Both the algebra and the vectors are necessary; however, singly or together, they provide an incomplete picture.

It is also a mistake to treat the algebra wholly as an abstraction. It is important to examine concrete examples. My own view is that application means nothing less than conception at an abstract level, design at a system level, calculation at a numerical level, and construction at a physical level of some actual apparatus intended to perform some useful function such as a microwave antenna to transmit and receive electromagnetic radiation for the purpose of communication.

For the committed beginner to the application of Clifford algebra, there are two obstacles to overcome. First, it is necessary to acquire a working knowledge of some of the abstract concepts (which I much prefer to call tools) in order to be able to formulate problems and their solutions. For anyone familiar with vector calculus, that serves as a good starting point.

Moving from vector calculus to Clifford algebra involves a paradigm shift in which problems in multiple dimensions are no longer treated as a system of partial differential equations in a multiplicity of single-dimensional scalar variables, but rather as a single ordinary differential equation in a single multi-dimensional, non-scalar variable.

It is not necessary to become familiar with all of the tools. For classical electromagnetism, Grassmann's central (now Clifford) multiplication plays the leading role, with most of the other kinds of multiplication seldom or never used. Combining central multiplication with the Cauchy integral, reincarnated in multiple dimensions using Clifford algebra, provides new tools which are sufficient in many practical cases to serve as effective replacements to the vector calculus, the Helmholtz equation, and the Green's functions.

Second is the knotty obstacle of numerical calculation. This has to be convenient (or at least not too inconvenient), efficient enough to allow the solution of meaningful problems within acceptable time frames, and most importantly, written largely by someone else.

The work here attempts to provide a bridge spanning both of these obstacles. Equal importance is given to fundamental theory, practical (geometrical) application, and software implementation. These three components are presented in the forms of written

---

3 Not dissimilar to the inverted 'u' symbol ∩ used earlier by Grassmann for the same purpose.

text, some equations, many figures, and extensive examples of computer code and the corresponding output.

Software implementations are based on computer program source code (written in the 'c' computer programming language) invoking individual routines, specifically constructed to do all of the actual calculation, from a suite[4] provided and distributed openly for that very purpose. When the user finds they need to do something not already provided, there is no significant obstacle to progress because the code is open for inspection, modification, and extension.

Chapters 1, 4, and 5 introduce the theory of Grassmann's multiplications and Clifford's algebra with a strong reliance on diagrams and pictograms to visually portray the various operations and instil a conceptual insight to the underlying mechanisms. Recourse to equations, which may well prove results but which can often fail to provide enlightenment, is avoided where possible.

Chapters 6–11 and 13 describe the mechanisms provided within the Clifford numerical suite (CNS) for representing different entities as Clifford numbers. These include vectors, quaternions, Pauli matrices, bicomplex numbers, electromagnetic fields, and arrays or matrices of any of those.

Chapters 12 and 18–25 present practical applications as examples showing how to implement numerical calculations for the various operations, on a variety of numerical data types. The applications progress from the simpler and perhaps somewhat contrived to the more advanced and realistic. The majority of the applications deal with problems having a geometrical nature or interpretation.

Chapter 12 demonstrates the evaluation of functions with a Clifford argument using power series expansions. Chapters 18 and 19 provide examples for finding the closest points on lines which may or may not intersect and the perspective projection of an object onto an oriented plane.

Chapters 13, 20, 21, and 25 deal with the solution of various linear systems of equations. Chapter 13 demonstrates the solution of linear systems in which the variables are bicomplex numbers using Gaussian elimination, whereas Chapter 25 does the same for variables which are extended electromagnetic fields. In Chapter 20, Grassmann's own method of elimination is used with scalars as variables. Grassmann's solution has a geometric interpretation as the ratio of hyper-volumes rather than the usual concept as the intersection of hyper-planes. Chapter 21 demonstrates a recursive implementation of the fast Fourier transform (FFT) for arbitrary Clifford numbers. The length of the transforms is limited to powers of 2.

Chapters 22–25 give examples involving electromagnetic fields. The simplest in Chapter 22 is the calculation of the electromagnetic field from a Hertzian dipole. More challenging in Chapter 23 is an implementation for the simulation of electromagnetic wave propagation using the finite difference time domain (FDTD) method. This is only in one dimension. The next application is in Chapter 24, where the Cauchy integral is used to extend the electromagnetic field measured on a closed surface to points away from the surface. This implementation is in three dimensions. Lastly, Chapter 25 demonstrates the effectiveness of solving electromagnetic scattering problems using Clifford

---

4 Referred to as the Clifford Numerical Suite, or CNS for short.

numbers representing those fields as single whole entities rather than separated into scalar components.

In addition to the implementations themselves and the examples of the output they produce, the chapter for each application reviews the theory or equations on which they are based.

Extensive examples for the implementation of software are interspersed where appropriate amongst the other material. Chapters 2 and 3 cover mechanisms provided for the input and output of numbers. Chapters 14–17 provide details on memory usage, errors and how to recover from them, methods for extending the software by writing your own small fragments of code and coupling them into the existing framework, and on Clifford identities which can be used to check the consistency of some of the different operations.

Chapters 26 and 27 give details of every individual routine[5] as a list organised by function, cross-referenced to detailed descriptions in alphabetical order.

As a consequence of providing the suite in the form of source code, all calculations compile to native machine code, and there is no need for any slow interpretive mechanisms or conversion of data types at the time of execution. This leads to comparatively little administrative overhead so that most of the computational time is spent directly on the numerical calculations.

The suite is able to handle Clifford numbers of up to $n = 31$ dimensions, storing only components which have non-zero value to make efficient use of memory. Calculations with dimensions in the double digits with full Clifford numbers (all non-zero components) can be slow, largely because there are many components which have to be multiplied in all 2^{2n} combinations. For 31 dimensions that can be more than 4×10^{18} complex multiplications. In contrast, there is no such issue when numbers are sparse, as with vectors, or of low dimension, as with electromagnetic fields.

As a tool, the CNS is well suited to the rapid development and testing of algorithms which rely on Clifford algebra or the algebra of Grassmann's other multiplication operations. The rapidity is a consequence of removing from the developer the need to delve into the minuscule detail, freeing them to operate at a higher level, much closer to the phenomenon of immediate interest. As an example, the electromagnetic field is treated as a single entity (Clifford number) rather than as two parts each with three components.

This work serves as both a resource for developing an understanding of the theory and principles behind the solution of problems within Clifford algebra, as well as a source of examples and a reference guide for the practical implementation of solutions in numerical code.

March 17, 2023

Andrew Seagar
Gold Coast

---

5 Over 300 routines.

Part I

Entities and Operations

1

Introduction

The purpose of the suite of routines described here, the Clifford numerical suite (CNS), is to provide a platform for the implementation of applications which require numerical evaluation of the various mathematical operations offered within the context of Clifford algebra. The routines are designed to allow the user to ignore most of the internal details of data structures and memory management if so desired.

The suite is constructed using the 'c' computer programming language (Kernighan & Ritchie 1978, King 2008) so that it can be embedded into, compiled with, and invoked directly from the user's code. This avoids the overheads of copying data associated with the call mechanisms involved in implementations underneath interpretive meta-languages[1] and consequently offers an overall higher speed of code execution. That in turn admits the possibility of solving problems which are more realistic and less trivial. The intention is to provide most users with a viable environment for the development of their own applications.

Providing the source code also allows the user to enhance and tailor it to best suit their own purposes. In cases where ultimate speed and memory efficiency is required, the suite can be used in the formative development phase, and substituted with the user's own hard-coded replacements in the end application phase.

1.1   Operations

The basic mathematical operations which are supported by the CNS are listed in Table 1.1 and described with examples in Chapters 4 and 5. Higher level and geometric operations such as inversion, reflection, projection, and rotation of vectors and other entities are also supported, as described in Chapters 6–10. The set of application programmer interfaces (API's) for accessing the suite is described in Chapters 26 and 27.

---

1 Such as Scilab.

Numerical Calculations in Clifford Algebra: A Practical Guide for Engineers and Scientists,
First Edition. Andrew Seagar.
© 2023 John Wiley & Sons Ltd. Published 2023 by John Wiley & Sons Ltd.

Table 1.1 Basic unary and binary operations.

Case	Unary operation	Notation	Binary operation	Notation
General	Negation	$-a$	Addition	$a + b$
	Involution	$\bar{a}$	Subtraction	$a - b$
	Reversion	$\tilde{a}$	Outer product	$a \wedge b$
	Clifford conjugation	$\bar{a}$	Inner product	$(a, b)\,e_{\emptyset}$
	Complex conjugation	a^*	Regressive product	$a \vee b$
			Central product	$a\,b$
			Left interior product	$a \lrcorner b$
			Right interior product	$a \llcorner b$
Special	Hodge dual	$* a \equiv \tilde{a} e_N$	Scalar product	(a, b)
	(Star, supplement)		Interior derivative	$i_a(b) \equiv a \lrcorner b$

1.2 History

In his books of 1844 and 1862, Grassmann (1995, 2000) introduces his anti-commutative outer multiplication $[a\,b]$ and unary supplement operation $\mid b$, using that pair to generate his commutative inner multiplication $[a \mid b]$, as listed in the central column of Table 1.2. In later times, notation for the outer multiplication has changed to $a \wedge b = [a\,b]$, the supplement operator has changed to the Hodge dual star operator $*b = \mid b$, and the inner multiplication has changed notation to $(a, b) = [a \mid b]$.

It is straightforward by simple substitution of notation to write three equivalent forms for Grassmann's inner multiplication $[a \mid b] = a \wedge *b = (a, b)$. Some care is needed in interpreting these 'equalities' because the product of the *scalar* inner multiplication (a, b) without the $e_{\emptyset}$ is a proper scalar, whereas the corresponding products of the other two inner multiplications as formulated are strictly pseudo-scalars.

Table 1.2 Equivalence between notations.

Exterior algebra various[a]	Extension theory Grassmann	Geometric algebra Clifford
Wedge product	Outer product	Outer product
$a \wedge b$	$[a\,b]$	$a \wedge b$
Hodge dual	Supplement	Hodge dual
$*b$	$\mid b$	$\tilde{b}e_N$
Inner product (Pseudo-scalar)	Inner product (Conflated)	Inner product (Grade 0)
$a \wedge *b$	$[a \mid b]$	$(a, b)\,e_{\emptyset}$

a) Peano, Cartan, Hodge.

However, in defining his inner multiplication [$a \mid b$], Grassmann added an extra provision to ensure that the product is a scalar rather than a pseudo-scalar. To do this, he conflated the unit pseudo-scalar e_N of dimension n, where found within the result, with the unit scalar 1^2. In that case, the product of his inner multiplication is by definition a scalar. Grassmann could do this because he did not admit the unit scalar 1 within the basis of his algebra. However, that means his inner multiplication gives products outside his algebra.

The inner multiplication within Clifford algebra is defined[3] without appealing to Hodge's star or Grassmann's supplement. The result is a scalar, not a pseudo-scalar, so there is no need to introduce any special provisions. Since Clifford's algebra admits scalars with unit $e_\emptyset$ as part of the lowest grade of the algebra (grade 0) it is possible, by placing the product in grade 0, to keep the algebra closed under inner multiplication. As a consequence, the form of the inner multiplication in Clifford algebra $(a, b)e_\emptyset$ adopted here shows the unit $e_\emptyset$, explicitly, and the conventional scalar product (a, b) – with the factor $e_\emptyset$ absent – produces only the scalar coefficient of the Clifford inner product. The numerical values (coefficients) of all three inner products in Table 1.2 are the same. They just have different units attached: e_N or 1 or $e_\emptyset$.

Grassmann further generalised his outer multiplication in two complementary forms: progressive multiplication, operating with the outer multiplication on two factors as given; and regressive multiplication, operating with the outer multiplication on the supplements of the factors, and taking the inverse supplement of that result as the final product.

With Grassmann's definition, non-zero components of the progressive and regressive multiplications never coincide. Using this property, Grassmann took the sum of the two to create a more general multiplication and dropped any formal distinction between progressive and regressive multiplications. Grassmann's progressive and regressive multiplications find application today within the context of computer graphics in the form of the join and meet operators.

1.3 Alternative Forms

The different unary and binary operations supported within Clifford algebra are not all independent. Table 1.3 lists each operation and gives where possible the formulæ needed to calculate it using other operations under the conditions listed in the leftmost column. The conditions refer to the sets A and B of units from which the numbers a and b are constructed. If the conditions are not met, the result of that particular operation is zero. See Section 5.3.2 for details.

Most of the operations are calculated in a straightforward manner within the context of Clifford's algebra, using only Clifford (central) multiplication, reversion, and scalar division. The Hodge star operator and regressive product, difficult to calculate within some algebraic frameworks, are particularly easy to calculate within Clifford algebra as the central multiplication of either two or three simple factors. The product $b\tilde{a}$ in both the

---

2 As well as scalars and pseudo-scalars, in vector calculus, vectors and bivectors are also conflated.
3 From the generalised signature in Section 5.3.2.

Table 1.3 Conversion of operations between Clifford and other algebras.

		Notation				
		Outer	Hodge	Regressive	Scalar	Clifford
Condition	**Operation**	$\wedge$	$*$	$\vee$	$(\,,)$	None
$A \cap B = \emptyset$	Outer product	$a \wedge b$	$*((*^{-1}a) \vee (*^{-1}b))$			$a\,b$
Any	Supplement		$*a$			$\tilde{a}\,e_N$
Any	Inverse of supplement		$*^{-1}b$			$\dfrac{e_N\,\tilde{b}}{e_N\tilde{e}_N}$
$A \cup B = e_N$	Regressive product[a]		$*^{-1}((*a) \wedge (*b))$	$a \vee b$		$b\,\tilde{e}_N\,a$
$A = B$	Grade 0 inner product		$*^{-1}(a \wedge (*b))$		$(a,b)e_\emptyset$	$b\,\tilde{a}$
$A = B$	Pseudo-scalar inner product	$a \wedge (*b)$			$(a,b)e_N$	$b\,\tilde{a}\,e_N$
$A = B$	Scalar inner product				(a,b)	
Any	Central product					$a\,b$

a) Grassmann imposed an additional condition $A \cap B \neq \emptyset$ which maintained closure of his regressive multiplication. Here, with support for scalars in grade 0, closure is maintained without the extra condition.

grade 0 and pseudo-scalar inner products can equally well be calculated[4] as $a\tilde{b}$, $\tilde{a}b$, or $\tilde{b}a$, because under the constraint $A = B$, only the products of those components in a and b which have the same units are taken into the result.

1.4 Naming

There is some ambiguity in the naming in English of Grassmann's first two kinds of multiplication operation. The concepts of 'inner' and 'outer' may be distinguished by two different distances in the same direction from a single reference point (called the centre), whereas 'interior' and 'exterior' may be distinguished by two equal distances in opposite directions from a single reference point (called the interface). Grassmann (1877) wrote in German **innere** (inner) and **äußere** (outer), and in French Grassmann (1855) wrote **extérieur** (exterior) and **intérieur** (interior). Here the alternatives used in Grassmann's native language are adopted. Following a suggestion by Sylvester, Clifford (1878) referred to Grassmann's inner and outer multiplications as 'scalar' and 'polar' noting the rationale that these terms describe the nature of the multiplications themselves under exchange of terms (commutative or anti-commutative) rather than the geometrical circumstances to which they apply. Clifford's usage has not been adopted. Grassmann also changed 'outer' to 'progressive', but that didn't take hold either.

Grassmann used the term 'product' for the result of a variety of operations, with 'multiplication' being just one of the variety. The outer product is the result of the outer multiplication of two numbers. The sum is the product of the addition of two numbers. Rather than adhering strictly to Grassmann's usage of the term 'multiplication' for the

---

4 Provided the coefficients commute.

operation and the term 'product' for the result, here 'the *xxx* product …' will often be used as a contraction for 'the product of the *xxx* multiplication …', because the latter is altogether too clumsy in many cases.

1.5 Structure

1.5.1 Algebraic

Clifford numbers are constructed as a sum of multiple linearly independent components, as shown in Table 1.4 (Seagar 2016). The components are formed by multiplying a numeric coefficient with a Clifford unit. A sum of different Clifford units has no algebraic simplification and thereby supports the linear independence.

Clifford units provide a mechanism to specify geometric entities of different types (such as points or scalars, lines or vectors, areas or bivectors, and volumes or trivectors) in any number of dimensions. Geometric entities of these different types are identified as being within different grades $\wedge$ of the algebra: grade 0 for scalars, grade 1 for vectors, and so on.

Notationally[5] Clifford units are written as the symbol 'e' with a subscript which is either a non-negative integer (such as 2), a lowercase letter (such as p) representing a non-negative integer, an uppercase letter (such as S) representing a set of non-negative integers, or the symbol '$\emptyset$' for the empty set.

The unit of grade 0, with the empty set as subscript, behaves much like the unit scalar '1' and is often omitted. The unit of scalars was called the 'absolute unit' by Grassmann. Units of grade 1, with single integers as subscripts, are called primal units. They cannot be reduced by factorisation into non-trivial products of other more elemental entities.

Units in grades higher than 1 can be constructed by multiplying together units in lower grades. These are called evolved units after Grassmann's abstract concept of evolution, through which higher- dimensional entities are constructed by extension from lower-dimensional entities. Grassmann's evolution is achieved using his outer multiplication for distinct vectors which are not orthogonal, or equally by central (Clifford) multiplication

Table 1.4 Examples of Clifford numbers.

	Grade				No. of components	
	$\wedge^0$	$\wedge^1$	$\wedge^2$	$\wedge^3$	$\wedge^0\wedge^1\wedge^2\wedge^3$	all
Dimension: 0	$2+3i$				1	1
1	$2+3i$	$+5e_0$			1 1	2
2	$2+3i$	$+5e_0+ze_1$	$+ie_0e_1$		1 2 1	4
3	$2+3i$	$+5e_0+ze_1+\pi e_2+ie_0e_1+ae_0e_2+be_1e_2+42.1e_0e_1e_2$			1 3 3 1	8
Coefficients	$(2+3i)$ 5 z π i a b 42.1					
Units: Absolute	e_0					
Primal	e_0 e_1 e_2					
Evolved	e_0e_1 e_0e_2 e_1e_2 $e_0e_1e_2$					
Components	$(2+3i)e_0$ $5e_0$ ze_1 πe_2 ie_0e_1 ae_0e_2 be_1e_2 $42.1e_0e_1e_2$					

---

5 Following Grassmann. Clifford used 'ι'.

in the case of orthogonal vectors. The primal Clifford units e_p are orthogonal[6] so that the product of distinct primal units by central or outer multiplication yields the same evolved result.

1.5.2 Numeric

Within the CNS numbers are constructed, and the basic operations listed in Table 1.1 are carried numerically, using various data types and structures defined according to the 'c' computer programming language (King 2008). There are only two data types which need be of any concern, described briefly here and with greater detail in Chapter 27.

At the lowest level is the data type 'cliff_comp' for Clifford components, which use the structure shown in Example 1.1:

Example 1.1

```
1:    /* single component of Clifford number */
2:    typedef struct{
3:        bitmap unit;        /* evolved unit */
4:        double coef[2];     /* real and imaginary parts */
5:    } cliff_comp;
```

---

to associate each complex numeric value with its own symbolic Clifford unit. The Clifford unit is stored as an unsigned integer with each bit representing one dimension. This limits the number of dimensions to the number of bits in the default integer data type for any particular compiler, often 32. A single full Clifford number of 32 dimensions would require sufficient memory to contain 2^{32} complex numbers (more than 64 G bytes).

At the highest level is the data type 'number' for Clifford numbers, which uses a type definition to provide a special name to the signed integer data type, as shown in Example 1.2:

Example 1.2

```
1:    /* token for a Clifford number */
2:    typedef int number;
```

Clifford numbers are constructed in the form of a tree from a pair of singly linked lists; one list for the branching internal skeletal structure of the tree and one list for the leaves at the extremities of the structure, where the actual data values reside. The numbers are referenced using the data type 'number', which is effectively an index to the location of the root of the tree in one or other of the linked lists. Management of the tree structure and linked lists is handled internally and is of little concern.

---

6 Grassmann's units may be chosen as orthogonal or not, according to the context.

The tree structure is sparse so that a Clifford number of 32 dimensions does not need 64 G bytes of memory unless every component is non-zero. For example, a Clifford vector of 32 dimensions would require sufficient memory to contain only 32 complex numbers. If some of the vectorial components are zero, it requires even less than that.

Clifford numbers which are no longer required may be recycled. This returns the memory allocated to their tree back to the linked lists so that it may be used to construct other numbers. Memory structure and management are described in detail in Chapter 14.

For many purposes, it is only necessary to deal with the data type 'number'. For more sophisticated effects, it may be necessary to operate directly on the data type 'cliff_comp'. For example, extension of the source code to incorporate custom-built functions is possible by writing sub-routines which operate numerically on one or two Clifford components using arguments consisting of pointers to the data type 'cliff_comp' and returning the result as the same data type. See Chapter 16 for details.

The code shown in Example 1.3a gives a minimal example of constructing a Clifford number and managing the associated memory. Line 4 declares a Clifford number 'u'. Line 6 establishes a trap handler for errors which prints some diagnostic information then cleanly terminates the code when errors are detected. Details on errors and more sophisticated usages of trap handlers are covered in Chapter 15.

Line 7 pre-allocates, from the 'c' program's operating environment to the internal linked lists, enough memory to contain 50 components. It doesn't matter too much if the amount allocated is insufficient, because memory expands automatically if required. However, when that occurs, the automatic expansion does temporarily pause the execution of code while data are copied from one range of memory addresses to another.

Example 1.3(a)

```
 1:   void demo0() /* minimal */
 2:   {
 3:       int lambda[3];
 4:       number u;
 5:
 6:       recovery_level(0,0,0); /* abort on all errors */
 7:       create_cliffmem(50);
 8:       declare_primal(3,pack_signature(3,lambda,-1));
 9:       print_signature();
10:
11:       u=scalar_number(1.0,2.0);
12:       print_number(u,8);
13:       recycle_number(u);
14:
15:       free_cliffmem(1);
16:   }
```

Line 8 establishes the Clifford context for numerical operations as being three-dimensional with negative signature for all three dimensions. If a mixed signature is required, that can be written into the array 'lambda [3]' manually rather than invoking the sub-routine `pack_signature()` to put three equal values in place.

Line 11 assigns memory, initialised to the (scalar) complex value $1 + 2i$, from the linked lists to the Clifford number. At this stage, there is one component used and 49 components free. Line 12 prints the number and line 13 returns the memory assigned to the number 'u', back to the linked lists for use in other Clifford numbers. As a consequence, there are once again 50 components free. Line 15 de-allocates all 50 components from the linked lists, returning the ownership of the memory back to the 'c' program's operating environment, and prints some information showing how much memory has been used and how much is currently in use. If the value zero rather than one is passed as the argument to the sub-routine `free_cliffmem()` on line 15, then the information is not printed. Printing the information can be useful in order to determine how to adjust the amount of memory allocated on line 7 so that the automatic expansion of memory is avoided. That improves the speed of execution.

Detailed descriptions of the functionality of all the sub-routines invoked in Example 1.3a are provided in Chapters 26 and 27.

The results produced by executing the code in Example 1.3a are shown in Example 1.3b. As a check, the signature associated with the numerical context is printed in line 2. The number of dimensions is determined by counting the number of values.

Example 1.3(b)

```
 1:    demo0()
 2:    [-1,-1,-1]
 3:    (1.00000e+00+i2.00000e+00)e{}
 4:    statistics for structural heap:
 5:    initial size (50)
 6:    peak size      (0), as % of initial size (0%)
 7:    full size      (50), expansion count (0)
 8:    free count     (50), free start (0)
 9:    statistics for data heap:
10:    initial size (50)
11:    peak size      (1), as % of initial size (2%)
12:    full size      (50), expansion count (0)
13:    free count     (50), free start (0)
```

The value of the Clifford number is printed in default format on line 3. In this case, it is simply a complex value, contained in the parentheses, post-multiplied by a Clifford unit. The Clifford unit is represented by the letter 'e' followed by a set in braces {...}. In this case, the Clifford number is a scalar so the set is the empty set, and the braces are empty. For non-scalar numbers, the braces contain integer values.

Memory usage information is printed on lines 4–8 for the skeletal structure of the tree, and lines 9–13 for the data values themselves. In this case, the 50 components allocated are established as two lists of 50 structures, one for the tree's skeletal structure (branches) and one for the data values (leaves). The trees for Clifford numbers usually require about the same number of each. Since the initial allocation here greatly exceeded the eventual usage, there was no need to expand the memory, and the number of times the memory was expanded is reported as zero. As a consequence, the full size of memory is equal to the initial size. If memory expansion had been needed, the full size would be greater than the initial size.

The amount of memory free (free count) at the time the information is printed is in this case the same as the full size. These values are the same because all memory used has been recycled. It is not necessary to do that; however, doing so systematically when numbers are no longer required should lead to equal values. If a strict recycling regime is followed, and these numbers are not the same, that may imply a programming error and can serve as a useful diagnostic indicator.

The value reported as the 'free start' is the index of the root of the linked list of unallocated memory. The value should be somewhere between 0 and 1 less than the full memory value. If it lies outside that range, then the internal memory structures have somehow become corrupt. Corruption can occur if the routines are used improperly. Usually, any corruption is detected and reported close to the place in the code where it happened.

The value reported as the peak memory size shows how much was actually used. In this case only one component was used, indicating that the initial allocation could tolerate being a lot less than 50. Rather than trying to predict the amount of memory required for any application, it is easiest to guess first, and use the values reported to revise the initial allocation either up or down at a later stage.

1.6 Entities

In addition to supporting scalars in grade 0, the use of higher grades and different dimensions in Clifford algebra also supports entities of more esoteric nature, as listed in Table 1.5.

The bicomplex numbers in the third row of the table differ from the simple complex scalars in the first row. The two-dimensional bicomplex numbers are a lower- dimensional version of the three-dimensional Clifford representation of quaternions.

Electromagnetic fields are stored as bivectors in four dimensions. The fourth dimension is associated with the non-spatial dimension of either time or frequency.

Table 1.5 Basic mathematical entities.

Grades	Dimension	Entity
0	Any	Scalars
1	1 or more	Vectors
0, 2	2	Bicomplex numbers
0, 2	3	Quaternions
0, 2	3	Pauli matrices
2	4	Electromagnetic fields
Any	Any	Arrays
Any/some	Any/some	Matrices

Vectors are described in Chapter 6. Quaternions and Pauli matrices, which are closely related to one another, are described in Chapters 7 and 8. Bicomplex numbers are described in Chapter 9 and the electromagnetic field in Chapter 10. Arrays of numbers are described in Chapter 11.

Chapter 13 describes matrices and matrix operations. Operations of addition, subtraction, and multiplication are supported for Clifford numbers with entities of any type in any number of dimensions. Support for matrix inverses is only provided for matrices whose elements are scalars, bicomplex numbers, quaternions, Pauli matrices, or electromagnetic fields. The user can extend that support to entities of other types by supplying sub-routines which take the inverses of those other types and declaring them as needed to the routines which handle the matrix inversion.

References

Clifford, W. K. (1878), 'Applications of Grassmann's extensive algebra', *American Journal of Mathematics* **1**(4), 350–358.

Grassmann, H. (1855), 'Sur les différents genres de multiplication', *Journal für die reine und angewandte Mathematik* **49**, 123–141.

Grassmann, H. (1877), 'Der ort der Hamilton'schen quaternionen in der Ausdehnungslehre', *Mathematische Annalen* **12**, 375–386.

Grassmann, H. (1995), The extension theory of 1844, *in* 'A New Branch of Mathematics: The Ausdehnungslehre of 1844, and Other Works', Open Court, Chicago, US-IL, pp. 2–295. (Translated by Lloyd C. Kannenberg).

Grassmann, H. (2000), *Extension Theory*, Vol. 19 of *History of Mathematics*, American Mathematical Society, Providence, US-RI (Translated by Lloyd C. Kannenberg).

Kernighan, B. W. and Ritchie, D. M. (1978), *The C Programming Language*, Prentice-Hall, Englewood Cliffs, US-NJ.

King, K. N. (2008), *C Programming: A Modern Approach*, 2nd edn, W. W. Norton, New York, US-NY.

Seagar, A. (2016), *Application of Geometric Algebra to Electromagnetic Scattering – the Clifford-Cauchy-Dirac Technique*, Springer, Singapore.

2

Input

The data type 'number' is used for all constants and variables. The coefficients of the individual components in numbers are complex values. There are no separate structures for integer 'int', real ('float', or 'double') non-complex values. Everything is complex with a Clifford unit attached.

The methods provided to set the values of constants are suitable for numbers which require initialisation without input from the keyboard. It is assumed that any syntactical errors in the format of the numbers are eliminated prior to compilation, and that there is not any need to check for errors at run time.

The methods provided to set the values of variables, given as input from the user at run time, allow for checking numeric syntax either explicitly, or implicitly via the error trapping and recovery mechanism. Examples of the implicit approach are given in Chapter 15.

2.1 Syntax

The syntax for a single component of a Clifford number in an input stream of characters is of the form:

```
(double,double)e{int,int,int,int...}
```

The two 'double' within parentheses (...) are the real and imaginary parts of the coefficient. These can be any valid literal constant floating point constructs.

The list (set) of 'int' within the braces {...} are the primal (grade 1) Clifford units, the multiplicative product of which gives the (higher grade) evolved unit for the particular Clifford number. There can be anywhere from 0 to 31[1] primal units. The primal units must be in numerical order from lowest (0) to highest (one less than the dimension). See Table 2.1 for examples of correct syntax.

---

1 Although in principle an unsigned integer can usually hold 32 bits, certain bit manipulations within the code as implemented are not guaranteed to avoid overflow and erroneous results may occur if all 32 dimensions are used.

Numerical Calculations in Clifford Algebra: A Practical Guide for Engineers and Scientists,
First Edition. Andrew Seagar.
© 2023 John Wiley & Sons Ltd. Published 2023 by John Wiley & Sons Ltd.

Table 2.1 Syntax for constants.

Code	Value	Grade, type	Description
(1.0,0.0)e{}	$1e_\emptyset$	0, Scalar	The real value one
(6e-3,0)e{}	$0.006e_\emptyset$	0, scalar	A small real number
(0.0,1.0)e{}	$ie_\emptyset$	0, Scalar	The complex square root of minus one
(0,9e3)e{}	$9000ie_\emptyset$	0, Scalar	A large imaginary number
(1,2)e{}	$(1+i2)e_\emptyset$	0, Scalar	A complex number
(7,0)e{2}	$7e_2$	1, Vector	A linear element, could be the z component of a vector
(3,4)e{0,2}	$(3+i4)e_0e_2$	2, bivector	An element of area could be the y component of an electric field[a]
(5,0)e{1,2,3}	$5e_1e_2e_3$	3, Trivector	A volume element

a) See Chapter 10.

2.2 Constants

Constants for some pre-defined entities can be set without explicitly specifying the Clifford units. To set values of more general numbers, with many and varied components, it is necessary to specify both the value of the coefficient and the unit. There are several alternative ways to do this, but none are especially convenient.

2.2.1 Specific Types

The values of the more common entities supported within the Clifford numerical suite can be set in a single invocation of certain dedicated sub-routines. This includes scalars, pseudo-scalars, vectors in three dimensions with real coefficients, quaternions, and Pauli matrices. When setting values using these sub-routines, there is no need to specify the Clifford units. Those are taken from the position the coefficients occupy in the list of arguments.

The code shown in Example 2.1a demonstrates in lines 12–18 mechanisms to create new numbers and set their values, and in lines 20–26, mechanisms to print them in forms appropriate to the entity each number represents.[2] The second argument in the sub-routine print_number() indicates the maximum number of components to print on one line before moving to the next line. The numeric format of the output is established as single precision real values with zero digits after the decimal point by the invocation of the sub-routine format_printing() on line 10. Any format string in the 'c' programming language (King 2008) which is compatible with the underlying double precision floating-point data type can be used.

---

2 Lines which are not numbered are a continuation of the previous numbered line.

Example 2.1(a)

```
 1:   void demo1() /* input -- specific numbers */
 2:   {
 3:       int lambda[3];
 4:       double a[3];
 5:       number z,s,u,x,v,q,p;
 6:
 7:       recovery_level(0,0,0);
 8:       create_cliffmem(50);
 9:       declare_primal(3,pack_signature(3,lambda,-1));
10:       format_printing("%.0f");
11:
12:       z=fresh_number();
13:       s=scalar_number(1.0,2.0);
14:       u=pseudo_number(1.0,2.0);
15:       x=simple_number(1.0,2.0,spack_unit("e{0,2}"));
16:       v=pack_R3vector(pack_R3array(1.0,2.0,3.0,a));
17:       q=pack_quaternion(1.0,2.0,3.0,4.0,5.0,6.0,7.0,8.0);
18:       p=pack_pauli(1.0,2.0,3.0,4.0,5.0,6.0,7.0,8.0);
19:
20:       printf("0 => ");              print_number(z,8);
                                        recycle_number(z);
21:       printf("1+i2 => ");           print_number(s,8);
                                        recycle_number(s);
22:       printf("(1+i2)e_N => ");      print_number(u,8);
                                        recycle_number(u);
23:       printf("(1+i2)e{0,2} => ");   print_number(x,8);
                                        recycle_number(x);
24:       printf("V => ");              print_vector(v);
                                        recycle_number(v);
25:       printf("Q => ");              print_quaternion(q);
                                        recycle_number(q);
26:       printf("[P] => \n");          print_pauli(p);
                                        recycle_number(p);
27:
28:       free_cliffmem(1);
29:   }
```

The sub-routine `fresh_number()` on line 12 allocates memory from the internal linked lists and creates a number with value zero. The sub-routines `scalar_number()` and `pseudo_number()` on lines 13 and 14 create numbers with the complex values as specified, in grade 0 and in the highest grade for the dimension set in the invocation of the sub-routine `declare_primal()` on line 9. In this case, the dimension is three, so the pseudo-scalar is created in grade 3. The sub-routine `simple_number()` on line 15 creates a number having a single component with the Clifford unit as specified. The sub-routine `pack_R3vector()` on line 16 constructs a real valued vector in three dimensions, suitable for use in simple geometric operations.

The output produced by executing the code in Example 2.1a is shown in Example 2.1b. The output for the printing statements is listed in lines 2–10. Quaternions, as on line 7, are expressed in terms of the identity quaternion 1 and the three other unit quaternions I, J, K. Pauli matrices are expressed in terms of the 2-by-2 complex matrix which is the result of adding the unit Pauli matrices scaled by the values of their coefficients, on lines 9 and 10.

Example 2.1(b)

```
 1:   demo1()
 2:   0 => (0+i0)e{}
 3:   1+i2 => (1+i2)e{}
 4:   (1+i2)e_N => (1+i2)e{0,1,2}
 5:   (1+i2)e{0,2} => (1+i2)e{0,2}
 6:   V => [1+i0, 2+i0, 3+i0]
 7:   Q => (1+i2)1+(3+i4)I+(5+i6)J+(7+i8)K
 8:   [P] =>
 9:   [(8+i10) (9-i1)]
10:   [(-3+i9) (-6-i6)]
11:   statistics for structural heap:
12:   initial size (50)
13:   peak size    (8), as % of initial size (16%)
14:   full size    (50), expansion count (0)
15:   free count   (50), free start (6)
16:   statistics for data heap:
17:   initial size (50)
18:   peak size    (15), as % of initial size (30%)
19:   full size    (50), expansion count (0)
20:   free count   (50), free start (12)
```

Vectors, quaternions, and Pauli matrices can be printed as for any other Clifford number using the sub-routine `print_number()` instead of their entity-specific printing mechanisms. Doing that shows the underlying structure of the Clifford numbers used to represent these entities. Details of vectors, quaternions, and Pauli matrices are given in Chapters 6–8.

2.2.2 General

The values of complicated entities must be set in a piece-wise manner. Several different mechanisms are provided to do so, any one of which may suit a particular situation better than the others. For these mechanisms, it is necessary to specify both the value of the coefficient and the Clifford unit associated with that value. Example 2.2a demonstrates four mechanisms used to set four different three-dimensional Clifford numbers to the same values. The latter three mechanisms ultimately reduce to the first mechanism internally, using the lower-level worker sub-routine `merge_comp()` to perform manipulations on the underlying tree.

The first mechanism on lines 16–24 merges new components (data type 'cliff_comp') into an existing number (data type 'number'). A new component is placed (using the sub-routine `identity()`) in a new location in the number's tree if no component with the same unit is already present. If there is already a component with the same unit, the two values are combined (using the sub-routine `addition_to()`). The sub-routine `spack_unit()` converts the Clifford unit in the form of a string of characters into a bit pattern stored as an unsigned integer.

The second mechanism, on lines 28–36, first constructs an array of components and then uses those to construct a new number. The third mechanism, on lines 40–48, first constructs two arrays: a complex array for the coefficients and a separate unsigned integer array for the units. The unsigned integer datatype 'bitmap' on line 5 is defined as a convenient abbreviation in one of the header files by the statement:

$$\texttt{typedef unsigned int bitmap;}$$

The two arrays are then used to construct the new number. This avoids the need to use the data type 'cliff_comp' for components.

The fourth mechanism, on lines 52–59, starts by defining a number with one component and repeatedly adds other components one at a time.

Example 2.2(a)

```
1:   void demo2() /* input -- general numbers */
2:   {
3:       double u_complex[8][2];
4:       cliff_comp comp[8];
5:       bitmap u_unit[8];
6:       int lambda[3];
7:       cliff_comp u_comp;
8:       number u,x;
9:
10:      recovery_level(0,0,0);
11:      create_cliffmem(50);
12:      declare_primal(3,pack_signature(3,lambda,-1));
13:      format_printing("%.0f");
14:
15:      /* method 1 */
16:      u=fresh_number();
17:      u=merge_comp(u,pack_comp(&u_comp,1.0,2.0,spack_unit("e{}")),
                    addition_to,identity);
18:      u=merge_comp(u,pack_comp(&u_comp,3.0,4.0,spack_unit("e{0}")),
                    addition_to,identity);
19:      u=merge_comp(u,pack_comp(&u_comp,5.0,6.0,spack_unit("e{1}")),
                    addition_to,identity);
20:      u=merge_comp(u,pack_comp(&u_comp,7.0,8.0,spack_unit("e{0,1}")),
                    addition_to,identity);
21:      u=merge_comp(u,pack_comp(&u_comp,9.0,10.0,spack_unit("e{2}")),
                    addition_to,identity);
22:      u=merge_comp(u,pack_comp(&u_comp,11.0,12.0,spack_unit("e{0,2}")),
                    addition_to,identity);
23:      u=merge_comp(u,pack_comp(&u_comp,13.0,14.0,spack_unit("e{1,2}")),
                    addition_to,identity);
24:      u=merge_comp(u,pack_comp(&u_comp,15.0,16.0,spack_unit("e{0,1,2}")),
                    addition_to,identity);
25:      printf("N => ");          print_number(u,4);      recycle_number(u);
26:
27:      /* method 2 */
28:      pack_comp(&comp[0],1.0,2.0,spack_unit("e{}"));
29:      pack_comp(&comp[1],3.0,4.0,spack_unit("e{0}"));
```

(Continued)

Example 2.2(a) (Continued)

```
30:     pack_comp(&comp[2],5.0,6.0,spack_unit("e{1}"));
31:     pack_comp(&comp[3],7.0,8.0,spack_unit("e{0,1}"));
32:     pack_comp(&comp[4],9.0,10.0,spack_unit("e{2}"));
33:     pack_comp(&comp[5],11.0,12.0,spack_unit("e{0,2}"));
34:     pack_comp(&comp[6],13.0,14.0,spack_unit("e{1,2}"));
35:     pack_comp(&comp[7],15.0,16.0,spack_unit("e{0,1,2}"));
36:     u=pack_number(8,comp);
37:     printf("N => ");           print_number(u,4);     recycle_number(u);
38:
39:     /* method 3 */
40:     u_complex[0][0]=1.0;        u_complex[0][1]=2.0;
                                    u_unit[0]=spack_unit("e{}");
41:     u_complex[1][0]=3.0;        u_complex[1][1]=4.0;
                                    u_unit[1]=spack_unit("e{0}");
42:     u_complex[2][0]=5.0;        u_complex[2][1]=6.0;
                                    u_unit[2]=spack_unit("e{1}");
43:     u_complex[3][0]=7.0;        u_complex[3][1]=8.0;
                                    u_unit[3]=spack_unit("e{0,1}");
44:     u_complex[4][0]=9.0;        u_complex[4][1]=10.0;
                                    u_unit[4]=spack_unit("e{2}");
45:     u_complex[5][0]=11.0;       u_complex[5][1]=12.0;
                                    u_unit[5]=spack_unit("e{0,2}");
46:     u_complex[6][0]=13.0;       u_complex[6][1]=14.0;
                                    u_unit[6]=spack_unit("e{1,2}");
47:     u_complex[7][0]=15.0;       u_complex[7][1]=16.0;
                                    u_unit[7]=spack_unit("e{0,1,2}");
48:     u=pack_components(8,u_complex,u_unit);
49:     printf("N => ");           print_number(u,4);     recycle_number(u);
50:
51:     /* method 4 */
52:     u=simple_number(1.0,2.0,spack_unit("e{}"));
53:     u=addto_number(u,x=simple_number(3.0,4.0,spack_unit("e{0}")));
                        recycle_number(x);
54:     u=addto_number(u,x=simple_number(5.0,6.0,spack_unit("e{1}")));
                        recycle_number(x);
55:     u=addto_number(u,x=simple_number(7.0,8.0,spack_unit("e{0,1}")));
                        recycle_number(x);
56:     u=addto_number(u,x=simple_number(9.0,10.0,spack_unit("e{2}")));
                        recycle_number(x);
57:     u=addto_number(u,x=simple_number(11.0,12.0,spack_unit("e{0,2}")));
                        recycle_number(x);
58:     u=addto_number(u,x=simple_number(13.0,14.0,spack_unit("e{1,2}")));
                        recycle_number(x);
59:     u=addto_number(u,x=simple_number(15.0,16.0,spack_unit
                        ("e{0,1,2}"))); recycle_number(x);
60:     printf("N => ");           print_number(u,4);     recycle_number(u);
61:
62:     free_cliffmem(1);
63:  }
```

Executing the code in Example 2.2a produces the results in Example 2.2b. Printing the numbers set using the four different mechanisms produces the same results in each case in lines 2–9. The numbers are printed over two lines because the second argument in the invocation of the sub-routine `print_number()` on lines 25, 37, 49, and 60 of Example 2.2a is set to the value 4. This instructs that no more than four components should be printed on any one line.

Note in the memory usage information on lines 12 and 17 that the total memory used, for the tree structure as well as for the data values themselves, is roughly twice the number of data values. This is typical.

Example 2.2(b)

```
 1:   demo2()
 2:   N => (1+i2)e{}+(3+i4)e{0}+(5+i6)e{1}+(7+i8)e{0,1}
 3:     +(9+i10)e{2}+(11+i12)e{0,2}+(13+i14)e{1,2}+(15+i16)e{0,1,2}
 4:   N => (1+i2)e{}+(3+i4)e{0}+(5+i6)e{1}+(7+i8)e{0,1}
 5:     +(9+i10)e{2}+(11+i12)e{0,2}+(13+i14)e{1,2}+(15+i16)e{0,1,2}
 6:   N => (1+i2)e{}+(3+i4)e{0}+(5+i6)e{1}+(7+i8)e{0,1}
 7:     +(9+i10)e{2}+(11+i12)e{0,2}+(13+i14)e{1,2}+(15+i16)e{0,1,2}
 8:   N => (1+i2)e{}+(3+i4)e{0}+(5+i6)e{1}+(7+i8)e{0,1}
 9:     +(9+i10)e{2}+(11+i12)e{0,2}+(13+i14)e{1,2}+(15+i16)e{0,1,2}
10:   statistics for structural heap:
11:   initial size (50)
12:   peak size      (7), as % of initial size (14%)
13:   full size      (50), expansion count (0)
14:   free count     (50), free start (5)
15:   statistics for data heap:
16:   initial size (50)
17:   peak size      (9), as % of initial size (18%)
18:   full size      (50), expansion count (0)
19:   free count     (50), free start (8)
```

2.3 Variables

Setting variables to values which are known only at the time code is executed, such as may be input from the keyboard, requires two stages. The first stage is to check that the value as given can be interpreted as a valid construct for a Clifford number. This involves checking that units, passed as strings of characters, adhere to the syntax required. Once that has been established, the second stage involves converting the strings into the equivalent internal (unsigned integer) form of units.

2.3.1 Checking and Converting

The code in Examples 2.3–2.5a provides three sub-routines which together prompt the user for input and read a Clifford number typed on the keyboard, one component at a time.

The code in Example 2.3 is a small worker sub-routine to read a line of text representing a Clifford number from the keyboard into a buffer so that the syntax can be checked before attempting to convert to a numeric form.

Example 2.3

```
1:   int keyboard_input(n,buffer) /* read line from keyboard into
                                      buffer */
2:        int n;
3:        char *buffer;
4:   {
5:      int count,end,status;
6:      char byte;
7:
8:      end=0;
9:      count=0;
10:     status=scanf("%c",&byte);
11:     if(status==1)
12:        {
13:         if(count<n)
14:            {
15:             buffer[end]=byte;
16:             end=end+1;
17:            }
18:         count=count+1; /* a character has been read */
19:        }
20:     while(byte!='\n') /* line ends when find "newline" */
21:        {
22:         status=scanf("%c",&byte);
23:         if(status==1)
24:            {
25:             if(count<n) /* save only as many as can fit */
26:                {
27:                 buffer[end]=byte;
28:                 end=end+1;
29:                }
30:             count=count+1;
31:            }
32:        }
33:     buffer[end-1]='\0';
34:     return(count);  /* number of characters read */
35:   }
```

The code in Example 2.4 is another worker sub-routine, which provides the functionality to accept and convert valid typed input. This sub-routine reads character data typed in on the keyboard, checks the syntax and either accepts the data and converts it to the data type 'number' or rejects the data and repeats the process.

Data are read in on line 11. The syntax of the coefficient is checked on lines 19 and 20, and the syntax of the unit is checked on line 28 and 29 using the sub-routine sscan_unit(). If the syntax is acceptable, the data input is converted to a 'number' on lines 31 and 32. Otherwise, an error message is printed on either line 22 or 36, and another attempt to read a number is initiated recursively on line 24 or 38.

Example 2.4

```
 1:   number read_cliffnum_check() /* keyboard entry of Clifford
                                      numbers */
 2:   {                             /* with explicit syntax check      */
 3:      char buffer[64];
 4:      char *message;
 5:      cliff_comp u_comp;
 6:      double x,y;
 7:      int i,n,k;
 8:      number u;
 9:
10:      printf("enter a single Clifford component: ");
11:      k=keyboard_input(64,buffer); /* max 63+NULL, discard extra */
12:      if(k>64)
13:         {
14:            printf("input buffer (%d>64) overflow '%s'",k,buffer);
15:            printf(" in routine: read_cliffnum_check()\n");
16:            brexit(0); /* user restart */
17:         }
18:      printf("%.63s\n",buffer);
19:      i=sscanf(buffer," (%lf,%lf)%n",&x,&y,&n);
20:      if(i!=2)
21:         {
22:            printf("complex number '(x,y)' not found at start of string
                     '%s'\n",buffer);
23:            printf("to finish enter: (0,0)e{}\n");
24:            u=read_cliffnum_check();
25:         }
26:      else
27:         {
28:            message=sscan_unit(buffer+n); /* check syntax here */
29:            if(strlen(message)==0)
30:               {
31:                  pack_comp(&u_comp,x,y,spack_unit(buffer+n));
32:                  u=pack_number(1,&u_comp);
33:               }
34:            else
35:               {
36:                  printf("%s",message);    printf(" in string '%s'\n",
                           buffer+n);
37:                  printf("to finish enter: (0,0)e{}\n");
38:                  u=read_cliffnum_check();
39:               }
40:         }
41:      return(u);
42:   }
```

The code listed in Example 2.5a operates a loop between lines 16 and 25. The first two lines set up the initial conditions. The main body of the loop repeatedly collects components typed at the keyboard using the sub-routine `read_cliffnum_check()` on line 24, and adds them to a running total using sub-routine `addto_number()` on line 20. The loop is terminated when the user types zero (as a Clifford number), detected by the sub-routine `is_nonzero()` on line 18.

Example 2.5(a)

```
 1:   void demo3() /* keyboard entry of Clifford numbers with explicit
                       checking */
 2:   {
 3:       int lambda[3]; ·
 4:       number u, total;
 5:
 6:       recovery_level(0,0,0);
 7:       create_cliffmem(50);
 8:       declare_primal(3,pack_signature(3,lambda,-1));
 9:       format_printing("%.1f");
10:
11:       /* give some instructions */
12:       printf("you can use 3-dimensional Clifford numbers\n");
13:       printf("to finish enter: (0,0)e{}\n");
14:
15:       /* run main loop */
16:       total=fresh_number();
17:       u=read_cliffnum_check();
18:       while(is_nonzero(u))
19:         {
20:            total=addto_number(total,u);
21:            printf("number added was: ");
22:            print_number(u,8);
23:            recycle_number(u);
24:            u=read_cliffnum_check();
25:         }
26:       recycle_number(u);
27:       printf("finished\n");
28:       printf("your total is:\n");
29:       print_number(total,8);
30:       recycle_number(total);
31:
32:       /* tidy up Clifford maths */
33:       free_cliffmem(1);
34:   }
```

Executing the code in Example 2.5a with one sequence of keyboard input produces the results shown in Example 2.5b. Numbers entered with a valid syntax on lines 4, 6, and 14 are added to the running total. Numbers with an invalid syntax are entered on lines 8 and 11. In those cases, error messages are issued on lines 9 and 12. After zero is entered on line 16, the total is printed and the code terminates.

Example 2.5(b)

```
 1:   demo3()
 2:   you can use 3-dimensional Clifford numbers
 3:   to finish enter:  (0,0)e{}
 4:   enter a single Clifford component:  (1,2)e{0}
 5:   number added was:  (1.0+i2.0)e{0}
 6:   enter a single Clifford component:  (3,4)e{1}
 7:   number added was:  (3.0+i4.0)e{1}
 8:   enter a single Clifford component:  xxxe{2}
 9:   complex number '(x,y)' not found at start of string 'xxxe{2}'
10:   to finish enter:  (0,0)e{}
11:   enter a single Clifford component:  (5,6)e{4}
12:   (err.5) primal unit (4) not < Clifford dimension (3) in string 'e{4}'
13:   to finish enter:  (0,0)e{}
14:   enter a single Clifford component:  (5,6)e{2}
15:   number added was:  (5.0+i6.0)e{2}
16:   enter a single Clifford component:  (0,0)e{}
17:   finished
18:   your total is:
19:   (1.0+i2.0)e{0}+(3.0+i4.0)e{1}+(5.0+i6.0)e{2}
20:   statistics for structural heap:
21:   initial size (50)
22:   peak size     (2), as % of initial size (4%)
23:   full size     (50), expansion count (0)
24:   free count    (50), free start (0)
25:   statistics for data heap:
26:   initial size (50)
27:   peak size     (4), as % of initial size (8%)
28:   full size     (50), expansion count (0)
29:   free count    (50), free start (3)
```

It is easy to make mistakes typing numbers at the keyboard using the code shown in Example 2.5a, which may lead to frustration. One method to avoid that is to prepare the data in advance in a file using a text editor (eliminating all syntax errors at that time) and then directing the file to the standard input stream, 'stdin'. An alternative method is to construct some new sub-routine similar to Examples 2.3 and 2.4, using the same concepts but reading the data directly from file on disc using the 'c' language (King 2008) standard input–output library routine fscanf() instead of scanf().

Reference

King, K. N. (2008), *C Programming: A Modern Approach*, 2nd edn, W. W. Norton, New York, US-NY.

3

Output

A variety of methods are provided to display formatted character strings representing the numeric values of the Clifford numbers contained in the data type 'number'. Most of the methods write the values to the standard output stream 'stdout', allowing some control over what information is printed and how the numerical values are formatted. When these methods produce unsatisfactory results, the best approach is to recover the values of coefficients and units in numeric form and format those according to the need of the particular application.

All of the examples in this chapter use a single three-dimensional number, generated by the code shown in Example 3.1. That way the outputs of all of the examples can be sensibly compared to one another.

Example 3.1

```
 1:   number make_data()
 2:   {
 3:     cliff_comp u_comp;
 4:     number u;
 5:
 6:     u=fresh_number();
 7:     u=merge_comp(u,pack_comp(&u_comp,1.11,2.22,spack_unit("e{}")),
                   addition_to,identity);
 8:     u=merge_comp(u,pack_comp(&u_comp,3.33,4.44,spack_unit("e{0}")),
                   addition_to,identity);
 9:     u=merge_comp(u,pack_comp(&u_comp,5.55,6.66,spack_unit("e{1}")),
                   addition_to,identity);
10:     u=merge_comp(u,pack_comp(&u_comp,7.77,8.88,spack_unit("e{0,1}")),
                   addition_to,identity);
11:     u=merge_comp(u,pack_comp(&u_comp,9.99,10.0,spack_unit("e{2}")),
                   addition_to,identity);
12:     u=merge_comp(u,pack_comp(&u_comp,11.1,12.2,spack_unit("e{0,2}")),
                   addition_to,identity);
13:     u=merge_comp(u,pack_comp(&u_comp,13.3,14.4,spack_unit("e{1,2}")),
                   addition_to,identity);
14:     u=merge_comp(u,pack_comp(&u_comp,15.5,16.6,spack_unit("e{0,1,2}")),
                   addition_to,identity);
15:     return(u);
16:   }
```

Numerical Calculations in Clifford Algebra: A Practical Guide for Engineers and Scientists,
First Edition. Andrew Seagar.
© 2023 John Wiley & Sons Ltd. Published 2023 by John Wiley & Sons Ltd.

3.1 Tree Format

Tree format produces output which shows in addition to the values of the number itself, the structure of the underlying tree. Example 3.2a prints the test number created by the sub-routine listed in Example 3.1. The printing format of the numeric values themselves is set on line 9 using the sub-routine `format_printing()`. The value of the test number (which occupies a full tree) is set on line 11 using the sub-routine `make_data()`, and the complete tree for the number is printed on line 13 using `print_tree()`.

Example 3.2(a)

```
 1:   void demo4() /* output -- tree format */
 2:   {
 3:       int lambda[3];
 4:       number u;
 5:
 6:       recovery_level(0,0,0);
 7:       create_cliffmem(50);
 8:       declare_primal(3,pack_signature(3,lambda,-1));
 9:       format_printing("%.2f");
10:
11:       u=make_data();
12:       printf("full tree\n");
13:       printf("T = \n");     print_tree(u);     /* binary tree format */
14:
15:       recycle_number(u);
16:       free_cliffmem(1);
17:   }
```

Results of executing the code in Example 3.2a are shown in Example 3.2b. The tree structure is listed from lines 4 to 27. The individual components are listed on the right-hand side with the complex coefficients in parentheses '(...)', followed by the unit as a list of integers within braces 'e { ... }'. The components are preceded by a single integer in brackets '[...]'. The integer represents the unit for the corresponding components in the form of a binary bitmap. For unit $e_0 e_2 = e$ { 0,2 }, the binary bitmap is $2^0 + 2^2 = 5$. For the absolute unit $e_\emptyset = e$ {}, the binary bitmap is 0.

Example 3.2(b)

```
 1:   demo4()
 2:   full tree
 3:   T =
 4:   (
 5:   |     [0-7] (
 6:   |     |    [0-3] (
 7:   |     |    |    [0-1] (
 8:   |     |    |    |    [0] (1.11+i2.22) e{}
 9:   |     |    |    |    [1] (3.33+i4.44) e{0}
10:   |     |    |    )
```

```
11:  |   |   |     [2-3] (
12:  |   |   |     |    [2] (5.55+i6.66) e{1}
13:  |   |   |     |    [3] (7.77+i8.88) e{0,1}
14:  |   |   | )
15:  |   | )
16:  |   |     [4-7] (
17:  |   |   |     [4-5] (
18:  |   |   |     |    [4] (9.99+i10.00) e{2}
19:  |   |   |     |    [5] (11.10+i12.20) e{0,2}
20:  |   |   | )
21:  |   |   |     [6-7] (
22:  |   |   |     |    [6] (13.30+i14.40) e{1,2}
23:  |   |   |     |    [7] (15.50+i16.60) e{0,1,2}
24:  |   |   | )
25:  |   | )
26:  | )
27:  )
28:  statistics for structural heap:
29:  initial size (50)
30:  peak size    (7), as % of initial size (14%)
31:  full size    (50), expansion count (0)
32:  free count   (50), free start (6)
33:  statistics for data heap:
34:  initial size (50)
35:  peak size    (8), as % of initial size (16%)
36:  full size    (50), expansion count (0)
37:  free count   (50), free start (7)
```

The node at the root of tree structure is indicated by the upper leftmost brackets '[0-7]' on line 5. The pair of integers inside the brackets indicate the range of units which in principle can be found beyond the root node. If any units within the range indicated are missing from the tree it means that the value of the corresponding coefficients is identically zero. In this case, all coefficients are non-zero so that in lines 8–23 all units in the range '[0-7]' are present.

The total amount of memory consumed can be calculated by counting up the number of lines containing brackets. For this case, there are 15 lines, seven with ranges '[i-j]' and eight with components '[k]'. Inspection of the statistics for the memory usage on line 30 shows that seven structural nodes (containing the ranges) have been used, and line 35 shows that eight data elements have been used for components. Note that the total amount of memory (number of nodes) consumed is about twice the number of components.

Nodes at each level of the tree beyond the root, aligned with the columns of vertical lines progressively further to the right, have two branches. Either of the branches may contain another branching node or an element of data. In this example, the tree extends three levels to support $2^3 = 8$ elements of data.

For comparison, Example 3.3a prints the same number as in Example 3.2a after all components not in grade 1 are set to 0. The test number is first defined on line 11 and then redefined on line 12 to contain only the grade 1 elements of its former self, using the sub-routine onegradeof_number(). That number is then printed on line 14. As a second demonstration, Example 3.3a also defines a number containing zero on line 17 and prints that on line 19.

Example 3.3(a)

```
 1:   void demo5() /* output - tree format */
 2:   {
 3:       int lambda[3];
 4:       number u;
 5:
 6:       recovery_level(0,0,0);
 7:       create_cliffmem(50);
 8:       declare_primal(3,pack_signature(3,lambda,-1));
 9:       format_printing("%.2f");
10:
11:       u=make_data();
12:       u=onegradeof_number(u,1,1);
13:       printf("tree with grade 1 elements only; vector\n");
14:       printf("T = \n");     print_tree(u);    /* binary tree format */
15:
16:       recycle_number(u);
17:       u=fresh_number();
18:       printf("tree containing only the number zero\n");
19:       printf("T = \n");     print_tree(u);    /* binary tree format */
20:
21:       recycle_number(u);
22:       free_cliffmem(1);
23:   }
```

The result of printing the number with only non-zero components in grade 1 is listed from lines 4 to 12 of Example 3.3b. In lines 7–10, the only components found with units in the range '[0-7]' on line 5 are those with non-zero coefficients. Less memory is used if the components which have zero value are not stored. Numerical calculations are more efficient because the zeros never play a part in any arithmetic operation. Here the total amount of memory used is two structural nodes ('[0-7]', '[0-3]') and three data elements ('[1]', '[2]', and '[4]'). As for the full tree in Example 3.2b, for this reduced tree, the total amount of memory consumed is about twice the number of components. That is typical of all numbers.

Example 3.3(b)

```
 1:   demo5()
 2:   tree with grade 1 elements only; vector
 3:   T =
 4:   (
 5:   |     [0-7] (
 6:   |     |     [0-3] (
 7:   |     |     |     [1] (3.33+i4.44)e{0}
 8:   |     |     |     [2] (5.55+i6.66)e{1}
 9:   |     |     )
```

```
10:   |    |     [4](9.99+i10.00)e{2}
11:   |    )
12:   )
13:   tree containing only the number zero
14:   T =
15:   (
16:   |     [0](0.00+i0.00)e{}
17:   )
18:   statistics for structural heap:
19:   initial size (50)
20:   peak size     (7), as % of initial size (14%)
21:   full size     (50), expansion count (0)
22:   free count    (50), free start (1)
23:   statistics for data heap:
24:   initial size (50)
25:   peak size     (8), as % of initial size (16%)
26:   full size     (50), expansion count (0)
27:   free count    (50), free start (4)
```

In this case, the tree structure has branches at the first-level node '[0-7]' which are of different types. One, '[0-3]', is another branching node and the other, '[4]', is an element of data.

The tree for the number zero is listed from line 15 to 17. The root node has no branches and holds the data element directly. This is true for any number which has only a single unit.

Inspection of the statistics for memory usage shows the same values as for Example 3.2. This is because the test number is created in full on line 11 of Example 3.3a, establishing the peak memory usage in advance of setting some components to zero.

3.2   Numeric Formats

Numbers may be printed in any valid format for the underlying double precision floating data type. There are two versions of most of the methods used for printing. One version formats numbers according to the default format, as specified by invoking the sub-routine format_printing(). The other version contains an extra argument used to carry the format specification, which is used instead of the default value.

3.2.1   Default Format

Example 3.4a sets the default format on line 13 as single precision floating with two digits following the decimal point and prints the test number, created on line 10, from line 16 through to 19. The names of the sub-routines used to invoke the printing all start with the characters 'print_'.

All components of the number are printed on line 16 using the sub-routine print_number(). The second argument in the list indicates that no more than four components should be printed on any one line. If the number contains more than four components, it will be printed on two or more lines.

In lines 17–19, the same number is printed as a vector (Chapter 6), quaternion (Chapter 7), and Pauli matrix (Chapter 8). Only those components of the number which are consistent with those particular types of data are printed. Some components are ignored. For example, the sub-routine `print_vector()` prints only those components occupying grade 1.

Example 3.4(a)

```
 1:  void demo6() /* output - default formats */
 2:  {
 3:      int lambda[3];
 4:      number u;
 5:
 6:      recovery_level(0,0,0);
 7:      create_cliffmem(50);
 8:      declare_primal(3,pack_signature(3,lambda,-1));
 9:
10:      u=make_data();
11:
12:      /* set default format */
13:      format_printing("%.2f");
14:
15:      /* print */
16:      printf("N = ");        print_number(u,4);
17:      printf("V = ");        print_vector(u);
18:      printf("Q = ");        print_quaternion(u);
19:      printf(" [P] =\n");    print_pauli(u);
20:
21:      recycle_number(u);
22:      free_cliffmem(0);
23:  }
```

The results of executing Example 3.4a are listed in Example 3.4b. Inspection of lines 2–8 shows that all of the entities are printed in the default format, as specified. If no default format is specified by omitting line 13 in Example 3.4a, the format `"%.5e"` is used.

Example 3.4(b)

```
 1:  demo6()
 2:  N = (1.11+i2.22)e{}+(3.33+i4.44)e{0}+(5.55+i6.66)e{1}
            +(7.77+i8.88)e{0,1}
 3:      +(9.99+i10.00)e{2}+(11.10+i12.20)e{0,2}+(13.30+i14.40)e{1,2}
            +(15.50+i16.60)e{0,1,2}
 4:  V = [3.33+i4.44,  5.55+i6.66,  9.99+i10.00]
 5:  Q = (1.11+i2.22)1+(13.30+i14.40)I+(-11.10-i12.20)J+(7.77+i8.88)K
 6:  [P] =
 7:  [ (9.99-i5.55)  (25.50-i1.10)]
 8:  [(3.30-i25.50)  (-7.77+i9.99)]
```

3.2.2 Defined Format

Lines 11–14 and 16–19 of Example 3.5a invoke sub-routines to print the test number in the format defined in the first argument of those sub-routines. The names of the sub-routines are all the same as in the corresponding cases for the default format of Example 3.4a, except that they start with the characters 'printf_' instead of 'print_'.

Lines 11–14 format the number in the form of a floating point mantissa with one digit after the decimal point and an integer exponent for radix 10. Lines 16–19 format the same number in the form of a floating point value with no digits showing after the decimal point and no exponent.

Example 3.5(a)

```
 1:   void demo7() /* output - defined formats */
 2:   {
 3:       int lambda[3];
 4:       number u;
 5:
 6:       recovery_level(0,0,0);
 7:       create_cliffmem(50);
 8:       declare_primal(3,pack_signature(3,lambda,-1));
 9:
10:       u=make_data();
11:       printf("N = ");        printf_number("%.1e",u,3);
12:       printf("V = ");        printf_vector("%.1e",u);
13:       printf("Q = ");        printf_quaternion("%.1e",u);
14:       printf(" [P] = \n");   printf_pauli("%.1e",u);
15:
16:       printf("\nN = ");      printf_number("%.0f",u,3);
17:       printf("V = ");        printf_vector("%.0f",u);
18:       printf("Q = ");        printf_quaternion("%.0f",u);
19:       printf(" [P] = \n");   printf_pauli("%.0f",u);
20:
21:       recycle_number(u);
22:       free_cliffmem(0);
23:   }
```

Example 3.5b lists the results of executing the code in Example 3.5a. Inspection of lines 2–9 show that they are all printed as requested with mantissa and exponent, and that lines 11–18 are also printed in the floating format as requested without fractional parts.

Example 3.5(b)

```
1:   demo7()
2:   N = (1.1e+00+i2.2e+00)e{}+(3.3e+00+i4.4e+00)e{0}
          +(5.5e+00+i6.7e+00)e{1}
3:       +(7.8e+00+i8.9e+00)e{0,1}+(1.0e+01+i1.0e+01)e{2}
          +(1.1e+01+i1.2e+01)e{0,2}
```

(Continued)

Example 3.5(b) (Continued)

```
 4:    +(1.3e+01+i1.4e+01)e{1,2}+(1.6e+01+i1.7e+01)e{0,1,2}
 5:    V = [3.3e+00+i4.4e+00,  5.5e+00+i6.7e+00,  1.0e+01+i1.0e+01
 6:    Q = (1.1e+00+i2.2e+00)1+(1.3e+01+i1.4e+01)I+(-1.1e+01-i1.2e+01)J
          +(7.8e+00+i8.9e+00)K
 7:    [P] =
 8:    [(1.0e+01-i5.5e+00)  (2.6e+01-i1.1e+00)]
 9:    [(3.3e+00-i2.6e+01)  (-7.8e+00+i1.0e+01)]
10:
11:    N = (1+i2)e{}+(3+i4)e{0}+(6+i7)e{1}
12:      +(8+i9)e{0,1}+(10+i10)e{2}+(11+i12)e{0,2}
13:      +(13+i14)e{1,2}+(16+i17)e{0,1,2}
14:    V = [3+i4,  6+i7,  10+i10]
15:    Q = (1+i2)1+(13+i14)I+(-11-i12)J+(8+i9)K
16:    [P] =
17:    [(10-i6)  (26-i1)]
18:    [(3-i26)  (-8+i10)]
```

3.3 Extended Formats

Printing formats may be extended to perform additional functions by adding two characters to the end of the standard format strings. The first character (#, +, 0, or -) specifies the type of rounding performed. The second character (c, r, i, m, or p) specifies which particular aspect of the complex coefficient should be reported. See Chapter 27 for a detailed description of the syntax.

3.3.1 Rounding

Rounding may be useful when working within the floating-point data types with data that are properly integral and which are used in operations that should in principle produce only integral results. Example 3.6a demonstrates the formats required to print complex coefficients with or without integer rounding. No rounding is specified by the character '#', as in line 11. Rounding up, to the nearest integer and down are specified by the characters '+', '0', and '-', respectively, as in lines 12–14.

Example 3.6(a)

```
 1:    void demo8() /* output - rounding */
 2:    {
 3:        int lambda[3];
 4:        number u;
 5:
 6:        recovery_level(0,0,0);
 7:        create_cliffmem(50);
 8:        declare_primal(3,pack_signature(3,lambda,-1));
```

```
 9:
10:    u=make_data();
11:    printf("N (no rounding) = ");       printf_number("%.2f#c",u,4);
                                            printf("\n");
12:    printf("N (round up) = ");          printf_number("%.2f+c",u,4);
                                            printf("\n");
13:    printf("N (round nearest) = ");     printf_number("%.2f0c",u,4);
                                            printf("\n");
14:    printf("N (round down) = ");        printf_number("%.2f-c",u,4);
15:
16:    recycle_number(u);
17:    free_cliffmem(0);
18:   }
```

Example 3.6b lists the results of executing the code in Example 3.6a. Careful inspection of lines 5–12 in comparison to lines 2 and 3 shows that they are all rounded as requested according to the extended format notation.

Example 3.6(b)

```
 1:    demo8()
 2:    N (no rounding) = (1.11+i2.22)e{}+(3.33+i4.44)e{0}+(5.55+i6.66)e{1}
          +(7.77+i8.88)e{0,1}
 3:       +(9.99+i10.00)e{2}+(11.10+i12.20)e{0,2}+(13.30+i14.40)e{1,2}
          +(15.50+i16.60)e{0,1,2}
 4:
 5:    N (round up) = (2.00+i3.00)e{}+(4.00+i5.00)e{0}+(6.00+i7.00)e{1}
          +(8.00+i9.00)e{0,1}
 6:       +(10.00+i10.00)e{2}+(12.00+i13.00)e{0,2}+(14.00+i15.00)e{1,2}
          +(16.00+i17.00)e{0,1,2}
 7:
 8:    N (round nearest) = (1.00+i2.00)e{}+(3.00+i4.00)e{0}
          +(6.00+i7.00)e{1}+(8.00+i9.00)e{0,1}
 9:       +(10.00+i10.00)e{2}+(11.00+i12.00)e{0,2}+(13.00+i14.00)e{1,2}
          +(16.00+i17.00)e{0,1,2}
10:
11:    N (round down) = (1.00+i2.00)e{}+(3.00+i4.00)e{0}+(5.00+i6.00)e{1}
          +(7.00+i8.00)e{0,1}
12:       +(9.00+i10.00)e{2}+(11.00+i12.00)e{0,2}+(13.00+i14.00)e{1,2}
          +(15.00+i16.00)e{0,1,2}
```

3.3.2 Parts of Coefficients

The second character of the extended printing format specifies whether the whole complex coefficient is printed, or only some particular aspect of its complex value. Example 3.7a demonstrates the formats required to print complex coefficients in whole or in part. The whole complex value is specified by the character 'c', as in line 11. The real part, imaginary part, magnitude, and phase are specified by the characters 'r', 'i', 'm', and 'p', respectively, as in lines 12–15. Specifying the extended format '#c' as in line 11 has the same effect as not adding any extension.

Example 3.7(a)

```
1:   void demo9() /* output - parts */
2:   {
3:       int lambda[3];
4:       number u;
5:
6:       recovery_level(0,0,0);
7:       create_cliffmem(50);
8:       declare_primal(3,pack_signature(3,lambda,-1));
9:
10:      u=make_data();
11:      printf("N (complex) = ");    printf_number("%.2f#c",u,4);
                                       printf("\n");
12:      printf("N (real) = ");       printf_number("%.2f#r",u,4);
                                       printf("\n");
13:      printf("N (imaginary) = ");  printf_number("%.2f#i",u,4);
                                       printf("\n");
14:      printf("N (magnitude) = ");  printf_number("%.2f#m",u,4);
                                       printf("\n");
15:      printf("N (phase) = ");      printf_number("%.2f#p",u,4);
16:
17:      recycle_number(u);
18:      free_cliffmem(0);
19:  }
```

Specifying only the real or imaginary part may be useful in certain cases where the other part is known to be zero. In other cases, the magnitude may be of primary importance, while the phase is of only passing interest.

Example 3.7b lists the results of executing the code in Example 3.7a. Careful inspection of lines 5–15 in comparison to lines 2 and 3 shows that they are printed as requested according to the extended format notation. Note that the phase is reported in radians not degrees, and that the imaginary part of the complex number by itself does not show the imaginary unit i.

Example 3.7(b)

```
1:   demo9()
2:   N (complex) = (1.11+i2.22)e{}+(3.33+i4.44)e{0}+(5.55+i6.66)e{1}
        +(7.77+i8.88)e{0,1}
3:       +(9.99+i10.00)e{2}+(11.10+i12.20)e{0,2}+(13.30+i14.40)e{1,2}
        +(15.50+i16.60)e{0,1,2}
4:
5:   N (real) = (1.11)e{}+(3.33)e{0}+(5.55)e{1}+(7.77)e{0,1}
6:       +(9.99)e{2}+(11.10)e{0,2}+(13.30)e{1,2}+(15.50)e{0,1,2}
```

```
 7:
 8:   N (imaginary) = (2.22)e{}+(4.44)e{0}+(6.66)e{1}+(8.88)e{0,1}
 9:     +(10.00)e{2}+(12.20)e{0,2}+(14.40)e{1,2}+(16.60)e{0,1,2}
10:
11:   N (magnitude) = (2.48)e{}+(5.55)e{0}+(8.67)e{1}+(11.80)e{0,1}
12:     +(14.14)e{2}+(16.49)e{0,2}+(19.60)e{1,2}+(22.71)e{0,1,2}
13:
14:   N (phase) = (1.11)e{}+(0.93)e{0}+(0.88)e{1}+(0.85)e{0,1}
15:     +(0.79)e{2}+(0.83)e{0,2}+(0.83)e{1,2}+(0.82)e{0,1,2}
```

3.4 Selected Components

Rather than printing all components of a number, sometimes it is useful to print only a selected component, such as one component of a vector. Example 3.8a demonstrates the mechanism provided to print the coefficient for a selected component using the sub-routine `print_single_coefficient()` in lines 15–18 and using `printf_single_coefficient()` in lines 21–25. For the form 'printf_', any floating point formats, including the extended formats in Section 3.3, can be used.

Example 3.8(a)

```
 1:   void demo10() /* output - selected coefficients */
 2:   {
 3:      int lambda[3];
 4:      number u;
 5:
 6:      recovery_level(0,0,0);
 7:      create_cliffmem(50);
 8:      declare_primal(3,pack_signature(3,lambda,-1));
 9:      format_printing("%.2f");
10:
11:      u=make_data();
12:      printf("N = ");                    print_number(u,4);
13:
14:      /* default format only */
15:      printf("N.e{} = ");   print_single_coefficient(u,spack_unit
                                   ("e{}"));
16:      printf("N.e{1} = ");  print_single_coefficient(u,spack_unit
                                   ("e{1}"));
17:      printf("N.e{1,2} = ");  print_single_coefficient(u,spack_unit
                                   ("e{1,2}"));
18:      printf("N.e{0,2} = ");  print_single_coefficient(u,spack_unit
                                   ("e{0,2}"));
19:
20:      /* specified format */
21:      printf("N.e{} = ");   printf_single_coefficient("%.2e",u,spack_unit
                                   ("e{}"));
```

(Continued)

Example 3.8(a) (Continued)

```
22:     printf("N.e{1} = ");   printf_single_coefficient("%.3f",u,spack_unit
                            ("e{1}"));
23:     printf("N.e{1,2}(real) = ");   printf_single_coefficient("%.2e#r",
                            u,spack_unit("e{1,2}"));
24:     printf("N.e{0,2}(round nearest, phase) = ");
25:     printf_single_coefficient("%.2e0p",u,spack_unit("e{0,2}"));
26:
27:     recycle_number(u);
28:     free_cliffmem(0);
29:   }
```

Example 3.8b lists the results of executing the code in Example 3.8a. Inspection of lines 4–11 in comparison to lines 2 and 3 shows that the coefficients for each of the selected components are printed in the format as specified. The units are not printed.

Example 3.8(b)

```
1:    demo10()
2:    N = (1.11+i2.22)e{}+(3.33+i4.44)e{0}+(5.55+i6.66)e{1}
        +(7.77+i8.88)e{0,1}
3:      +(9.99+i10.00)e{2}+(11.10+i12.20)e{0,2}+(13.30+i14.40)e{1,2}
        +(15.50+i16.60)e{0,1,2}
4:    N.e{} = (1.11+i2.22)
5:    N.e{1} = (5.55+i6.66)
6:    N.e{1,2} = (13.30+i14.40)
7:    N.e{0,2} = (11.10+i12.20)
8:    N.e{} = (1.11e+00+i2.22e+00)
9:    N.e{1} = (5.550+i6.660)
10:   N.e{1,2}(real) = (1.33e+01)
11:   N.e{0,2}(round nearest, phase) = (1.00e+00)
```

3.5 Primitive Formats

Two primitive formats are provided for printing coefficients and units as separate lists. For these formats, a coefficient at one position in one list corresponds to the unit in the same position in the other list. Units are in the form of a binary integer bitmap.

Example 3.9a demonstrates the first format on lines 15 and 18, which print plain lists in the order of the units as bitmaps from low to high. Lines 21 and 24 prints lists ordered first by grade from low to high, and within grade, by unit as bitmap from low to high.

Example 3.9(a)

```
1:    void demo11() /* output - raw formats */
2:    {
3:      int lambda[3];
```

```
 4:     number u;
 5:
 6:     recovery_level(0,0,0);
 7:     create_cliffmem(50);
 8:     declare_primal(3,pack_signature(3,lambda,-1));
 9:     format_printing("%.2f");
10:
11:     u=make_data();
12:     printf("N = ");                                   print_number(u,4);
13:
14:     /* coefficients only */
15:     printf("coefficients...\nN = ");                  print_listof_
                                                          coefficients(u,4);
16:
17:     /* integer units only */
18:     printf("units...\nN = ");                         print_listof_units(u,8);
19:
20:     /* coefficients in grades */
21:     printf("coefficients in grades...\nN = "); print_gradedlistof_
                                                          coefficients(u,4);
22:
23:     /* units in grades */
24:     printf("units in grades...\nN = ");               print_gradedlistof_
                                                          units(u,8);
25:
26:     recycle_number(u);
27:     free_cliffmem(0);
28:     }
```

Example 3.9b lists the results of executing the code in Example 3.9a. Inspection of lines 5–8 in comparison to lines 2 and 3 shows that the elements (coefficients and units) in plain list form are in the same order as when numbers are printed by the print_number() mechanism. The elements in the plain lists are separated by commas.

Inspection of lines 10–13 shows that the order of the elements is different to that for the plain list. Grades are separated by semicolons, and within each grade, the elements are separated by commas. The binary bit patterns for the units in each grade have the same number of bits: zero in grade 0 (0), one in grade 1 (1,2,4), two in grade 2 (3,5,6), and three in grade 3 (7).

Example 3.9(b)

```
1:  demo11()
2:  N = (1.11+i2.22)e{}+(3.33+i4.44)e{0}+(5.55+i6.66)e{1}
       +(7.77+i8.88)e{0,1}
3:      +(9.99+i10.00)e{2}+(11.10+i12.20)e{0,2}+(13.30+i14.40)e{1,2}
       +(15.50+i16.60)e{0,1,2}
4:  coefficients...
5:  N = [(1.11+i2.22),(3.33+i4.44),(5.55+i6.66),(7.77+i8.88),
```

(Continued)

Example 3.9(b) (Continued)

```
 6:    (9.99+i10.00),(11.10+i12.20),(13.30+i14.40),(15.50+i16.60)
 7:    units...
 8:    N = [0,1,2,3,4,5,6,7]
 9:    coefficients in grades...
10:    N = [(1.11+i2.22);(3.33+i4.44),(5.55+i6.66),(9.99+i10.00);
11:    (7.77+i8.88),(11.10+i12.20),(13.30+i14.40);(15.50+i16.60)]
12:    units in grades...
13:    N = [0;1,2,4;3,5,6;7]
```

3.6   Recovered Values

When none of the formats for printing described here are suitable, the final method is to do it yourself. This approach involves recovering the data from the tree in the form of two plain arrays, one with pairs of real numbers for the complex coefficients, and the other with integer values for the corresponding units as bitmaps.

Example 3.10a demonstrates the mechanism provided to recover the coefficients and units on lines 17 and 20 using sub-routines count_components() and unpack _components(). The sub-routine count_components() returns the number of components in the array. This is the count of only the non-zero components, or the single zero component if the number has no non-zero components. Counting the components can be used to determine how large to make the arrays into which they are unpacked.

The sub-routine unpack_components() copies the real, imaginary, and unit values into an array of pairs of double precision data values, and an array of single unsigned integer values. If the arrays are not large enough some kind of memory corruption is likely to occur.

Once the values have been copied into the two arrays, they are printed from those in whatever format one might choose. The loop from lines 25 to 31 prints the values in much the same as the standard format using the two sub-routines print_coef() and print_unit(). The loop from lines 35 to 41 prints the values in a much plainer format using the standard input–output library routine printf().

Example 3.10(a)

```
 1:    void demo12() /* output - recovered values */
 2:    {
 3:       double u_complex[8][2];
 4:       bitmap u_unit[8];
 5:       int lambda[3];
 6:       number u;
 7:       int i,c;
 8:
 9:       recovery_level(0,0,0);
10:       create_cliffmem(50);
11:       declare_primal(3,pack_signature(3,lambda,-1));
```

```
12:     format_printing("%.2f");
13:
14:     u=make_data();
15:
16:     /* recover values */
17:     c=count_components(u);
18:     printf("%d components to unpack\n",c);
19:
20:     c=unpack_components(u,u_complex,u_unit);
21:     printf("%d unpacked components to print\n",c);
22:
23:     /* print like standard format */
24:     printf("print like standard format\n");
25:     for(i=0;i<c;i++)
26:       {
27:         printf("%d> ",i+1);
28:         print_coef(u_complex[i][0],u_complex[i][1]);
29:         print_unit(u_unit[i]);
30:         printf("\n");
31:       }
32:
33:     /* print my format */
34:     printf("now print in my special plain format\n");
35:     for(i=0;i<c;i++)
36:       {
37:         printf("%d> ",i+1);
38:         printf("%5.2f    %5.2f",u_complex[i][0],u_complex[i][1]);
39:         printf("  %3d",u_unit[i]);
40:         printf("\n");
41:       }
42:
43:     recycle_number(u);
44:     free_cliffmem(0);
45:   }
```

Example 3.10b lists the results of executing the code in Example 3.10a. Lines 5–12 show the recovered values printed in the standard way, and lines 14–21 show the same values printed in the wholly user defined way.

Example 3.10(b)

```
1:   demo12()
2:   8 components to unpack
3:   8 unpacked components to print
4:   print like standard format
5:   1> (1.11+i2.22)e{}
6:   2> (3.33+i4.44)e{0}
7:   3> (5.55+i6.66)e{1}
8:   4> (7.77+i8.88)e{0,1}
9:   5> (9.99+i10.00)e{2}
```

(Continued)

Example 3.10(b) (Continued)

```
10:    6> (11.10+i12.20)e{0,2}
11:    7> (13.30+i14.40)e{1,2}
12:    8> (15.50+i16.60)e{0,1,2}
13:    now print in my special plain format
14:    1>  1.11      2.22      0
15:    2>  3.33      4.44      1
16:    3>  5.55      6.66      2
17:    4>  7.77      8.88      3
18:    5>  9.99     10.00      4
19:    6> 11.10     12.20      5
20:    7> 13.30     14.40      6
21:    8> 15.50     16.60      7
```

4

Unary Operations

4.1 Theory

Five elementary unary operations, as listed in the middle section of Table 4.1, are provided within the Clifford numerical suite. The first four operations change the signs of some of the components, without ever affecting the units. The Hodge star operation changes some signs and always modifies units. It is not elementary in the sense that it is composed from simpler operations (reversion and central multiplication).

4.1.1 Negation

Negation of Clifford numbers involves changing the sign of every coefficient. Figure 4.1 depicts on the left-hand side a numerical operation involving three primal units e_2, e_3, e_5 shown as boxes with two Clifford (central) multiplication operations and one negation operation. The central multiplication operations, for which there is no symbol, are located between the boxes.

The negation operation has the effect of multiplying the entity on its right-hand side by the value -1. The result is shown on the right-hand side of Figure 4.1. The three prime numbers '2 3 5' joined by dots denotes the single Clifford number produced by multiplying the three primal units, and the minus sign is an ordinary minus sign.

The same operation is written in the form of Clifford algebra in Eq. (4.1):

$$a = e_2 e_3 e_5$$
$$-a = -e_2 e_3 e_5 \qquad\qquad (4.1)$$

Applying the negation operator twice reproduces the original number.

4.1.2 Involution

The involution[1] operation involves introducing a change of sign for every primal unit. Figure 4.2 depicts on the left-hand side, a numerical operation involving three primal

---

1 Strictly speaking the *main* involution operation.

Numerical Calculations in Clifford Algebra: A Practical Guide for Engineers and Scientists, First Edition. Andrew Seagar.
© 2023 John Wiley & Sons Ltd. Published 2023 by John Wiley & Sons Ltd.

Table 4.1 Unary operations.

Case	Unary operation	Notation
General	Negation	$-a$
	Involution	$\neg a$
	Reversion	$\tilde{a}$
	Clifford conjugation	$\bar{a}$
	Complex conjugation	a^*
Special	Hodge dual	$*a \equiv \tilde{a}\, e_N$
	(Star, supplement)	

$(-)\;\boxed{2}\;\boxed{3}\;\boxed{5}\;\;\rightarrow\;-2.3.5$ **Figure 4.1** Negation.

$(\neg)\boxed{\boxed{2}\;\boxed{3}\;\boxed{5}} \rightarrow (-)\boxed{2}\,(-)\boxed{3}\,(-)\boxed{5}$ **Figure 4.2** Involution.
$\qquad\rightarrow (-)\,(-)\,(-)\boxed{2}\;\boxed{3}\;\boxed{5}$
$\qquad\rightarrow (-)\boxed{2}\;\boxed{3}\;\boxed{5} \qquad \rightarrow\;-2.3.5$

units e_2, e_3, e_5 with two Clifford (central) multiplication operations and one involution operation. The involution operation applies to all three units, not just the first one.

In this case, the involution introduces an overall change of sign in the final result, shown on the right-hand side of Figure 4.2. If an additional primal unit is introduced, the result has no change of sign.

The same operation is written in the form of Clifford algebra in Eq. (4.2). Here the symbol k takes the value of the grade of the unit to which the involution operator is applied.

$$a = e_2 e_3 e_5$$
$$\neg a = (-1)^k\, a$$
$$= (-1)^3\, e_2 e_3 e_5$$
$$= -e_2 e_3 e_5 \tag{4.2}$$

Applying the involution operator twice reproduces the original number.

4.1.3 Pair Exchange

If we do some kind of operation and end up with a result in which the primal units are out of order, then a sequence of pair exchange operations is required to re-establish the proper order from low to high. Multiplication of two primal Clifford units is not commutative. Each time a pair of different primal units is exchanged, a change of sign is introduced. See Section 5.3.1 for details.

Figure 4.3   Pair exchange.

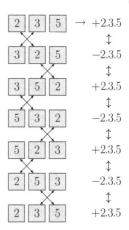

Figure 4.3 demonstrates a sequence of pair exchange operations working down the page, to produce all combinations of the three primal units, and returning at the end to the starting value.

The operations invoked in the pair exchanges are written in the form of Clifford algebra in Eq. (4.3).

$$e_r e_s = -e_s e_r$$
$$e_2 e_3 = -e_3 e_2$$
$$e_2 e_5 = -e_5 e_2$$
$$e_3 e_5 = -e_5 e_3 \tag{4.3}$$

4.1.4   Reversion

The reversion operation involves the reversal of the order of all primal units. Figure 4.4 depicts on the left-hand side, the reversion operator applied to three primal units. The reversion is achieved simply, without any calculation, by writing the three primal units in the opposite order. Pair exchange working down the page is then required to re-establish the order from low to high.

In this case, three exchanges are used. Working back up the page from the bottom right to the top with the corresponding three changes of sign produces the final result at the top right. In this case, the final result has the opposite sign to the initial value.

Figure 4.4   Reversion.

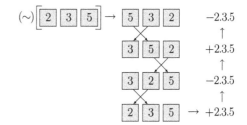

The operation of reversion is written in the form of Clifford algebra in Eq. (4.4). Again, the symbol k takes the value of the grade of the unit to which the operation is applied.

$$a = e_2 e_3 e_5$$
$$\tilde{a} = (-1)^{k(k-1)/2} \, a$$
$$= (-1)^{3(2)/2} \, e_2 e_3 e_5$$
$$= (-1)^3 \, e_2 e_3 e_5$$
$$= -e_2 e_3 e_5 \tag{4.4}$$

Applying the reversion operator twice reproduces the original number.

4.1.5 Clifford Conjugation

The operation of Clifford conjugation, as shown in Eq. (4.5), is the combination of both involution and reversion. The result is the same regardless of the order of those operations.

$$a = e_2 e_3 e_5$$
$$\bar{a} = \tilde{\overset{\rightharpoonup}{a}} = \overset{\rightharpoonup}{\tilde{a}} = (-1)^{k(k+1)/2} \, a$$
$$= (-1)^{3(4)/2} \, e_2 e_3 e_5$$
$$= (-1)^6 \, e_2 e_3 e_5$$
$$= e_2 e_3 e_5 \tag{4.5}$$

In this case, the result has the same sign as the initial value.

Applying the Clifford conjugation operator twice reproduces the original number.

4.1.6 Supplementation and Pseudo-scalar

The operation of supplementation involves replacing from the entire set of primal units for any particular dimension, those present in a unit by those which are missing. Figure 4.5 depicts the operation of supplementation in three dimensions where two primal units indicated by solid blocks are present, and one indicated by the dashed block is missing. The number of dimensions is indicated by the size of the container, which is large enough to hold a maximum of three blocks.

The lower part of the figure depicts the result of the operation, being everything which is missing from the upper part of the figure. If the number of dimensions was four instead of three, the containers would need to be big enough to contain four blocks. In that case if the upper container contained two blocks, the result would also contain two blocks ($2 + 2 = 4$).

Performing the operation of supplementation in Clifford algebra of dimension n requires use of the corresponding pseudo-scalar e_N, to form the product shown in the bottom row

Figure 4.5 Supplement.

Figure 4.6   Pseudo-scalar e_N for dimension $n = 3$.

of Table 4.1. The pseudo-scalar can be envisaged as a container filled with all of the primal units for that dimension, as in Figure 4.6 for dimension $n = 3$.

The big N is a symbolic representation of the entire set of primal units $\{0, 1, 2, \ldots, n-1\}$, pictorially represented as the container filled completely with the small n number of blocks.

The pseudo-scalar is written in terms of Clifford algebra in Eq. (4.6), as the product obtained by central multiplication of all n primal units, left to right from lowest to highest:

$$e_N = e_0 e_1 e_2 \ldots e_{n-1} \tag{4.6}$$

Using the pseudo-scalar, reversion, and central multiplication, the operation of supplementation for Figure 4.5 is written in the form of Clifford algebra as follows:

$$a = e_0 e_1$$
$$* a = \tilde{a} e_N$$
$$= (e_1 e_0)(e_0 e_1 e_2)$$
$$= e_0^2 e_1^2 e_2$$
$$= e_2 \tag{4.7}$$

The Clifford squares of e_0 and e_1 can be ± 1 depending on choice of signature. In this case, the result is the same regardless of the signature because there are an even number of squares, and it is assumed that the signatures are all uniform, i.e. all $+1$ or all -1. That is not always the case.

Applying the Hodge star (supplementation) operator twice does not necessarily reproduce the original number. Applying four times does.

It is worthwhile noting that calculating the supplement (or Hodge star) within the framework of Clifford algebra is perfectly straightforward.

4.2   Practice

4.2.1   Example Code

The code in Example 4.1a demonstrates the use of the unary mathematical operations. A two-dimensional number context with all negative signature is established on line 8. A test number is created in lines 11–14 and printed on line 24. The different unary operations are invoked on lines 16–22. The results of those operations are printed on lines 25–38.

Example 4.1(a)

```
1:   void demo13() /* unary arithmetic operators */
2:   {
3:       int lambda[2];
4:       number u,n,c,r,i,x,s,h,h2,h3,h4;
```

(Continued)

Example 4.1(a) (Continued)

```
5:
6:      recovery_level(0,0,0);
7:      create_cliffmem(50);
8:      declare_primal(2,pack_signature(2,lambda,-1));
9:      format_printing("%.0f");
10:
11:     u=scalar_number(1.0,2.0);
12:     u=addto_number(u,simple_number(3.0,4.0,spack_unit("e{0}")));
13:     u=addto_number(u,simple_number(5.0,6.0,spack_unit("e{1}")));
14:     u=addto_number(u,pseudo_number(7.0,8.0));
15:
16:     n=negativeof_number(u,0);
17:     i=involutionof_number(u,0);
18:     r=reversionof_number(u,0);
19:     x=cliffordconjugateof_number(u,0);
20:     c=complexconjugateof_number(u,0);
21:     s=supplementof_number(u,0);
22:     h=hodgestarof_number(u,0);
23:
24:     printf(" N  = ");                    print_number(u,8);
                                             recycle_number(u);
25:     printf("-N  = negative of N\n");
26:     printf("    = ");                    print_number(n,8);
                                             recycle_number(n);
27:     printf("N^\" = involution of N\n");
28:     printf("    = ");                    print_number(i,8);
                                             recycle_number(i);
29:     printf("N^~ = reversion of N\n");
30:     printf("    = ");                    print_number(r,8);
                                             recycle_number(r);
31:     printf("N^- = Clifford conjugate of N\n");
32:     printf("    = ");                    print_number(x,8);
                                             recycle_number(x);
33:     printf("N^* = complex conjugate of N\n");
34:     printf("    = ");                    print_number(c,8);
                                             recycle_number(c);
35:     printf("|N  = supplement of N\n");
36:     printf("    = ");                    print_number(s,8);
                                             recycle_number(s);
37:     printf("*N  = Hodge star of N\n");
38:     printf("    = ");                    print_number(h,8);
39:
40:     h2=hodgestarof_number(h,1);
41:     h3=hodgestarof_number(h2,0);
42:     h4=hodgestarof_number(h3,0);
43:
44:     printf("**N = double Hodge star of N\n");
45:     printf("    = ");                    print_number(h2,8);
                                             recycle_number(h2);
46:     printf("***N = triple Hodge star of N\n");
47:     printf("    = ");                    print_number(h3,8);
                                             recycle_number(h3);
48:     printf("****N = quadruple Hodge star of N\n");
```

```
49:     printf("      = ");              print_number(h4,8);
                                         recycle_number(h4);
50:
51:     free_cliffmem(0);
52: }
```

The Hodge star operator (supplement) is applied three more times in lines 40–42, and the results are printed on lines 44–49.

4.2.2 Example Output

The results of running the code in Example 4.1a are listed in Example 4.1b on lines 2–22. Comparison to line 2 shows that negation changes the sign of all components, and involution changes components in only odd numbered grades. Reversion changes the sign on grades which give remainders of 2 or 3 on integer division by 4, and Clifford conjugation changes the sign on grades which give remainders of 1 or 2 on integer division by 4. The Hodge star and supplement operations are identical and give exactly the same results. Notice with the supplement (Hodge star), the original coefficients are now running from back to front instead of front to back.

All of the operations except for the Hodge star (supplement) reproduce the original number if applied twice. The output on lines 17–22 verify that it is necessary to apply the Hodge star four times before reproducing the original number.

Example 4.1(b)

```
1:   demo13()
2:    N   = (1+i2)e{}+(3+i4)e{0}+(5+i6)e{1}+(7+i8)e{0,1}
3:   -N   = negative of N
4:        = (-1-i2)e{}+(-3-i4)e{0}+(-5-i6)e{1}+(-7-i8)e{0,1}
5:    N^" = involution of N
6:        = (1+i2)e{}+(-3-i4)e{0}+(-5-i6)e{1}+(7+i8)e{0,1}
7:    N^~ = reversion of N
8:        = (1+i2)e{}+(3+i4)e{0}+(5+i6)e{1}+(-7-i8)e{0,1}
9:    N^- = Clifford conjugate of N
10:       = (1+i2)e{}+(-3-i4)e{0}+(-5-i6)e{1}+(-7-i8)e{0,1}
11:   N^* = complex conjugate of N
12:       = (1-i2)e{}+(3-i4)e{0}+(5-i6)e{1}+(7-i8)e{0,1}
13:   |N  = supplement of N
14:       = (7+i8)e{}+(5+i6)e{0}+(-3-i4)e{1}+(1+i2)e{0,1}
15:   *N  = Hodge star of N
16:       = (7+i8)e{}+(5+i6)e{0}+(-3-i4)e{1}+(1+i2)e{0,1}
17:   **N = double Hodge star of N
18:       = (1+i2)e{}+(-3-i4)e{0}+(-5-i6)e{1}+(7+i8)e{0,1}
19:  ***N = triple Hodge star of N
20:       = (7+i8)e{}+(-5-i6)e{0}+(3+i4)e{1}+(1+i2)e{0,1}
21: ****N = quadruple Hodge star of N
22:       = (1+i2)e{}+(3+i4)e{0}+(5+i6)e{1}+(7+i8)e{0,1}
```

5

Binary Operations

Eight elementary binary operations, as listed in the middle section of Table 5.1, are provided within the Clifford numerical suite. These are not all independent, and some can be defined in terms of others. The scalar multiplication returns the complex coefficient of the inner multiplication, with no unit $e_\emptyset$ attached. The interior derivative is a special case of left interior multiplication when the multiplier is a vector (i.e. in grade 1).

5.1 Geometric Origins

5.1.1 Outer Multiplication

Outer multiplication is an operation which calculates the value produced by an abstract process Grassmann called evolution, when applied in a geometric context. For the evolution of two vectors a, b in Figure 5.1, the vector a is translated in the direction of the vector b.

Grassmann interpreted the process of evolution as generating the area swept over by vector a during the act of the process. The result of the evolution is the two-dimensional entity, shaded in grey in Figure 5.1, known as a bivector. Whereas a vector plays the geometric role of an oriented line, a bivector plays the geometric role of an oriented area.

The evolution of vectors by other vectors produces bivectors. From lines, evolution produces areas. Similarly, the evolution of a bivector by a vector produces a trivector, an element of volume created by sweeping the bivector (area) through space in the direction of the vector. Grassmann considered higher-dimensional entities as evolved from lower-dimensional entities. Areas are evolved from line and line, volumes are evolved from area and line.

The numerical value produced by the process of evolution, as calculated using outer multiplication, is given for vectors in Eq. (5.1).

$$\begin{array}{ll} \text{Parallelogram} & \text{Rectangle} \\ a \wedge b = |a||b| \sin\theta & = |a||b| \sin 90° \\ & = |a||b| \end{array} \tag{5.1}$$

If the angle between the vectors is zero, as in Figure 5.2, then the area generated is zero, and from Eq. (5.1), the numerical value in Eq. (5.2) is zero.

$$a \wedge a = |a||a| \sin 0° = 0 \tag{5.2}$$

Numerical Calculations in Clifford Algebra: A Practical Guide for Engineers and Scientists,
First Edition. Andrew Seagar.
© 2023 John Wiley & Sons Ltd. Published 2023 by John Wiley & Sons Ltd.

Table 5.1   Elementary binary operations.

Case	Binary operation	Notation
General	Addition	$a + b$
	Subtraction	$a - b$
	Outer multiplication	$a \wedge b$
	Inner multiplication	$(a, b) e_\emptyset$
	Regressive multiplication	$a \vee b$
	Central multiplication	$a\, b$
	Left interior multiplication	$a \lrcorner b$
	Right interior multiplication	$a \llcorner b$
Special	Scalar multiplication	(a, b)
	Interior derivative	$i_a(b) \equiv a \lrcorner b$

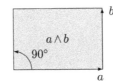

Figure 5.1   Outer multiplication of unequal vectors.

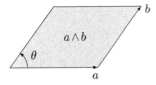

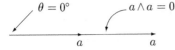

Figure 5.2   Outer multiplication of equal vectors.

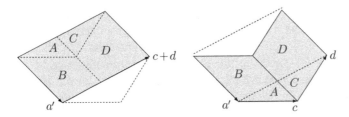

Figure 5.3   Left to right distribution of outer multiplication over addition.

Figure 5.3 demonstrates the distributive property of outer multiplication over addition. The left side of the figure shows, as the shaded parallelogram $A + B + C + D$, the product from multiplying vector a' with the vector that is the result of adding two vectors $c + d$. The right-hand side of the figure shows the sum of the products from multiplying vector a' with each of vectors c and d separately as the two shaded parallelograms $A + B$ and $C + D$. The results on the left-hand side and right-hand side are both composed of the sum of the same four areas A, B, C, D and therefore equal.

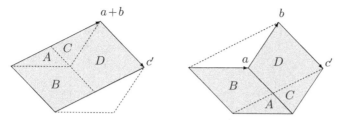

Figure 5.4   Right to left distribution of outer multiplication over addition.

The distributive property of outer multiplication over addition is expressed algebraically in Eq. (5.3). In this case, the distribution is from the left towards the right.

$$a' \wedge (c + d) = (A + B + C + D)$$
$$= (A + B) + (C + D) = (a' \wedge c) + (a' \wedge d) \tag{5.3}$$

Figure 5.4 demonstrates the distributive property of outer multiplication over addition again. This time the left side of the figure shows, as the shaded parallelogram $A + B + C + D$, the product from multiplying the vector that is the result of adding two vectors $a + b$, with the vector c'. The right-hand side of the figure shows the sum of the products from multiplying vectors a and b each separately with vector c' as the two shaded parallelograms $A + B$ and $C + D$. The results on the left-hand side and right-hand side are both composed of the sum of the same four areas A, B, C, D and therefore equal.

In this case, the distribution is from the right towards the left as expressed algebraically in Eq. (5.4).

$$(a + b) \wedge c' = (A + B + C + D)$$
$$= (A + B) + (C + D) = (a \wedge c') + (b \wedge c') \tag{5.4}$$

Substitution of $c+d$ for c' in Eq. (5.4) and then utilising Eq. (5.3) gives Eq. (5.5), which uses the distributive property from both left and right at the same time.

$$(a + b) \wedge (c + d) = a \wedge (c + d) + b \wedge (c + d)$$
$$= (a \wedge c) + (a \wedge d) + (b \wedge c) + (b \wedge d) \tag{5.5}$$

Figure 5.5 demonstrates the anti-commutativity of outer multiplication. The two areas formed by multiplying two vectors in opposite orders induce the circulations shown by the curved arrows[1]. The direction of the circulation is the same as that of tracing the parallelogram around its sides following the direction of the vectors from which it is formed. The two circulations are of opposite sign.

Figure 5.5   Anti-commutativity of outer multiplication.

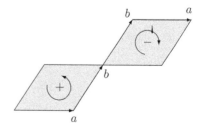

---

1 In two dimensions sign of the circulation is the same as the sign of the oriented area.

The result of performing the outer multiplication on the long diagonals, $a + b$ and $b + a$, of the two parallelograms in Figure 5.5 is zero because the two diagonals are the same in length and direction, as in Figure 5.2. The first line of Eq. (5.6) expresses that result in the form of algebra:

$$(a + b) \wedge (b + a) = 0$$
$$a \wedge b + a \wedge a + b \wedge b + b \wedge a = 0$$
$$a \wedge b + 0 + 0 + b \wedge a = 0 \qquad (5.6)$$
$$a \wedge b + b \wedge a = 0$$
$$b \wedge a = -a \wedge b$$

The second line follows from the distributive property, Eq. (5.5), and the third line follows from Figure 5.2. Both fourth and final lines state algebraically that outer multiplication of vectors is anti-commutative.

5.1.2 Orthogonal Components

It is often convenient to express vectors in terms of orthogonal components. Figure 5.6 shows in the top left corner, as the shaded parallelogram, the outer multiplication of two vectors a and b. The parallelogram is cut along lines parallel to the horizontal (x) and vertical (y) directions into four pieces A, B, C, D, which are rearranged in the top right corner of the figure. The area is re-interpreted in the bottom left as the remainder after cutting a small rectangle Q out of a larger rectangle P. The small rectangle is then moved away to expose the large rectangle fully.

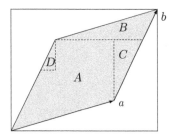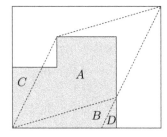

$$A + B + C + D = P - Q$$

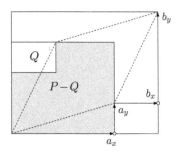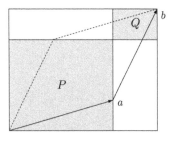

Figure 5.6 Outer multiplication using orthogonal components.

The outer multiplication for the two vectors a and b is calculated geometrically in terms of the difference in area of the large and small rectangle in the left-hand column of Eq. (5.7). The areas of those rectangles are given by the outer multiplication of the vectors which define their sides, being simply the orthogonal components of the vectors a and b.

$$
\begin{array}{ll}
\text{Geometry} & \text{Algebra} \\[4pt]
a \wedge b = P - Q & a \wedge b = (a_x + a_y) \wedge (b_x + b_y) \\[4pt]
\quad = a_x \wedge b_y - b_x \wedge a_y & \quad = a_x \wedge b_x + a_x \wedge b_y + a_y \wedge b_x + a_y \wedge b_y \qquad (5.7)\\[4pt]
& \quad = a_x \wedge b_y + a_y \wedge b_x \\[4pt]
& \quad = a_x \wedge b_y - b_x \wedge a_y
\end{array}
$$

The outer multiplication for the two vectors a and b is calculated algebraically in the right-hand column of Eq. (5.7) in terms of the distributive property of multiplication over addition from Eq. (5.5), appealing to Figure 5.2 for the product when two vectors are in the same direction and appealing to Figure 5.5 for the anti-commutativity of two vectors. The algebraic result reduces to the same value as the geometric result.

5.1.3 Inner Multiplication

Figure 5.7 shows for two dimensions, the outer multiplication of two vectors $a{\wedge}b$ and also the outer multiplication of the first vector with the supplement of the second $a{\wedge}{*}b$.

The outer multiplication of the first vector with the supplement of the second gives the product from the inner multiplication with pseudo-scalar as unit, as in Table 5.3. In two dimensions the pseudo-scalar is a bivector, so the result is an oriented area. As drawn, the supplement of the second vector is 90° anti-clockwise from the original, and the circulation of the product is in the anti-clockwise direction, as shown by the curved arrow.

From the areas of the shaded parallelograms in Figure 5.7, the numerical values produced by the outer and inner multiplications are related to the magnitude of the corresponding pairs of vectors and the angles between them, as given in Eq. (5.8). Whereas the outer multiplication is proportional to the sine of the angle, the inner multiplication is proportional to the cosine of the angle.

$$
\begin{cases}
\text{outer: } a \wedge b = |a||b| \sin \theta \\
\text{inner: } a \wedge {*}b = |a||{*}b| \sin(\theta + 90°) = |a||b| \cos \theta
\end{cases} \qquad (5.8)
$$

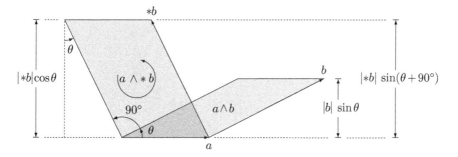

Figure 5.7 Outer and inner multiplication of vectors.

The case depicted in Figure 5.7 and Eq. (5.8) is for positive values of signature (see Sections 5.3.1 and 5.3.2). If the values of the signature are negative, the supplement of the second vector is in the opposite direction, the product of the multiplication is a parallelogram angled towards the right-hand side below the horizontal axis, and the circulation of the product is in the clockwise direction.

For example[2], if $a = 3e_0$ and $b = 4e_0 + 3e_1$, then the inner multiplication (with pseudo-scalar unit) of a and b is

$$
\begin{aligned}
a \wedge *b &= 3e_0 \wedge * (4e_0 + 3e_1) &&= 3e_0 \wedge (4\widetilde{e_0} + 3\widetilde{e_1})e_0 e_1 \\
&= 3e_0 \wedge (4e_0 e_0 e_1 + 3e_1 e_0 e_1) &&= 3e_0 \wedge (4e_1 e_0 e_0 - 3e_0 e_1 e_1) \\
&= 3e_0 \wedge (4e_1 \lambda - 3e_0 \lambda) &&= 3e_0 \wedge (4e_1 - 3e_0)\lambda &&= 12\lambda e_0 \wedge e_1
\end{aligned} \tag{5.9}
$$

The result is a bivector with positive (anti-clockwise) circulation if the signature λ of both primal units is $+1$ and a bivector with negative (clockwise) circulation if the signature is -1.

5.1.4 Names

Grassmann (1995, 1855) chose the names outer and inner multiplication based on the geometrical behaviour of their products for two vectors a and b as in Figure 5.8. Vector b_1 is the projection of a onto b, and c_1 is the projection of a onto c. Vectors b and c are perpendicular, and all three vectors are unit length.

Equation (5.10) asserts that the value of the product from the multiplication '$\wedge*$' depends only on b_1, the orthogonal component of a which is inside b,

$$
a \wedge *b = (b_1 + c_1) \wedge *b = b_1 \wedge *b \tag{5.10}
$$

and similarly, Eq. (5.11) asserts that the value of the product from the multiplication '$\wedge$' depends only on c_1, the orthogonal component of a which is outside b.

$$
a \wedge b = (b_1 + c_1) \wedge b = c_1 \wedge b \tag{5.11}
$$

If the assertions are true, there is justification to name the first of these multiplications *inner* and the second *outer*.

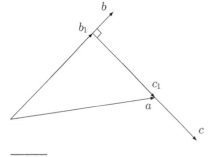

Figure 5.8 Geometric behaviour of inner and outer multiplication for vectors.

---

2 In doing this, the definition of the supplement in Clifford algebra from Table 5.3 and the rules for central (Clifford) multiplication of primal units in Eq. (5.19) are used. Calculation of the supplement using other methods is very clumsy. Central multiplication, which is used when there is no intervening symbol between factors, is explained in Sections 5.2.2, 5.2.3, 5.3.1, and 5.3.2.

One easy way to verify (not prove) the assertions is by using numerical values for the vectors drawn in Figure 5.8: $a = \frac{1}{\sqrt{50}}(7e_0 + 1e_1)$, $b = \frac{1}{\sqrt{50}}(5e_0 + 5e_1)$, and $c = \frac{1}{\sqrt{50}}(5e_0 - 5e_1)$. The components of vector a in the directions of b and c can be calculated using the projection operators in Table 6.1. Doing so produces the values $b_1 = \frac{1}{\sqrt{50}}(4e_0 + 4e_1)$ and $c_1 = \frac{1}{\sqrt{50}}(3e_0 - 3e_1)$, which on inspection are clearly parallel to b and c, respectively.

Substituting the numerical values into the left-hand side of Eq. (5.11) gives

$$a \wedge b = \frac{1}{50}(7e_0 + 1e_1) \wedge (5e_0 + 5e_1) = \frac{1}{50}(35e_0 \wedge e_1 + 5e_1 \wedge e_0)$$

$$= \frac{1}{50}30e_0 \wedge e_1 = \frac{3}{5}e_0 \wedge e_1 \tag{5.12}$$

and for the right-hand side:

$$c_1 \wedge b = \frac{1}{50}(3e_0 - 3e_1) \wedge (5e_0 + 5e_1) = \frac{1}{50}(15e_0 \wedge e_1 - 15e_1 \wedge e_0)$$

$$= \frac{1}{50}30e_0 \wedge e_1 = \frac{3}{5}e_0 \wedge e_1 \tag{5.13}$$

Similarly, for the left-hand side of Eq. (5.10):

$$a \wedge *b = \frac{1}{50}(7e_0 + 1e_1) \wedge * (5e_0 + 5e_1) \qquad = \frac{1}{50}(7e_0 + 1e_1) \wedge (5\tilde{e}_0 + 5\tilde{e}_1)e_0e_1$$

$$= \frac{1}{50}(7e_0 + 1e_1) \wedge (5e_0e_0e_1 + 5e_1e_0e_1) \qquad = \frac{1}{50}(7e_0 + 1e_1) \wedge (5\lambda e_1 - 5\lambda e_0)$$

$$= \frac{1}{50}(35\lambda e_0 \wedge e_1 - 5\lambda e_1 \wedge e_0) \qquad = \frac{1}{50}40\lambda e_0 \wedge e_1 \qquad = \frac{4}{5}\lambda e_0 \wedge e_1 \tag{5.14}$$

and for the right-hand side:

$$b_1 \wedge *b = \frac{1}{50}(4e_0 + 4e_1) \wedge * (5e_0 + 5e_1) \qquad = \frac{1}{50}(4e_0 + 4e_1) \wedge (5\lambda e_1 - 5\lambda e_0)$$

$$= \frac{1}{50}(20\lambda e_0 \wedge e_1 - 20\lambda e_1 \wedge e_0) \qquad = \frac{1}{50}40\lambda e_0 \wedge e_1 \qquad = \frac{4}{5}\lambda e_0 \wedge e_1 \tag{5.15}$$

Comparison of the value of Eq. (5.12) with Eq. (5.13), and the value of Eq. (5.14) with Eq. (5.15) supports the view that the assertions in both Eqs. (5.11) and (5.10) are valid, and that the names chosen by Grassmann do properly reflect the geometrical behaviour of the corresponding operations on vectors.

5.2 Multiplication of Units

5.2.1 Progressive and Regressive Multiplication

Equation (5.16) defines Grassmann's regressive multiplication in algebraic form. Regressive multiplication ($\vee$) is the inverse supplement ($*^{-1}$) of the outer ($\wedge$) multiplication of the supplement ($*$) of the two factors x, y:

$$x \vee y = *^{-1}((*x) \wedge (*y)) \tag{5.16}$$

or alternatively[3]:

$$\begin{cases} *(x \vee y) = (*x) \wedge (*y) \\ *^{-1}(x \wedge y) = (*^{-1}x) \vee (*^{-1}y) \end{cases} \tag{5.17}$$

Progressive multiplication is identical to outer multiplication ($\wedge$); it is just a new name.

---

3 Grassmann gave the definition as in the first line of Eq. (5.17).

Unless the result is completely zero, the progressive (outer) multiplication produces (by evolution) entities of equal or higher dimension, and the regressive multiplication produces (by devolution) entities of equal or lower dimension. See Tables 5.7 and 5.8 for examples.

Figure 5.9 shows for three dimensions Grassmann's progressive and regressive multiplication in the pictorial form. The numbered blocks represent primal units, the orthogonal units for vectors. The boxes containing the blocks represent higher-dimensional units constructed by evolution from the primal units therein. An empty box represents the unit for the empty set, $e_\emptyset$, the unit of scalars. The size of the box in relation to the size of the blocks represents the number of dimensions. Box x and box y represent two factors combined under some operation which produces results in Boxes 1 and 2.

When there are no units in common between the two factors combined by the operations of progressive and regressive multiplication, all units can fit into the first box, and the second box is empty, as in Figure 5.9a,c. When there are units in common between the two factors, one is placed in each box, as in Figure 5.9b,d.

If the second box is empty, the first box holds the value of the progressive product, as in Figure 5.9a,c. If the first box is full, the second box holds the value of the regressive product, as in Figure 5.9c,d. Be careful to note that the second empty box in Figure 5.9c does not represent zero value. It represents an empty set, which is the unit for scalars, and may take a zero or non-zero[4] value.

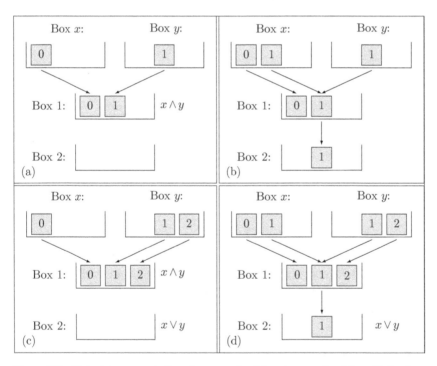

Figure 5.9 Pictorial representation of progressive ($\wedge$) and regressive ($\vee$) multiplication.

---

4 Grassmann excluded scalars from the basis of his algebra, so defined the result in this case to be zero. In this work, scalars are accommodated by accepting as result the value produced by Eq. (5.16) in grade 0.

Table 5.2 Rules for progressive (∧) and regressive (∨) multiplication.

		Box 2	
		Empty	**Not empty**
Box 1	Not full (a)	$x \wedge y = $ box 1 $x \vee y = $ zero	(b) $\quad x \wedge y = $ zero $x \vee y = $ zero
	Full (c)	$x \wedge y = $ box 1 $x \vee y = $ box 2	(d) $\quad x \wedge y = $ zero $x \vee y = $ box 2

The rules for progressive and regressive multiplication of units are summarised in Table 5.2.

5.2.2 Outer, Inner, and Central Multiplication

Central multiplication of two numbers p and q is written as pq with no intervening symbol. Central multiplication of two units pq and qr, which are evolved from primal units and which have a common factor q, can be expanded as inner and outer multiplications. The terms p, r which are not in common are combined by outer multiplication $p \wedge r$. The terms which are in common are combined by inner multiplication $(q, \tilde{q})e_\emptyset$ with reversion of one of the factors. Finally, the two products are multiplied together:

$$(pq)(qr) = p \wedge r \,(q, \tilde{q})e_\emptyset \tag{5.18}$$

The product of the inner multiplication is a scalar (in grade 0), so the final multiplication of the two products is the usual multiplication by a scalar as in Section 5.2.3.

Figure 5.10 shows for three dimensions Grassmann's central multiplication in the pictorial form. When there are no units in common between the two factors, all units can fit into the first box, and the second box is empty. Comparison of Figure 5.10a,c with the corresponding case for progressive and regressive multiplications in Figure 5.9a,c shows that when there are no units in common, central multiplication plays the role of outer multiplication.

When there are units in common between the two factors both units are placed in the second box, as in Figure 5.10b,d. This differs from the corresponding case for progressive and regressive multiplications in Figure 5.9b,d, where one unit is placed in each box.

For central multiplication, the units in the first box are multiplied using outer multiplication. The units in the second box are taken as two equal factors, the reversion operator is applied to one, and then they are multiplied using inner multiplication. The inner multiplication produces a scalar in grade 0. Finally, the first box is multiplied by the scalar from the second box.

When all units are in common, the first box is empty, and central multiplication plays the role of inner multiplication (with the reversion of one factor).

The rules for central multiplication of units are summarised in Table 5.3.

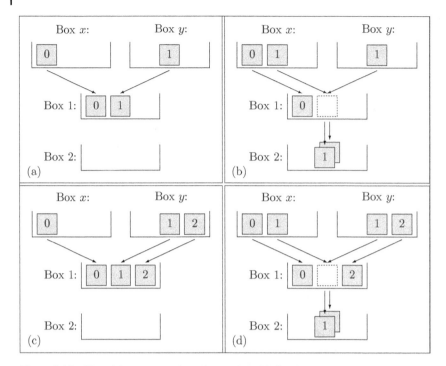

Figure 5.10 Pictorial representation of central multiplication.

Table 5.3 Rules for central multiplication between units.

			Box 2		
			Empty		**Not empty**
Box 1	Not full	(a)	$xy = \text{box } 1 \times \text{box } 2$ $pr = (p \wedge r)e_\emptyset$ $e_0 e_1 = (e_0 \wedge e_1)e_\emptyset$	(b)	$xy = \text{box } 1 \times \text{box } 2$ $(pq)q = p\,(q, \tilde{q})e_\emptyset$ $(e_0 e_1)e_1 = e_0\,(e_1, \tilde{e_1})e_\emptyset$
	Full	(c)	$xy = \text{box } 1 \times \text{box } 2$ $p(qr) = (p \wedge q \wedge r)e_\emptyset$ $e_0(e_1 e_2) = (e_0 \wedge e_1 \wedge e_2)e_\emptyset$	(d)	$xy = \text{box } 1 \times \text{box } 2$ $(pq)(qr) = (p \wedge r)\,(q, \tilde{q})e_\emptyset$ $(e_0 e_1)(e_1 e_2) = (e_0 \wedge e_2)(e_1, \tilde{e_1})e_\emptyset$

5.2.3 Multiplication By Scalars

Within Clifford algebra, complex scalars are accommodated in grade 0, with unit[5] $e_\emptyset$. The unit $e_\emptyset$ is not a primal unit because it plays no role in the evolution of higher-dimensional entities from ones of lower dimension. Any other unit multiplied by $e_\emptyset$ alone remains unchanged. However, if a non-unit value of coefficient is attached to the unit $e_\emptyset$ that will affect the value of the coefficient of the product.

---

5 Grassmann called the unit scalar '1', without the $e_\emptyset$ attached, the absolute unit.

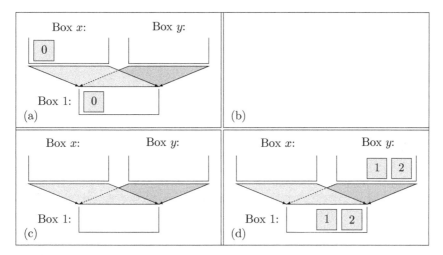

Figure 5.11 Pictorial representation of multiplication by a scalar.

Table 5.4 Rules for multiplication by a scalar.

		Box y	
		Empty	Not empty
Box x	Not empty	(a) $pe_\emptyset = p$ $xy = \text{box 1}$ $e_0 e_\emptyset = e_0$	(b) Not multiplication by scalar
	Empty	(c) $e_\emptyset e_\emptyset = e_\emptyset$ $xy = \text{box 1}$	(d) $xy = \text{box 1}$ $e_\emptyset q = q$ $e_\emptyset(e_1 e_2) = e_1 e_2$

Figure 5.11 shows for three dimensions multiplication by a scalar[6] within Clifford algebra. One of the factors is represented by an empty box, standing for the unit of scalars, the empty set unit $e_\emptyset$. The result in box 1 contains whatever units are in the other factor.

The rules for multiplication by scalars are summarised in Table 5.4.

5.3 Central Multiplication

With central multiplication, some units are multiplied by outer multiplication and other units are multiplied by inner multiplication. If outer and inner multiplication are two extremes, as at the opposite ends of a bridge, then central multiplication may be thought of as the central span, forming a connection as in Figure 5.12 between the others, and encompassing the properties of both. As the vector b sweeps from left to right, the behaviour of

---

6 Multiplication by a scalar, in which one of the factors is a scalar, should not be confused with scalar multiplication, for which the result is a scalar.

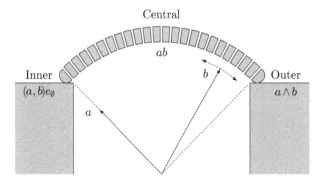

Figure 5.12 Multiplication of fixed vector *a* and variable vector *b* supported as inner[7] and outer multiplication at two ends of central 90° span.

the central multiplication varies continuously from more like inner to more like outer. In some circumstances, central multiplication behaves like outer multiplication, and in some circumstances, it behaves like inner multiplication, and in other circumstances, it may be thought of as doing something in between.

5.3.1 Primal Units

The rules for central multiplication amongst scalars and primal units can be expressed algebraically in the form of the equations:

$$\begin{cases} e_\emptyset e_\emptyset = & e_\emptyset \\ e_\emptyset e_q = & e_q \\ e_p e_\emptyset = & e_p \\ e_p e_p = & \lambda_p e_\emptyset \\ e_p e_q = & -e_q e_p \end{cases} \tag{5.19}$$

The rules for the first three lines come from Figure 5.11. The rule in the final line comes from Figure 5.5. Although in Eq. (5.19), the multiplication is central and in Figure 5.5, it is outer, here the two are the same because the units e_p and e_q are vectors which are both primal and orthogonal. That means Figure 5.10a and Table 5.3a are in effect. Note in the final line of Eq. (5.19) that the multiplication of primal units is anti-commutative. Commutativity does not apply in Clifford algebra.

The rule in line 1 was used by Grassmann, with the unit scalar 1 in place of the unit $e_\emptyset$ of grade 0, to define the unit scalar as the absolute unit. Here in the Clifford context, by virtue of the rule in line 1, it is $1e_\emptyset = e_\emptyset$ which plays the role of absolute unit.

The rule in line 4 defines the product from multiplying a primal unit with itself as a scalar value $\lambda_p e_\emptyset$. The integer coefficient λ_p is known as the signature. The signature can be ± 1 or zero. The value of the signature for one primal unit is not necessarily the same as the value for another.

More generally, the signature is the scalar coefficient for the inner multiplication of the unit with itself $\lambda_p e_\emptyset = (e_p, e_p)e_\emptyset = e_p \tilde{e}_p$. Here the units are primal and applying the reversion operator has no effect so that $e_p \tilde{e}_p = e_p e_p$ as in line 4.

---

7 For vectors $b = \tilde{b}$, so inner multiplication $(a, b)e_\emptyset = (a, \tilde{b})e_\emptyset$, as for the scalar part of central multiplication.

Table 5.5 Rules for central multiplication xy between primal units and scalars.

xy		y	
	$e_\emptyset$	e_p	e_q
$e_\emptyset$	$e_\emptyset$	e_p	e_q
$x \quad e_p$	e_p	$\lambda_p e_\emptyset$	$e_p e_q$
e_q	e_q	$-e_p e_q$	$\lambda_q e_\emptyset$

All elementary operations within the Clifford numerical suite are supported for any signatures, zero, positive, or negative, or any combination of those. In many applications, it is possible to adopt the same value of signature for all primal units. In that case, it is possible to embed the signature into the rules and make them simpler (although different for opposite choice of signatures). Here no choices are made, and the signature is maintained explicitly, in plain sight.

The rules for central multiplication of primal units and scalars are summarised in Table 5.5.

5.3.2 Evolved and Other Units

Table 5.6 defines all seven types of multiplication in Table 5.1 in terms of central (Clifford) multiplication for two factors x, y. The factors are any scalar, primal, or

Table 5.6 Multiplication of evolved and other Clifford units x and y.

			Evolved and other units									
	x		$e_\emptyset$	$e_\emptyset$	$e_\emptyset$	e_A	e_A	e_A	e_B	e_B	e_B	
	y		$e_\emptyset$	e_A	e_B	$e_\emptyset$	e_A	e_B	$e_\emptyset$	e_A	e_B	
Multiplication	$x \diamond y$	Constraints	*If specified constraints are honoured then*: value is below									*Else*:
Outer (wedge)	$x \wedge y$	$A \cap B = \emptyset$	0	e_A	e_B	e_A	0	$e_A e_B$	e_B	$e_B e_A$	0	0
Inner (grade 0)	$(x,y)e_\emptyset$	$A = B$	$e_\emptyset$	0	0	0	$\lambda_A e_\emptyset$	0	0	0	$\lambda_B e_\emptyset$	0
Regressive	$x \vee y$	$A \cup B = N$	0	0	0	0	$e_A \widetilde{e}_N e_A$	$e_B \widetilde{e}_N e_A$	0	$e_A \widetilde{e}_N e_B$	$e_B \widetilde{e}_N e_B$	0
Central (Clifford)	xy	Always	$e_\emptyset$	e_A	e_B	e_A	$e_A e_A$	$e_A e_B$	e_B	$e_B e_A$	$e_B e_B$	Never
Left interior	$x \lrcorner y$	$A \subseteq B$	$e_\emptyset$	e_A	e_B	0	$\lambda_A e_\emptyset$	$\widetilde{e}_A e_B$	0	0	$\lambda_B e_\emptyset$	0
Right interior	$x \llcorner y$	$A \subseteq B$	$e_\emptyset$	0	0	e_A	$\lambda_A e_\emptyset$	0	e_B	$e_B \widetilde{e}_A$	$\lambda_B e_\emptyset$	0
Inner (scalar)	(x,y)	$A = B$	1	0	0	0	λ_A	0	0	0	λ_B	0

evolved units. The products of the multiplication are zero unless the constraints in the second column of Table 5.6 are honoured for the sets of integers A, B which identify the primal units from which the factors are composed. The pseudo-scalar in n dimensions $e_N = e_0 e_1 e_2 \ldots e_{n-1}$ is represented in the table by the set $N = \{0, 1, 2, \ldots, n-1\}$.

The signature of primal units λ_p is generalised for evolved units $e_S = e_{s_1} e_{s_2} \ldots e_{s_m}$ as the product from multiplying all of the signatures of the primal units:

$$\lambda_S = \prod_{k=1}^{m} \lambda_{s_k} \tag{5.20}$$

The value of the generalised signature appears as the coefficient of the product from the multiplication of an evolved (or any) unit with its own reversion. This is identical to the product calculated by inner multiplication and is indeed the way the inner product is most easily calculated:

$$\lambda_S e_\emptyset = \tilde{e}_S e_S = e_S \tilde{e}_S = (e_S, e_S) e_\emptyset \tag{5.21}$$

As a consequence, the signature $\lambda_\emptyset$ for the unit of scalars $e_\emptyset$ is always $+1$.

5.3.3 Numbers

Clifford algebra is not commutative. However, it is associative[8] and does support the distributive property[9] of Clifford (central) multiplication over addition. Using the distributive property, the central multiplication of two Clifford numbers:

$$\begin{cases} x = \sum_i a_i e_{A_i} \\ y = \sum_j b_j e_{B_j} \end{cases} \tag{5.22}$$

can be effected by multiplying every component in the first number by every component in the second number, and then adding all of the results together:

$$xy = \sum_i \sum_j a_i b_j e_{A_i} e_{B_j} \tag{5.23}$$

The multiplication of the units e_{A_i}, e_{B_j} in Eq. (5.23) is calculated according to the rules for central multiplication in Table 5.6. It is important to keep the units from the first number ahead of those from the second number, because they are not commutative. Multiplication of the coefficients a_i, b_j is the same as for any complex numbers.

If, for example:

$$\begin{cases} x = 3e_0 + 4e_1 e_2 \\ y = 5e_1 + 6e_0 e_2 \end{cases} \tag{5.24}$$

---

8 Vector calculus is not associative.
9 Grassmann took the distributive property as a necessary condition for defining any kind of multiplication operation.

then,

$$xy = (3e_0 + 4e_1e_2)(5e_1 + 6e_0e_2)$$
$$= 3 \times 5e_0e_1 + 3 \times 6e_0e_0e_2 + 4 \times 5e_1e_2e_1 + 4 \times 6e_1e_2e_0e_2$$
$$= 15e_0e_1 + 18e_0e_0e_2 + 20e_1e_2e_1 + 24e_1e_2e_0e_2$$
$$= 15e_0e_1 + 18e_0e_0e_2 - 20e_1e_1e_2 + 24e_0e_1e_2e_2$$
$$= 15e_0e_1 + 18\lambda_0e_2 - 20\lambda_1e_2 + 24e_0e_1\lambda_2$$
$$= (15 + 24\lambda_2)e_0e_1 + (18\lambda_0 - 20\lambda_1)e_2 \tag{5.25}$$

The final value of the numerical result depends on the choice of the signatures $\lambda_0, \lambda_1, \lambda_2$.

Each of the other kinds of multiplication can be effected in a similar way by following the corresponding rules from Table 5.6. For example, outer multiplication is given by

$$x \wedge y = \sum_i \sum_j a_i b_j e_{A_i} \wedge e_{B_j} = \sum_i \sum_{A_i \cap B_j = 0} a_i b_j e_{A_i} e_{B_j} \tag{5.26}$$

and inner multiplication is given by

$$(x, y)e_\emptyset = \sum_i \sum_j a_i b_j (e_{A_i}, e_{B_j})e_\emptyset = \sum_i \sum_{A_i = B_j} a_i b_j e_{A_i} \widetilde{e_{B_j}} \tag{5.27}$$

The difference by comparison to Eq. (5.23) is the presence of the constraint equation from the second column in Table 5.6 limiting the scope of the inner summation, and the use of different expressions for the multiplication kernel from the body of Table 5.6, either reordering factors or introducing additional factors or operators.

5.4 Practice

5.4.1 Example Code

The code shown in Example 5.1a establishes a three-dimensional Clifford context, for Clifford arithmetic with all negative signature, from lines 7 to 10. Two numbers are set to prescribed values in lines 12–16 and printed on lines 27 and 28. The results produced by adding, subtracting, and multiplying the numbers using the first six elementary types of multiplication in Table 5.1 are calculated in lines 18–25. The results are printed from lines 30 to 43.

Example 5.1(a)

```
1:   void demo14 () /* binary arithmetic operators */
2:   {
3:       int lambda [3];
4:       cliff_comp y_comp;
5:       number u,v,a,s,i,c,o,l,r,m;
6:
7:       recovery_level (0,0,0);
8:       create_cliffmem (50);
```

(Continued)

Example 5.1(a) (Continued)

```
 9:     declare_primal(3,pack_signature(3,lambda,-1));
10:     format_printing("%.0f");
11:
12:     u=simple_number(1.0,2.0,spack_unit("e{0}"));
13:     u=merge_comp(u,pack_comp(&y_comp,3.0,4.0,spack_unit("e{1}")),
                    addition_to,identity);
14:     u=merge_comp(u,pack_comp(&y_comp,5.0,6.0,spack_unit("e{0,1}")),
                    addition_to,identity);
15:     v=simple_number(7.0,8.0,spack_unit("e{0}"));
16:     v=merge_comp(v,pack_comp(&y_comp,9.0,10.0,spack_unit("e{0,1}")),
                    addition_to,identity);
17:
18:     a=add_number(u,v,0);
19:     s=sub_number(u,v,0);
20:     o=outer_mul(u,v,0);
21:     i=inner_mul(u,v,0);
22:     m=regressive_mul(u,v,0);
23:     c=central_mul(u,v,0);
24:     l=left_mul(u,v,0);
25:     r=right_mul(u,v,0);
26:
27:     printf("    U    = ");    print_number(u,8);    recycle_number(u);
28:     printf("    V    = ");    print_number(v,8);    recycle_number(v);
29:
30:     printf("    U+V  = ");    print_number(a,8);    recycle_number(a);
31:     printf("    U-V  = ");    print_number(s,8);    recycle_number(s);
32:     printf("  U/\\V  = outer (wedge) multiplication\n          = ");
33:                              print_number(o,8);    recycle_number(o);
34:     printf("(U,V)e{} = inner multiplication\n          = ");
35:                              print_number(i,8);    recycle_number(i);
36:     printf("  U\\/V  = regressive multiplication\n          = ");
37:                              print_number(m,8);    recycle_number(m);
38:     printf("    UV   = central (Clifford)
                              multiplication\n          = ");
39:                              print_number(c,8);    recycle_number(c);
40:     printf("  U_|V   = left interior multiplication\n          = ");
41:                              print_number(l,8);    recycle_number(l);
42:     printf("  U|_V   = right interior multiplication\n          = ");
43:                              print_number(r,8);    recycle_number(r);
44:     free_cliffmem(0);
45:  }
```

Although the multiplication operations are binary, the list of arguments in the sub-routines which are used to invoke them takes three values. The value of the third argument indicates whether or not either of the preceding two arguments should be recycled as part of the calculation. This is useful to eliminate intermediate temporary results cleanly when nesting operations, as explained in Section 14.2.3. A value of zero, as used here, indicates that nothing should be recycled because the numbers are needed for later calculations and printing.

5.4.2 Example Output

The results of running the code listed in Example 5.1a are shown in Example 5.1b. It is straightforward to check the results of addition and subtraction in lines 4 and 5, but less so to check the products of the different types of multiplications in lines 6–17. In Chapter 17, the multiplications are verified using identities.

Example 5.1(b)

```
 1:   demo14()
 2:        U      = (1+i2)e{0}+(3+i4)e{1}+(5+i6)e{0,1}
 3:        V      = (7+i8)e{0}+(9+i10)e{0,1}
 4:       U+V     = (8+i10)e{0}+(3+i4)e{1}+(14+i16)e{0,1}
 5:       U-V     = (-6-i6)e{0}+(3+i4)e{1}+(-4-i4)e{0,1}
 6:       U/\V    = outer (wedge) multiplication
 7:               = (11-i52)e{0,1}
 8:     (U,V)e{}  = inner multiplication
 9:               = (-6+i82)e{}
10:       U\/V    = regressive multiplication
11:               = (0+i0)e{}
12:        UV     = central (Clifford) multiplication
13:               = (24-i126)e{}+(-13+i66)e{0}+(-2+i54)e{1}+(11-i52)e{0,1}
14:       U_|V    = left interior multiplication
15:               = (-6+i82)e{}+(-13+i66)e{0}+(11-i28)e{1}
16:       U|_V    = right interior multiplication
17:               = (-6+i82)e{}+(-13+i82)e{1}
```

5.4.3 Multiplication Tables

Tables 5.7 and 5.8 show the products resulting from multiplying all units in three dimensions using the first six elementary multiplication operators, for positive and negative signatures, respectively.

The table for central multiplication is full, whereas the tables for all other types of multiplication are entirely zero above or below the leading or trailing diagonal. In the case of inner multiplication, the table is zero both above and below the diagonal.

Figure 5.13 depicts as shaded each of the territories within the tables where the six types of multiplication produce their non-zero values. Although the territories of the five types of multiplication other than central together do cover the whole table, the coverage is not complete because their shaded territories do contain some zero values. As a consequence, central multiplication cannot be constructed using some combination of the others. However, the others can be constructed from central multiplication by discarding some values. Which values should be discarded is determined by the column of constraints in Table 5.6.

If the first factor in the multiplications is replaced by a vector D of differentials along each of a set of orthogonal axes, such as $D = \frac{\partial}{\partial x}e_0 + \frac{\partial}{\partial y}e_1 + \frac{\partial}{\partial z}e_2$ in three dimensions, then the six types of multiplication behave as six different kinds of differential operators. Clifford (central) multiplication, which covers the whole territory, plays the role of an

Table 5.7 Multiplication tables for positive signatures.

Outer		b							
	$a \wedge b$	$e_\emptyset$	e_0	e_1	e_0e_1	e_2	e_0e_2	e_1e_2	$e_0e_1e_2$
	$e_\emptyset$	$e_\emptyset$	e_0	e_1	e_0e_1	e_2	e_0e_2	e_1e_2	$e_0e_1e_2$
	e_0	e_0	0	e_0e_1	0	e_0e_2	0	$e_0e_1e_2$	0
	e_1	e_1	$-e_0e_1$	0	0	e_1e_2	$-e_0e_1e_2$	0	0
a	e_0e_1	e_0e_1	0	0	0	$e_0e_1e_2$	0	0	0
	e_2	e_2	$-e_0e_2$	$-e_1e_2$	$e_0e_1e_2$	0	0	0	0
	e_0e_2	e_0e_2	0	$-e_0e_1e_2$	0	0	0	0	0
	e_1e_2	e_1e_2	$e_0e_1e_2$	0	0	0	0	0	0
	$e_0e_1e_2$	$e_0e_1e_2$	0	0	0	0	0	0	0

Inner		b							
	$(a,b)e_\emptyset$	$e_\emptyset$	e_0	e_1	e_0e_1	e_2	e_0e_2	e_1e_2	$e_0e_1e_2$
	$e_\emptyset$	$e_\emptyset$	0	0	0	0	0	0	0
	e_0	0	$e_\emptyset$	0	0	0	0	0	0
	e_1	0	0	$e_\emptyset$	0	0	0	0	0
a	e_0e_1	0	0	0	$e_\emptyset$	0	0	0	0
	e_2	0	0	0	0	$e_\emptyset$	0	0	0
	e_0e_2	0	0	0	0	0	$e_\emptyset$	0	0
	e_1e_2	0	0	0	0	0	0	$e_\emptyset$	0
	$e_0e_1e_2$	0	0	0	0	0	0	0	$e_\emptyset$

Regressive		b							
	$a \vee b$	$e_\emptyset$	e_0	e_1	e_0e_1	e_2	e_0e_2	e_1e_2	$e_0e_1e_2$
	$e_\emptyset$	0	0	0	0	0	0	0	$e_\emptyset$
	e_0	0	0	0	0	0	0	$e_\emptyset$	e_0
	e_1	0	0	0	0	0	$-e_\emptyset$	0	e_1
a	e_0e_1	0	0	0	0	$e_\emptyset$	e_0	e_1	e_0e_1
	e_2	0	0	0	$e_\emptyset$	0	0	0	e_2
	e_0e_2	0	0	$-e_\emptyset$	$-e_0$	0	0	e_2	e_0e_2
	e_1e_2	0	$e_\emptyset$	0	$-e_1$	0	$-e_2$	0	e_1e_2
	$e_0e_1e_2$	$e_\emptyset$	e_0	e_1	e_0e_1	e_2	e_0e_2	e_1e_2	$e_0e_1e_2$

(Continued)

Table 5.7 (Continued)

Central					b				
	ab	$e_\emptyset$	e_0	e_1	e_0e_1	e_2	e_0e_2	e_1e_2	$e_0e_1e_2$
	$e_\emptyset$	$e_\emptyset$	e_0	e_1	e_0e_1	e_2	e_0e_2	e_1e_2	$e_0e_1e_2$
	e_0	e_0	$e_\emptyset$	e_0e_1	e_1	e_0e_2	e_2	$e_0e_1e_2$	e_1e_2
	e_1	e_1	$-e_0e_1$	$e_\emptyset$	$-e_0$	e_1e_2	$-e_0e_1e_2$	e_2	$-e_0e_2$
a	e_0e_1	e_0e_1	$-e_1$	e_0	$-e_\emptyset$	$e_0e_1e_2$	$-e_1e_2$	e_0e_2	$-e_2$
	e_2	e_2	$-e_0e_2$	$-e_1e_2$	$e_0e_1e_2$	$e_\emptyset$	$-e_0$	$-e_1$	e_0e_1
	e_0e_2	e_0e_2	$-e_2$	$-e_0e_1e_2$	e_1e_2	e_0	$-e_\emptyset$	$-e_0e_1$	e_1
	e_1e_2	e_1e_2	$e_0e_1e_2$	$-e_2$	$-e_0e_2$	e_1	e_0e_1	$-e_\emptyset$	$-e_0$
	$e_0e_1e_2$	$e_0e_1e_2$	e_1e_2	$-e_0e_2$	$-e_2$	e_0e_1	e_1	$-e_0$	$-e_\emptyset$

Left					b				
	$a \lrcorner b$	$e_\emptyset$	e_0	e_1	e_0e_1	e_2	e_0e_2	e_1e_2	$e_0e_1e_2$
	$e_\emptyset$	$e_\emptyset$	e_0	e_1	e_0e_1	e_2	e_0e_2	e_1e_2	$e_0e_1e_2$
	e_0	0	$e_\emptyset$	0	e_1	0	e_2	0	e_1e_2
	e_1	0	0	$e_\emptyset$	$-e_0$	0	0	e_2	$-e_0e_2$
a	e_0e_1	0	0	0	$e_\emptyset$	0	0	0	e_2
	e_2	0	0	0	0	$e_\emptyset$	$-e_0$	$-e_1$	e_0e_1
	e_0e_2	0	0	0	0	0	$e_\emptyset$	0	$-e_1$
	e_1e_2	0	0	0	0	0	0	$e_\emptyset$	e_0
	$e_0e_1e_2$	0	0	0	0	0	0	0	$e_\emptyset$

Right					b				
	$a \llcorner b$	$e_\emptyset$	e_0	e_1	e_0e_1	e_2	e_0e_2	e_1e_2	$e_0e_1e_2$
	$e_\emptyset$	$e_\emptyset$	0	0	0	0	0	0	0
	e_0	e_0	$e_\emptyset$	0	0	0	0	0	0
	e_1	e_1	0	$e_\emptyset$	0	0	0	0	0
a	e_0e_1	e_0e_1	$-e_1$	e_0	$e_\emptyset$	0	0	0	0
	e_2	e_2	0	0	0	$e_\emptyset$	0	0	0
	e_0e_2	e_0e_2	$-e_2$	0	0	e_0	$e_\emptyset$	0	0
	e_1e_2	e_1e_2	0	$-e_2$	0	e_1	0	$e_\emptyset$	0
	$e_0e_1e_2$	$e_0e_1e_2$	e_1e_2	$-e_0e_2$	e_2	e_0e_1	$-e_1$	e_0	$e_\emptyset$

Table 5.8 Multiplication tables for negative signatures.

Outer	b							
$a \wedge b$	$e_\emptyset$	e_0	e_1	e_0e_1	e_2	e_0e_2	e_1e_2	$e_0e_1e_2$
$e_\emptyset$	$e_\emptyset$	e_0	e_1	e_0e_1	e_2	e_0e_2	e_1e_2	$e_0e_1e_2$
e_0	e_0	0	e_0e_1	0	e_0e_2	0	$e_0e_1e_2$	0
e_1	e_1	$-e_0e_1$	0	0	e_1e_2	$-e_0e_1e_2$	0	0
e_0e_1	e_0e_1	0	0	0	$e_0e_1e_2$	0	0	0
e_2	e_2	$-e_0e_2$	$-e_1e_2$	$e_0e_1e_2$	0	0	0	0
e_0e_2	e_0e_2	0	$-e_0e_1e_2$	0	0	0	0	0
e_1e_2	e_1e_2	$e_0e_1e_2$	0	0	0	0	0	0
$e_0e_1e_2$	$e_0e_1e_2$	0	0	0	0	0	0	0

Here a labels the rows.

Inner	b							
$(a,b)e_\emptyset$	$e_\emptyset$	e_0	e_1	e_0e_1	e_2	e_0e_2	e_1e_2	$e_0e_1e_2$
$e_\emptyset$	$e_\emptyset$	0	0	0	0	0	0	0
e_0	0	$-e_\emptyset$	0	0	0	0	0	0
e_1	0	0	$-e_\emptyset$	0	0	0	0	0
e_0e_1	0	0	0	$e_\emptyset$	0	0	0	0
e_2	0	0	0	0	$-e_\emptyset$	0	0	0
e_0e_2	0	0	0	0	0	$e_\emptyset$	0	0
e_1e_2	0	0	0	0	0	0	$e_\emptyset$	0
$e_0e_1e_2$	0	0	0	0	0	0	0	$-e_\emptyset$

Here a labels the rows.

Regressive	b							
$a \vee b$	$e_\emptyset$	e_0	e_1	e_0e_1	e_2	e_0e_2	e_1e_2	$e_0e_1e_2$
$e_\emptyset$	0	0	0	0	0	0	0	$-e_\emptyset$
e_0	0	0	0	0	0	0	$-e_\emptyset$	$-e_0$
e_1	0	0	0	0	0	$e_\emptyset$	0	$-e_1$
e_0e_1	0	0	0	0	$-e_\emptyset$	$-e_0$	$-e_1$	$-e_0e_1$
e_2	0	0	0	$-e_\emptyset$	0	0	0	$-e_2$
e_0e_2	0	0	$e_\emptyset$	e_0	0	0	$-e_2$	$-e_0e_2$
e_1e_2	0	$-e_\emptyset$	0	e_1	0	e_2	0	$-e_1e_2$
$e_0e_1e_2$	$-e_\emptyset$	$-e_0$	$-e_1$	$-e_0e_1$	$-e_2$	$-e_0e_2$	$-e_1e_2$	$-e_0e_1e_2$

Here a labels the rows.

(Continued)

Table 5.8 (Continued)

Central					b			
ab	$e_\emptyset$	e_0	e_1	e_0e_1	e_2	e_0e_2	e_1e_2	$e_0e_1e_2$
$e_\emptyset$	$e_\emptyset$	e_0	e_1	e_0e_1	e_2	e_0e_2	e_1e_2	$e_0e_1e_2$
e_0	e_0	$-e_\emptyset$	e_0e_1	$-e_1$	e_0e_2	$-e_2$	$e_0e_1e_2$	$-e_1e_2$
e_1	e_1	$-e_0e_1$	$-e_\emptyset$	e_0	e_1e_2	$-e_0e_1e_2$	$-e_2$	e_0e_2
e_0e_1	e_0e_1	e_1	$-e_0$	$-e_\emptyset$	$e_0e_1e_2$	e_1e_2	$-e_0e_2$	$-e_2$
e_2	e_2	$-e_0e_2$	$-e_1e_2$	$e_0e_1e_2$	$-e_\emptyset$	e_0	e_1	$-e_0e_1$
e_0e_2	e_0e_2	e_2	$-e_0e_1e_2$	$-e_1e_2$	$-e_0$	$-e_\emptyset$	e_0e_1	e_1
e_1e_2	e_1e_2	$e_0e_1e_2$	e_2	e_0e_2	$-e_1$	$-e_0e_1$	$-e_\emptyset$	$-e_0$
$e_0e_1e_2$	$e_0e_1e_2$	$-e_1e_2$	e_0e_2	$-e_2$	$-e_0e_1$	e_1	$-e_0$	$e_\emptyset$

Left					b			
$a \lrcorner b$	$e_\emptyset$	e_0	e_1	e_0e_1	e_2	e_0e_2	e_1e_2	$e_0e_1e_2$
$e_\emptyset$	$e_\emptyset$	e_0	e_1	e_0e_1	e_2	e_0e_2	e_1e_2	$e_0e_1e_2$
e_0	0	$-e_\emptyset$	0	$-e_1$	0	$-e_2$	0	$-e_1e_2$
e_1	0	0	$-e_\emptyset$	e_0	0	0	$-e_2$	e_0e_2
e_0e_1	0	0	0	$e_\emptyset$	0	0	0	e_2
e_2	0	0	0	0	$-e_\emptyset$	e_0	e_1	$-e_0e_1$
e_0e_2	0	0	0	0	0	$e_\emptyset$	0	$-e_1$
e_1e_2	0	0	0	0	0	0	$e_\emptyset$	e_0
$e_0e_1e_2$	0	0	0	0	0	0	0	$-e_\emptyset$

Right					b			
$a \llcorner b$	$e_\emptyset$	e_0	e_1	e_0e_1	e_2	e_0e_2	e_1e_2	$e_0e_1e_2$
$e_\emptyset$	$e_\emptyset$	0	0	0	0	0	0	0
e_0	e_0	$-e_\emptyset$	0	0	0	0	0	0
e_1	e_1	0	$-e_\emptyset$	0	0	0	0	0
e_0e_1	e_0e_1	e_1	$-e_0$	$e_\emptyset$	0	0	0	0
e_2	e_2	0	0	0	$-e_\emptyset$	0	0	0
e_0e_2	e_0e_2	e_2	0	0	$-e_0$	$e_\emptyset$	0	0
e_1e_2	e_1e_2	0	e_2	0	$-e_1$	0	$e_\emptyset$	0
$e_0e_1e_2$	$e_0e_1e_2$	$-e_1e_2$	e_0e_2	e_2	$-e_0e_1$	$-e_1$	e_0	$-e_\emptyset$

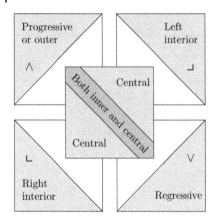

Figure 5.13 Identification of territory covered by different types of multiplication.

ordinary differential operator. The other multiplications, which only have partial coverage, play the roles of partial differential operators.

The ordinary differential operator formed by applying D using central multiplication was first introduced by Dirac (1928) in a different form, using a basis of Pauli matrices (see Chapter 8) instead of a basis of Grassmann's primal (vector) units. Examples of solutions to problems involving the Dirac operator (as it is now known) using Clifford algebra and the Clifford numerical suite are given in Chapters 23 and 24.

References

Dirac, P. A. M. (1928), 'The quantum theory of the electron', *Proceedings of the Royal Society. Series A, Containing Papers of a Mathematical and Physical Character* **117**(778), 610–624.

Grassmann, H. (1855), 'Sur les différents genres de multiplication', *Journal für die reine und angewandte Mathematik* **49**, 123–141.

Grassmann, H. (1995), On the various types of multiplication, *in* 'A New Branch of Mathematics: The Ausdehnungslehre of 1844, and Other Works', Open Court, Chicago, US-IL, pp. 451–468. (Translated by Lloyd C. Kannenberg).

6

Vectors and Geometry

6.1 Theory

Vectors $V = a\hat{\mathbf{x}} + b\hat{\mathbf{y}} + c\hat{\mathbf{z}}$ are represented in Clifford algebra by identifying the orthogonal unit vectors $\hat{\mathbf{x}}, \hat{\mathbf{y}}, \hat{\mathbf{z}}$ with particular Clifford units:

$$\begin{cases} \hat{\mathbf{x}} \leftrightarrow e_0 \\ \hat{\mathbf{y}} \leftrightarrow e_1 \\ \hat{\mathbf{z}} \leftrightarrow e_2 \end{cases} \tag{6.1}$$

in a context, where the signature λ is either all $+1$ or all -1. Vectors in three dimensions form a particular subset of three-dimensional Clifford numbers. Although three-dimensional Clifford numbers have four grades and eight components, vectors retain only one of the grades (grade 1) and three of the components. Vectors in dimensions higher than three require Clifford numbers of the corresponding dimension.

Table 6.1 lists the operations provided within the Clifford numerical suite for vectors of either signature. All multiplications are central (Clifford). Vectors n, n_1, and n_2 have real coefficients and unit length, and the symbol $\mathbb{I}$ stands for the identity operator so that $\mathbb{I}V = V$.

6.1.1 Magnitude

The magnitude of a real valued vector is the square root of the sum of the squares of all of its coefficients. The signature plays no role, so the magnitude of a vector having the same components with either signature is the same. If the coefficients are complex, the product from multiplying the coefficient with its complex conjugate is taken instead of the square.

For real valued vectors, the magnitude gives the length of the vector.

The square of the magnitude is almost the same as the inner product of the vector with its complex conjugate. For positive signatures, the two are the same; for negative signatures, they differ in sign.

The square of a unit vector $n^2 = \lambda e_\emptyset$ is the real valued unit scalar, with positive or negative sign depending on the signature. For real valued vectors, a unit vector may be constructed by dividing the vector by its magnitude. For vectors with complex coefficients, unit vectors should be constructed using the sub-routine `normalise_vector()` instead of division by the magnitude. That makes sure the individual complex squares of components (not

Numerical Calculations in Clifford Algebra: A Practical Guide for Engineers and Scientists, First Edition. Andrew Seagar.
© 2023 John Wiley & Sons Ltd. Published 2023 by John Wiley & Sons Ltd.

Table 6.1 Geometric operations on vectors.

			$\lambda = -1$	$\lambda = +1$
Magnitude	$\|V\|$	$=$	$\sqrt{-(V, V^*)}$	$\sqrt{(V, V^*)}$
Inversion (reciprocal)	$1/V$	$=$	V/V^2	
Reflection	$\mathbb{Q}V$	$=$	nVn	$-nVn$
Inner projection	$\mathbb{P}^-V$	$=$	$\frac{1}{2}(\mathbb{I} - \mathbb{Q})V$	
Outer projection	$\mathbb{P}^+V$	$=$	$\frac{1}{2}(\mathbb{I} + \mathbb{Q})V$	
Rotation	$\mathbb{R}V$	$=$	$\mathbb{Q}_2\mathbb{Q}_1 V = n_2 n_1 V n_1 n_2$	

products of values with their conjugates) deliver after summation the real valued unit scalar with the appropriate sign.

6.1.2 Inverse

The inverse of a vector is another vector. Multiplying a vector by its inverse from either side produces the unit scalar value $1e_\emptyset$.

In Clifford algebra, many vectors do have an inverse, but that is not guaranteed. All real vectors, other than those of zero length, have an inverse. Some complex vectors which have non-zero magnitude do not have any inverse. Take as example the vector $V = e_0 + ie_1$. The square of the magnitude is obtained from inner multiplication as the non-zero value:

$$(e_0 + ie_1, e_0 + i^*e_1) = e_0 \tilde{e}_0 + ie_1 i^* \tilde{e}_1 = e_0^2 + (ii^*)e_1^2 = \lambda(1 + 1)e_\emptyset = 2\lambda e_\emptyset \neq 0 \qquad (6.2)$$

However the denominator, when attempting to calculate the inverse, produces the value:

$$(e_0 + ie_1)(e_0 + ie_1) = e_0^2 + ie_0 e_1 + ie_1 e_0 + i^2 e_1^2 = (\lambda + 0 - \lambda)e_\emptyset = 0 \qquad (6.3)$$

The denominator is zero so that no inverse can be found.

6.1.3 Reflection

A unit vector n is used as a reference to determine what should happen when other vectors undergo a reflection $\mathbb{Q}$. In three dimensions, the reference vector plays the role of the unit normal to the mirror in which other vectors are reflected. If the reference vector is in the x direction, as in Figure 6.1, other vectors will be reflected in the yz plane.

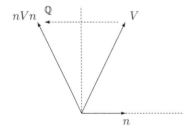

Figure 6.1 Reflection of vector, in context with negative signature $\lambda = -1$.

In any number of dimensions, a vector undergoing reflection is split into two parts, one in parallel with the reference vector and one perpendicular to it. The part in parallel with the reference vector is reversed and then recombined with the (unchanged) perpendicular part.

6.1.4 Projection

As with reflection, a unit reference vector n is used to determine what happens when a vector undergoes a projection operation. Within the Clifford numerical suite, two orthogonal projection operations are provided. The first, $\mathbb{P}^-$, is an inner projection, which calculates the part of the vector parallel to the reference vector. The second, $\mathbb{P}^+$, is an outer projection, which calculates the part of the vector perpendicular to the reference vector.

Together the projection operators $\mathbb{P}^-$ and $\mathbb{P}^+$ can be used to split vectors into two orthogonal components. The sum of the two projections reproduces the original vector:

$$\mathbb{P}^-V + \mathbb{P}^+V = (\mathbb{P}^- + \mathbb{P}^+)V = \mathbb{I}V = V \tag{6.4}$$

In three dimensions, if the reference vector is $n = \frac{1}{\sqrt{2}}(\hat{\mathbf{y}} - \hat{\mathbf{x}})$, as in Figure 6.2, the inner projection gives the component of the vector V along (in this case opposite to) the direction of $\hat{\mathbf{y}} - \hat{\mathbf{x}}$ and the outer projection gives the components of the vector along the directions of $\hat{\mathbf{x}} + \hat{\mathbf{y}}$ and $\hat{\mathbf{z}}$.

6.1.5 Rotation

Rotation is achieved by a sequence of two different reflections. The two reference unit vectors for the reflections define the plane of the rotation. In three dimensions, the axis of rotation is perpendicular to that plane.

The angle and direction of rotation are determined by the angle between the two reference vectors, and the order in which their reflections are applied. The angle of rotation is twice the smallest positive angle between the reference vectors. Two vectors $45°$ apart give a rotation of $90°$. The direction of rotation is in the same sense (clockwise or anticlockwise) as the direction of the smallest angle from the reference vector for the first reflection towards the reference vector for the second reflection.

For example, in three dimensions, if the two reference vectors are $n_1 = \frac{1}{\sqrt{2}}(\hat{\mathbf{x}} - \hat{\mathbf{y}})$ and $n_2 = \hat{\mathbf{x}}$, as in Figure 6.3, the vector $1\hat{\mathbf{x}} + 2\hat{\mathbf{y}}$ is rotated to $-2\hat{\mathbf{x}} + 1\hat{\mathbf{y}}$, and any components in the z direction remain unchanged. If the two reference vectors are taken in the opposite order,

Figure 6.2 Orthogonal projections of vector, in context with negative signature $\lambda = -1$.

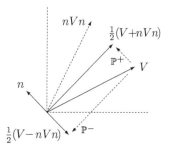

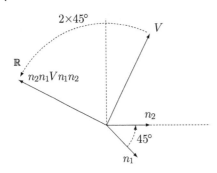

Figure 6.3 Rotation of vector.

the angle between them is −45°, and the direction of rotation is opposite to that shown in the figure.

6.2 Practice

6.2.1 Example Code

The code shown in Example 6.1a demonstrates use of the operations listed in Table 6.1. A three-dimensional context with negative signature is established in lines 7–10. A Clifford number is initialised to a three-dimensional real valued vector $1\hat{x} + 2\hat{y} + 3\hat{z}$ on line 13, and printed on line 20.

Example 6.1(a)

```
 1:   void demo15() /* vectors */
 2:   {
 3:       int lambda[3];
 4:       double a[3];
 5:       number v,w,x,d,m,n,i,r,o;
 6:
 7:       recovery_level(0,0,0);
 8:       create_cliffmem(50);
 9:       declare_primal(3,pack_signature(3,lambda,-1));
10:       format_printing("%.5f");
11:
12:       /* magnitude and inverse */
13:       v=pack_R3vector(pack_R3array(1.0,2.0,3.0,a));
14:       m=magnitudeof_vector(v,0);
15:       n=normalise_vector(v,0);
16:       i=inverseof_vector(v,0);
17:       x=central_mul(v,i,0);
18:       w=magnitudeof_number(x,0);
19:       d=magnitudeof_number(sub_number(x,scalar_number(1.0,0.0),2),1);
20:       printf("V          = ");        print_vector(v);
21:       printf("|V|        = ");        print_number(m,8);
                                          recycle_number(m);
22:       printf("V/|V|      = ");        print_vector(n);
                                          recycle_number(n);
```

```
23:     printf("1/V        = ");        print_vector(i);
                                        recycle_number(i);
24:     printf("V*1/V      = ");        print_number(x,8);
                                        recycle_number(x);
25:     printf("|V*1/V|    = ");        print_number(w,8);
                                        recycle_number(w);
26:     printf("|V*1/V-1| = ");         printf_number("%e",d,8);
                                        recycle_number(d);
27:
28:     /* reflection */
29:     n=pack_R3vector(pack_R3array(1.0,0.0,0.0,a));
30:     r=reflect_vector(v,n,0);
31:     printf("\nV             = ");        print_vector(v);
                                             recycle_number(v);
32:     printf("unit vector n = ");          print_vector(n);
                                             recycle_number(n);
33:     printf("n*V*n = reflection of V along direction n\n");
34:     printf("             = ");            print_vector(r);
                                             recycle_number(r);
35:
36:     /* projection */
37:     v=normalise_vector(pack_R3vector(pack_R3array(7.0,1.0,0.0,a)),1);
38:     n=normalise_vector(pack_R3vector(pack_R3array(1.0,1.0,0.0,a)),1);
39:     i=innerspace:vector(v,n,0);
40:     o=outerspace:vector(v,n,0);
41:     printf("\nunit vector V = ");         print_vector(v);
                                             recycle_number(v);
42:     printf("unit vector n = ");          print_vector(n);
43:     printf("0.5(V-n*V*n)  = component of V in direction of n\n");
44:     printf("             = ");            print_vector(i);
                                             recycle_number(i);
45:     printf("0.5(V+n*V*n)  = component of V not in direction of n\n");
46:     printf("             = ");            print_vector(o);
                                             recycle_number(o);
47:
48:     /* rotation */
49:     v=pack_R3vector(pack_R3array(1.0,2.0,3.0,a));
50:     n=normalise_vector(pack_R3vector(pack_R3array(1.0,-1.0,0.0,
                                         a)),1);
51:     x=pack_R3vector(pack_R3array(1.0,0.0,0.0,a));
52:     r=rotate_number(v,n,x,0);
53:     printf("\nV              = ");        print_vector(v);
                                             recycle_number(v);
54:     printf("unit vector n1 = ");         print_vector(n);
                                             recycle_number(n);
55:     printf("unit vector n2 = ");         print_vector(x);
                                             recycle_number(x);
56:     printf("n2*n1*V*n1*n2  = rotation of V in plane n1,n2; 2*angle
                                 between n1,n2\n");
57:     printf("              = ");           print_vector(r);
                                             recycle_number(r);
58:
59:     free_cliffmem(0);
60:   }
```

On line 14, the magnitude of the vector is calculated. On line 15, the vector is normalised to unit length The inverse of the vector is taken on line 16 and multiplied by the vector itself on line 17. The magnitude of that number, which should be unity, is calculated on line 18, and the magnitude of the difference between unity and the product of vector and inverse, which should be zero, is calculated on line 19.

The final argument in the list of parameters for the sub-routines invoked to perform all operations on lines 14–19 other than `scalar_number()`, indicates whether the numbers in the preceding arguments are to be recycled. A value of zero indicates no action should be taken. A non-zero value indicates by the bits in the value those arguments which should be recycled; '1' for the first argument, '2' for the second, and '3' for both first and second. This mechanism is used in line 19 to dispose of the temporary results created by invocations to `scalar_number()` and `sub_number()`, while preserving the number 'x' for later use. For details on recycling see Section 14.2.3.

Results of the calculations of magnitude and inverse are printed on lines 21–26.

Examples of calculations involving reflection, projection, and rotation, along with the statements which print the results, are given on lines 29–34, 37–46, and 49–57, respectively.

6.2.2 Example Output

The results of running the code listed in Example 6.1a are shown in Example 6.1b. Lines 3–8 give the results for various mixed inverse and magnitude operations. The vector on line 2 has non-unit magnitude on line 3. The result of normalising the vector to unit length is given on line 4. That can be checked by squaring the components and adding them together.

Example 6.1(b)

```
 1:  demo15()
 2:  V           = [1.00000+i0.00000, 2.00000+i0.00000, 3.00000+i0.00000]
 3:  |V|         = (3.74166+i0.00000)e{}
 4:  V/|V|       = [0.26726+i0.00000, 0.53452+i0.00000, 0.80178+i0.00000]
 5:  1/V         = [-0.07143+i0.00000, -0.14286+i0.00000, -0.21429
                   +i0.00000]
 6:  V*1/V       = (1.00000+i0.00000)e{}
 7:  |V*1/V|     = (1.00000+i0.00000)e{}
 8:  |V*1/V-1|   = (1.110223e-16+i0.000000e+00)e{}
 9:
10:  V                = [1.00000+i0.00000, 2.00000+i0.00000, 3.00000
                        +i0.00000]
11:  unit vector n = [1.00000+i0.00000, 0.00000+i0.00000, 0.00000
                      +i0.00000]
12:  n*V*n = reflection of V along direction n
13:                   = [-1.00000+i0.00000, 2.00000+i0.00000, 3.00000
                        +i0.00000]
14:
15:  unit vector V = [0.98995+i0.00000, 0.14142+i0.00000, 0.00000
                      +i0.00000]
16:  unit vector n = [0.70711+i0.00000, 0.70711+i0.00000, 0.00000
                      +i0.00000]
17:  0.5(V-n*V*n)  = component of V in direction of n
```

```
18:                    = [0.56569+i0.00000, 0.56569+i0.00000, 0.00000
                         +i0.00000]
19:   0.5(V+n*V*n)     = component of V not in direction of n
20:                    = [0.42426+i0.00000, -0.42426+i0.00000, 0.00000
                         +i0.00000]

21:
22:   V                = [1.00000+i0.00000, 2.00000+i0.00000, 3.00000
                         +i0.00000]
23:   unit vector n1   = [0.70711+i0.00000, -0.70711+i0.00000, 0.00000
                         +i0.00000
24:   unit vector n2   = [1.00000+i0.00000, 0.00000+i0.00000, 0.00000
                         +i0.00000]
25:   n2*n1*V*n1*n2    = rotation of V in plane n1,n2; 2*angle
                         between n1,n2
26:                    = [-2.00000+i0.00000, 1.00000+i0.00000, 3.00000
                         +i0.00000]
```

The product from multiplying the vector on line 2 with its inverse on line 5 is the unit scalar on line 6, which has magnitude equal to one on line 7. The accuracy of the inverse is given in line 8 by calculating the magnitude of the value obtained after subtracting one from the product of vector and its inverse. The underlying machine precision is of type 'double' with around 15–16 decimal places.

Lines 10–13 give the results of reflecting a vector. The example chosen is that shown in Figure 6.1, and the result is verified as seen there.

Lines 15–20 give the projections of a vector in two perpendicular directions. The vector $V = a$ chosen in this example is the unit vector a from Figure 5.8, and the reference vector $n = b$ is the vector b from that figure. Lines 17 and 18 give the inner projection, the vector b_1 in Figure 5.8, and lines 19 and 20 give the outer projection, vector c_1 in Figure 5.8. As a check for b_1, $0.565692 \times \sqrt{50} = 4.00003$, and for c_1, $0.424262 \times 2\sqrt{50} = 2.99997$. These values are accurate to the actual values 4 and 3 within the precision of the printing format chosen.

Lines 22–26 demonstrate the 90° rotation for the case shown in Figure 6.3. The numerical result is as depicted in the figure.

7

Quaternions

Quaternions were constructed by Hamilton (1844) as a generalisation of complex numbers from two to three dimensions. They were used by Maxwell (1873a, 1873b) to write his equations of electromagnetism.

7.1 Theory

Quaternions $Q = s1 + pI + qJ + rK$ are represented in Clifford algebra by identifying the unit quaternions $1, I, J, K$ with particular Clifford units:

$$\begin{cases} 1 \leftrightarrow e_\emptyset \\ I \leftrightarrow e_1 e_2 \\ J \leftrightarrow e_2 e_0 \\ K \leftrightarrow e_0 e_1 \end{cases} \tag{7.1}$$

in a context where the signature λ, as deduced by both Grassmann (1995) and Clifford (1878), is all -1. Quaternions form a particular subset of three-dimensional Clifford numbers. Whereas three-dimensional Clifford numbers have four grades and eight components, quaternions retain only two of the grades (0 and 2) and four of the components.

The unit quaternions obey the rules for multiplication (Hamilton 1844):

$$\begin{cases} 1 = -I^2 = -J^2 = -K^2 \\ I = JK \\ J = KI \\ K = IJ \end{cases} \tag{7.2}$$

These rules are honoured by the Clifford algebra when Grassmann's central multiplication is used.

Table 6.1 lists the operations provided within the Clifford numerical suite for quaternions. All multiplications are central (Clifford). Vectors $n, n_1,$ and n_2 have real coefficients and unit lengths.

Numerical Calculations in Clifford Algebra: A Practical Guide for Engineers and Scientists,
First Edition. Andrew Seagar.
© 2023 John Wiley & Sons Ltd. Published 2023 by John Wiley & Sons Ltd.

7.1.1 Magnitude

The magnitude of a real valued quaternion is the square root of the sum of the squares of all of its coefficients. If the coefficients are complex, the product from multiplying the coefficient with its complex conjugate is taken instead of the square.

For quaternions, the square of the magnitude is the same as the inner product of the quaternion with its complex conjugate.

7.1.2 Inverse

The inverse of a quaternion is another quaternion. Multiplying a quaternion by its inverse from either side produces the unit quaternion 1.

In Clifford algebra, many quaternions do have an inverse, but that is not guaranteed. Some quaternions which have non-zero magnitude do not have any inverse. Take as example the quaternion $Q = I + iJ$, which has $s = r = 0$, $p = 1$ and $q = i$. The square of the magnitude is obtained from inner multiplication as the non-zero value:

$$(I + iJ, I + i^*J) = I\tilde{I} + iJi^*\tilde{J} = -I^2 - (ii^*)J^2 = 1 + 1 = (2)1 \neq 0 \tag{7.3}$$

However the denominator, when attempting to calculate the inverse, produces the value:

$$(I + iJ)(\tilde{I} + i\tilde{J}) = (I + iJ)(-I - iJ) = -I^2 - iIJ - iJI - i^2J^2 = (1 + 0 - 1)1 = 0 \tag{7.4}$$

The denominator is zero so that no inverse can be found.

7.1.3 Reflection and Projection

Reflection and projection for quaternions can be achieved as for vectors with a reflection operator Q. However, comparison of the formulæ for the operator for vectors with negative signature in Table 6.1, to the quaternion operator in Table 7.1 shows that the quaternion operator involves a different sign than for vectors. This is because the number of primal units making up the quaternion units is either one more or fewer than for vectors. Pair exchanges of the primal units in the calculation of the reflection, therefore, involve either one more or fewer than for vectors. In either case, the difference of one unit changes the sign of the product.

The form of quaternion reflection operator as in Table 7.1 is appropriate for any numbers with negative signature which have non-zero coefficients only in the even grades. That can

Table 7.1 Geometric operations on quaternions.

Magnitude	$\lvert Q\rvert = \sqrt{(Q, Q^*)}$
Inversion (reciprocal)	$1/Q = \tilde{Q}/(Q\tilde{Q})$
Reflection	$\mathbb{Q}\,Q = -nQn$
Inner projection	$\mathbb{P}^-\,Q = \frac{1}{2}(1 - \mathbb{Q})Q$
Outer projection	$\mathbb{P}^+\,Q = \frac{1}{2}(1 + \mathbb{Q})Q$
Rotation	$\mathbb{R}\,Q = n_2 n_1 Q n_1 n_2$

further be extended to any number u of either positive or negative signature with components in any grades by exploiting the involution operator to introduce the extra negation appearing in even grades as compared to odd grades:

$$\mathbb{Q}u = \begin{cases} \overrightarrow{n\hat{u}n} & \text{for uniform signature } \lambda = +1 \\ -\overrightarrow{n\hat{u}n} & \text{for uniform signature } \lambda = -1 \end{cases} \tag{7.5}$$

The projection operators $\mathbb{P}^+$ and $\mathbb{P}^-$ are then calculated as in Table 7.1 from the extended (new) reflection operator in the normal way.

Within the Clifford numerical suite, the reflection operator `reflect_number()` is provided for all numbers of non-zero uniform signature using the formulæ from Eq. (7.5), and the two projection operators are similarly provided as the sub-routines `innerspace_number()` and `outerspace_number()`. The equivalent vector operators ending in 'vector' rather than 'number' are maintained for cases where it is desired to treat even and odd grades differently. Such is the case for electromagnetism (in Chapter 10) where entities associated with space need to be treated differently from those associated with time.

7.1.4 Rotation

The geometric operation of rotation applies independently to the scalar and bivector parts of quaternions. After the operation, the scalar part remains a scalar, and the bivector part remains a bivector. The result is therefore another quaternion.

With the identification between unit quaternions and Clifford units as adopted[1] in Eq. (7.1), the unit quaternion I plays the role of the unit bivector occupying the yz plane, as shown in Figure 7.1. Similarly, the unit quaternions J and K play the roles of the unit bivectors occupying the zx and xy planes, respectively. Note that unit vectors $\hat{x}$, $\hat{y}$, and $\hat{z}$ play the roles of the normals to the planes identified with the quaternions I, J, and K in that order. The unit quaternion 1 plays the role of a point at the origin, O.

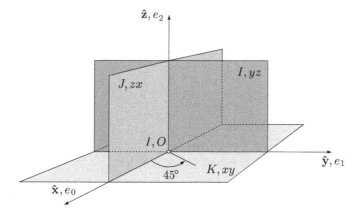

Figure 7.1 Geometric interpretation of unit quaternions I, J, K as bivectors.

---

1 Other identifications could equally well be adopted, as in Chapter 10.

Rotation of the quaternion I by 90° can be achieved using two unit reference vectors separated by 45°, such as $n_1 = e_0$ and $n_2 = \frac{1}{\sqrt{2}}(e_0 + e_1)$. Inspection of Figure 7.1 suggests the result should coincide with the quaternion J. That is confirmed utilising Clifford algebra in Eq. (7.6) applying the rotation operator given in Table 7.1.

$$
\begin{aligned}
\mathbb{R}\,I = n_2 n_1 I n_1 n_2 \quad &= \frac{1}{\sqrt{2}}(e_0 + e_1)e_0 e_1 e_2 e_0 \frac{1}{\sqrt{2}}(e_0 + e_1) = -\frac{1}{2}(e_0 + e_1)e_1 e_2(e_0 + e_1) \\
= \frac{1}{2}(e_0 e_1 - 1)(e_0 + e_1)e_2 &= \frac{1}{2}(e_0 e_1 e_0 + e_0 e_1 e_1 - e_0 - e_1)e_2 = \frac{1}{2}(e_1 - e_0 - e_0 - e_1)e_2 \\
= -e_0 e_2 \quad &= e_2 e_0 = J
\end{aligned}
$$

$$(7.6)$$

7.1.5 Intersection

The intersection of planes gives lines. For example, intersecting the yz plane with the zx plane gives the z axis. Planes through the origin can be represented by quaternions $Q = pI + qJ + rK$ where the scalar part s is zero. The intersection of these planes is effected by using Grassmann's regressive multiplication. Within Clifford algebra, using the definition for regressive product in Tables 1.3 and 5.6, the calculation is accomplished using central multiplication as in Eq. (7.7):

$$
\begin{aligned}
I \vee J = *^{-1}((* I) \wedge (* J)) &= J\widehat{(e_0 e_1 e_2)}I = e_2 e_0 (e_2 e_1 e_0)e_1 e_2 = e_2(-e_2 e_0)e_1(-e_1 e_0)e_2 \\
= e_\emptyset e_0 e_\emptyset e_0 e_2 &= e_0 e_0 e_2 = -e_\emptyset e_2 = -e_2
\end{aligned}
$$

$$(7.7)$$

Within a sign, the result is the vectorial unit e_2, which is identified with the unit vector $\hat{z}$.

7.1.6 Factorisation

A quaternion Q can be constructed as the central product ab of a pair of vectors a, b in a three-dimensional context:

$$Q = s1 + pI + qJ + rK = s + V\sigma = ab = a \wedge b + (a, b) \tag{7.8}$$

where s is a scalar, V is a vector in three dimensions, and $\sigma = -e_0 e_1 e_2$ is the negatively signed unit of volume for the three spatial dimensions[2] with square $\sigma^2 = +1$.

Alternatively, if the value of a quaternion is known, two factors a, b can be determined from the equations:

$$
\begin{cases}
(a, b) = s \\
(a \wedge b)\sigma = V
\end{cases}
\tag{7.9}
$$

The first line of Eq. (7.9) seeks two vectors with known value of inner (dot) product, and the second line seeks two vectors in the plane perpendicular to V with known value of outer (cross) product.

A solution can be obtained by geometrical construction as in Figure 7.2. Firstly the basis vectors for the axes e_0, e_1, e_2 are projected onto the plane through the origin

---

2 For electromagnetic fields in addition to the three spatial dimensions, there is an extra dimension of time or frequency.

Figure 7.2 Construction of vector factors of quaternion in plane perpendicular to vector V.

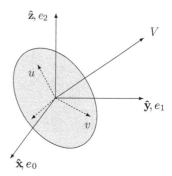

perpendicular to the vector V. The two largest of the projections are identified as vectors u and v, and the linear equations

$$\begin{cases} (u, \alpha u + \beta v) & = s \\ u \wedge (\alpha u + \beta v) \sigma & = V \end{cases} \tag{7.10}$$

are solved for the two complex scalars α and β

$$\begin{cases} \beta = V[(u \wedge v)\sigma]^{-1} \\ \alpha = [s - \beta(u, v)]/(u, u) \end{cases} \tag{7.11}$$

The two vectorial factors are then

$$\begin{cases} a = u \\ b = \alpha u + \beta v \end{cases} \tag{7.12}$$

The factorisation is not unique because there are six unknown coefficients in the two vectors, and only four known coefficients in the quaternion.

Within the Clifford numerical suite factorisation of quaternions is implemented, as described here, by the sub-routine `factorsof_quaternion()`.

7.2 Practice

7.2.1 Example Code

Example 7.1a provides examples using central and regressive multiplication of quaternions as well as other operations from Table 7.1. A three-dimensional Clifford context with all negative signature is established in lines 8–11. Lines 14–17 initialise four numbers to the four unit quaternions, which are then printed in lines 19–22.

Example 7.1(a)

```
1:  void demo16() /* quaternions */
2:  {
3:      double a[3];
4:      int lambda[3];
```

(Continued)

Example 7.1(a) (Continued)

```
 5:    number q[7],f[2];
 6:    number o,x,y,z,n;
 7:
 8:    recovery_level(0,0,0);
 9:    create_cliffmem(50);
10:    declare_primal(3,pack_signature(3,lambda,-1));
11:    format_printing("%.0f");
12:
13:    /* unit quaternions*/
14:    o=pack_quaternion(1.0,0.0,0.0,0.0,0.0,0.0,0.0,0.0);
15:    x=pack_quaternion(0.0,0.0,1.0,0.0,0.0,0.0,0.0,0.0);
16:    y=pack_quaternion(0.0,0.0,0.0,0.0,1.0,0.0,0.0,0.0);
17:    z=pack_quaternion(0.0,0.0,0.0,0.0,0.0,0.0,1.0,0.0);
18:
19:    printf("1 = ");   print_quaternion(o);
20:    printf("I = ");   print_quaternion(x);
21:    printf("J = ");   print_quaternion(y);
22:    printf("K = ");   print_quaternion(z);
23:    recycle_number(o);
24:
25:    /* quaternion central multiplication */
26:    q[0]=central_mul(x,x,0);
27:    q[1]=central_mul(y,y,0);
28:    q[2]=central_mul(z,z,0);
29:    q[3]=central_mul(x,y,0);
30:    q[4]=central_mul(y,z,0);
31:    q[5]=central_mul(z,x,0);
32:    q[6]=central_mul(central_mul(x,y,0),z,1);
33:
34:    printf("\nI*I = ");  print_quaternion(q[0]);
35:    printf("J*J = ");    print_quaternion(q[1]);
36:    printf("K*K = ");    print_quaternion(q[2]);
37:    printf("I*J = ");    print_quaternion(q[3]);
38:    printf("J*K = ");    print_quaternion(q[4]);
39:    printf("K*I = ");    print_quaternion(q[5]);
40:    printf("I*J*K = ");  print_quaternion(q[6]);
41:    recycle_array(7,q);
42:
43:    /* quaternion regressive multiplication */
44:    q[0]=regressive_mul(x,x,0);
45:    q[1]=regressive_mul(y,y,0);
46:    q[2]=regressive_mul(z,z,0);
47:    q[3]=regressive_mul(x,y,0);
48:    q[4]=regressive_mul(y,z,0);
49:    q[5]=regressive_mul(z,x,0);
50:    q[6]=regressive_mul(regressive_mul(x,y,3),z,3);
51:
52:    printf("\nI\\/I = ");    print_number(q[0],8);
53:    printf("J\\/J = ");      print_number(q[1],8);
54:    printf("K\\/K = ");      print_number(q[2],8);
55:    printf("I\\/J = ");      print_vector(q[3]);
56:    printf("J\\/K = ");      print_vector(q[4]);
57:    printf("K\\/I = ");      print_vector(q[5]);
58:    printf("I\\/J\\/K = "); print_number(q[6],8);
```

```
59:      recycle_array(7,q);
60:
61:      /* magnitude and inverse */
62:      q[0]=pack_quaternion(1.0,2.0,3.0,4.0,5.0,6.0,7.0,8.0);
63:      x=magnitudeof_quaternion(q[0],0);
64:      q[1]=inverseof_quaternion(q[0],0);
65:      q[2]=central_mul(q[0],q[1],0);
66:      y=magnitudeof_quaternion(q[2],0);
67:      o=scalar_number(1.0,0.0);
68:      z=magnitudeof_number(sub_number(q[2],o,2),1);
69:
70:      format_printing("%.3f");
71:      printf("\nQ = ");         print_quaternion(q[0]);
72:      printf("|Q|  = ");        print_number(x,8);
73:      printf("1/Q = ");         print_quaternion(q[1]);
74:      printf("Q*1/Q = ");       printf_quaternion("%.1e",q[2]);
75:      printf("|Q*1/Q|  = ");    print_number(y,8);
76:      printf("|Q*1/Q-1| = ");   printf_number("%e",z,8);
77:      recycle_number(x);        recycle_number(y);
78:      recycle_number(z);
79:      recycle_array(3,q);
80:
81:      /* rotation */
82:      n=pack_R3vector(pack_R3array(1.0,0.0,0.0,a));
83:      x=normalise_vector(pack_R3vector(pack_R3array(1.0,1.0,0.0,a)),1);
84:      y=pack_quaternion(0.0,0.0,1.0,0.0,0.0,0.0,0.0,0.0);
85:      z=rotate_number(y,n,x,0);
86:
87:      printf("\nunit vector n1 = "); print_vector(n);
88:      printf("unit vector n2 = ");   print_vector(x);
89:      printf("Q = ");                print_quaternion(y);
90:      printf("n2*n1*Q*n1*n2  = rotation of Q in plane n1,n2;
                               2*angle between n1,n2\n");
91:      printf("               = ");   print_quaternion(z);
92:      recycle_number(x);        recycle_number(y);
93:      recycle_number(z);        recycle_number(n);
94:
95:      random_seed(0xABCDEF);
96:      q[0]=random_quaternion();
97:      factorsof_quaternion(q[0],f,0);
98:      x=central_mul(f[0],f[1],0);
99:      y=magnitudeof_number(sub_number(q[0],x,0),1);
100:
101:     printf("\nquaternion Q        = ");  print_quaternion(q[0]);
102:     printf("factor f1 (vector)  = ");    print_number(f[0],8);
103:     printf("factor f2 (vector)  = ");    print_number(f[1],8);
104:     printf("f1*f2 as quaternion = ");    print_quaternion(x);
105:     printf("f1*f2 as number\n = ");      print_number(x,8);
106:     printf("error |Q-f1*f2|     = ");    printf_number("%.2e",y,8);
107:     recycle_number(x);        recycle_number(y);
108:     recycle_number(q[0]);     recycle_array(2,f);
109:
110:     free_cliffmem(0);
111: }
```

The unit quaternions are multiplied in various combinations on lines 26–32 using central multiplication, with the results stored in an array rather than in separate variables. The results are printed in lines 34–40, and the whole array is recycled on line 41. Using the array cuts down on the number of times individual numbers need to be recycled.

Lines 44–50 demonstrate the use of regressive multiplication, with results printed on lines 52–58. The magnitude and inverse of quaternions are used in different ways on lines 62–68, with results printed on lines 71–76. The unit quaternion I is subject to a rotation of 90° in lines 82–85, with results printed on lines 87–91.

A random quaternion generated on line 96 is factorised as two vectors on line 97. The two factors are multiplied together on line 98, and the magnitude of the error between that product and the original quaternion is calculated on line 99. Results of these operations are printed in lines 101–106.

7.2.2 Example Output

The results from running the code listed in Example 7.1a are given in Example 7.1b. Lines 2–5 show the unit quaternions, and lines 7–13 give the results produced using central multiplication. Note that the values produced are consistent with the rules for the behaviour of quaternions in Eq. (7.2).

Example 7.2(b)

```
 1:    demo16 ()
 2:    1 = (1+i0)1
 3:    I = (1+i0)I
 4:    J = (1+i0)J
 5:    K = (1+i0)K
 6:
 7:    I*I = (-1+i0)1
 8:    J*J = (-1+i0)1
 9:    K*K = (-1+i0)1
10:    I*J = (1+i0)K
11:    J*K = (1+i0)I
12:    K*I = (1+i0)J
13:    I*J*K = (-1+i0)1
14:
15:    I\/I = (0+i0)e{}
16:    J\/J = (0+i0)e{}
17:    K\/K = (0+i0)e{}
18:    I\/J = [0+i0, 0+i0, -1+i0]
19:    J\/K = [-1+i0, 0+i0, 0+i0]
20:    K\/I = [0+i0, -1+i0, 0+i0]
21:    I\/J\/K = (1+i0)e{}
22:
23:    Q = (1.000+i2.000)1+(3.000+i4.000)I+(5.000+i6.000)J+(7.000+i8.000)K
24:    |Q| = (14.283+i0.000)e{}
25:    1/Q = (0.009-i0.007)1+(-0.017+i0.018)I+(-0.025+i0.029)J
                                         +(-0.033+i0.041)K
26:    Q*1/Q = (1.0e+00+i5.9e-17)1+(2.8e-17-i3.5e-18)I+(-0.0e+00+i1.0e-17)J
                                         +(0.0e+00-i1.4e-17)K
```

```
27:  |Q*1/Q|  = (1.000+i0.000)e{}
28:  |Q*1/Q-1| = (6.754292e-17+i0.000000e+00)e{}
29:
30:  unit vector n1 = [1.000+i0.000, 0.000+i0.000, 0.000+i0.000]
31:  unit vector n2 = [0.707+i0.000, 0.707+i0.000, 0.000+i0.000]
32:  Q = (1.000+i0.000)I
33:  n2*n1*Q*n1*n2  = rotation of Q in plane n1,n2; 2*angle between
                       n1,n2
34:                 = (1.000+i0.000)J
35:
36:  quaternion Q         = (0.083-i0.128)1+(0.405+i0.796)I
                            +(0.366-i0.271)J+(-0.029-i0.819)K
37:  factor f1 (vector)   = (0.216+i0.267)e{0}+(1.116-i0.131)e{1}
                            +(-0.076-i0.306)e{2}
38:  factor f2 (vector)   = (-0.106+i0.727)e{0}+(-0.083+i0.128)e{1}
                            +(0.314+i0.764)e{2}
39:  f1*f2 as quaternion = (0.083-i0.128)1+(0.405+i0.796)I
                            +(0.366-i0.271)J+(-0.029-i0.819)K
40:  f1*f2 as number
41:    = (0.083-i0.128)e{}+(-0.029-i0.819)e{0,1}+(-0.366+i0.271)e{0,2}
                            +(0.405+i0.796)e{1,2}
42:  error |Q-f1*f2|      = (2.88e-16+i0.00e+00)e{}
```

Lines 15–21 give the results when regressive multiplication is used instead of central multiplication. Observe that the product multiplying unit quaternions I and J gives the unit vector $-\hat{z}$ (or $-e_2$) as calculated in Eq. (7.7). The results for the calculations involving magnitude and inverse behave as expected in lines 23–28 within the accuracy of the underlying double precision data type.

Lines 30–34 show that the result produced by rotating the unit quaternion I by 90° in the xy plane is the unit quaternion J. This is in agreement with the value calculated in Eq. (7.6).

Lastly, the quaternion on line 36 produces the two factors on lines 37 and 38. The factors are not a unique pair; there are other possibilities. Comparing the product of the two factors on line 39 with the original on line 36 verifies that this pair of factors do behave as intended. The error in the product of the factors is seen on line 42 as zero within machine precision.

References

Clifford, W. K. (1878), 'Applications of Grassmann's extensive algebra', *American Journal of Mathematics* **1**(4), 350–358.

Grassmann, H. (1995), The position of the Hamiltonian quaternions in extension theory, *in* 'A New Branch of Mathematics: The Ausdehnungslehre of 1844, and Other Works', Open Court, Chicago, US-IL, pp. 525–538. (Translated by Lloyd C. Kannenberg).

Hamilton, Sir W. R. (1844), 'On quaternions; or on a new system of imaginaries in algebra', *Philosophical Magazine and Journal of Science* **25**, 10–13.

Maxwell, J. C. (1873a), *A Treatise on Electricity and Magnetism*, Vol. 1, Clarendon Press, Oxford, UK.

Maxwell, J. C. (1873b), *A Treatise on Electricity and Magnetism*, Vol. 2, Clarendon Press, Oxford, UK.

8

Pauli Matrices

Pauli (1927) matrices were constructed in the context of quantum mechanics involving electrons.

8.1 Theory

Pauli matrices $P = sI + pX + qY + rZ$ are represented in Clifford algebra by identifying the unit Pauli matrices[1] I, X, Y, Z with particular Clifford units:

$$
\begin{cases}
\begin{pmatrix} 1 & 0 \\ 0 & 1 \end{pmatrix} = I \leftrightarrow e_\emptyset \\[2mm]
\begin{pmatrix} 0 & 1 \\ 1 & 0 \end{pmatrix} = X \leftrightarrow ie_1 e_2 \\[2mm]
\begin{pmatrix} 0 & -i \\ i & 0 \end{pmatrix} = Y \leftrightarrow ie_2 e_0 \\[2mm]
\begin{pmatrix} 1 & 0 \\ 0 & -1 \end{pmatrix} = Z \leftrightarrow ie_0 e_1
\end{cases}
\tag{8.1}
$$

in a context, where the signature λ is all -1. Pauli matrices form a particular subset of three-dimensional Clifford numbers. Although three-dimensional Clifford numbers have four grades and eight components, Pauli matrices retain only two of the grades (zero and two) and four of the components.

The unit Pauli matrices obey the rules for multiplication:

$$
\begin{cases}
I = X^2 = Y^2 = Z^2 \\
X = -iYZ \\
Y = -iZX \\
Z = -iXY
\end{cases}
\tag{8.2}
$$

These rules are honoured by the Clifford algebra when Grassmann's central multiplication is used.

---

1 Pauli used S_x, S_y, S_z. Dirac later renamed these as $\sigma_1, \sigma_2, \sigma_3$.

Numerical Calculations in Clifford Algebra: A Practical Guide for Engineers and Scientists,
First Edition. Andrew Seagar.
© 2023 John Wiley & Sons Ltd. Published 2023 by John Wiley & Sons Ltd.

Table 8.1 Geometric operations on Pauli matrices.

Magnitude	$\lvert P \rvert \;=\; \sqrt{(P, P^{*})}$
Inversion (reciprocal)	$1/P \;=\; \check{P}/(P\check{P})$
Reflection	$\mathbb{Q}\,P \;=\; -nPn$
Inner projection	$\mathbb{P}^{-}\,P \;=\; \frac{1}{2}(\mathbb{I} - \mathbb{Q})\,P$
Outer projection	$\mathbb{P}^{+}\,P \;=\; \frac{1}{2}(\mathbb{I} + \mathbb{Q})\,P$
Rotation	$\mathbb{R}\,P \;=\; n_2 n_1 P n_1 n_2$

Comparison of Eq. (8.1) to the equivalent identification for quaternions in Eq. (7.1) indicates that the Pauli matrices and quaternions are closely related. The non-scalar Pauli units X, Y, Z are effectively the same as the quaternion units multiplied by the imaginary unit i. They behave under central multiplication most like imaginary[2] bivectors and could be interpreted geometrically as such. If proper bivectors were required, quaternions may be a better choice.

Table 8.1 lists the operations provided within the Clifford numerical suite for Pauli matrices. All multiplications are central (Clifford). Vectors n, n_1, and n_2 have real coefficients and unit length.

8.1.1 Recovery of Components

Unlike quaternions and vectors, the individual Pauli unit matrices do not prevent the different components from being combined into a single matrix entity. The underlying structure of units is then not readily visible. That may or may not be important, depending on the particular application.

When individual components are combined into a single matrix entity, nothing is lost since the original four components can always be recovered from the four independent values within the matrix. Recovery of an individual component P' which has Pauli unit matrix U (I, X, Y, or Z) from a Pauli matrix P can be achieved using inner multiplication and inverse operations as in Eq. (8.3):

$$P' = (P, U)/(U, U) \tag{8.3}$$

8.1.2 Magnitude

The magnitude of a Pauli matrix can be calculated directly from the coefficients of the individual components. If the coefficients are real, the magnitude is the square root of the sum of the squares of all of its coefficients. If the coefficients are complex, the product from multiplying the coefficient with its complex conjugate is taken instead of the square.

---

2 Because of the imaginary value i which emerges in their products.

The matrix itself cannot be used directly to calculate the magnitude. The sum of the matrix values multiplied by their complex conjugates gives a different result. It would be necessary to recover the coefficients from the matrix as in Section 8.1.1 first.

For Pauli matrices within the Clifford formalism, the square of the magnitude is the same as the inner product of the matrix with its complex conjugate.

8.1.3 Inverse

The inverse of a Pauli matrix is another Pauli matrix. Multiplying a Pauli matrix by its inverse from either side produces the unit Pauli matrix I.

In Clifford algebra, many Pauli matrices do have an inverse, but that is not guaranteed. Some Pauli matrices which have non-zero magnitude do not have any inverse. Take as example the Pauli matrix $P = X + iY = \begin{pmatrix} 0 & 2 \\ 0 & 0 \end{pmatrix}$, which has $s = r = 0$, $p = 1$, and $q = i$. The square of the magnitude is obtained from inner multiplication as the non-zero value:

$$(X + iY, X^* + i^*Y^*) = X\tilde{X}^* + iYi^*\tilde{Y}^* = -X\tilde{X} - iYi^*\tilde{Y} = X^2 + (ii^*)Y^2 = I + I = 2I \neq 0$$

$$(8.4)$$

However, the matrix $\begin{pmatrix} 0 & 2 \\ 0 & 0 \end{pmatrix}$ is singular by inspection and clearly has no inverse.

8.1.4 Reflection, Projection, and Rotation

Reflection and projection of Pauli matrices behave in exactly the same way as for quaternions in Section 7.1.3. The rotation of Pauli matrices with no scalar component behaves in the same way as for quaternions in Section 7.1.4. In all cases, there is simply an extra factor of i in all calculations carried in Clifford form.

8.2 Practice

8.2.1 Example Code

The code listed in Example 8.1a demonstrates the use of operations on Pauli matrices from Table 8.1. Lines 8–11 establish a three-dimensional Clifford context with all negative signature. The unit Pauli matrices are created in lines 14–17 and printed in lines 19–22. The rules for multiplication from Eq. (8.2) are tested in lines 25–30 and printed in lines 33–38.

A non-unit Pauli matrix with value

$$P = (1 + 2i)I + (3 + 4i)X + (5 + 6i)Y + (7 + 8i)Z = \begin{pmatrix} 8 + 10i & 9 - i \\ -3 + 9i & -6 - 6i \end{pmatrix} \qquad (8.5)$$

is created on line 43 and printed on line 51. The four components are extracted in lines 44–47, and the results are printed in lines 52–55.

Calculations using the magnitude and inverse operations are executed in lines 59–64 and printed in lines 67–72. The Pauli unit matrix X is rotated by 90° in the xy plane in lines 78–81, with results printed in lines 83–87.

Example 8.1(a)

```
1:   void demo17() /* pauli matrices */
2:   {
3:      double a[3];
4:      int lambda[3];
5:      number p[7];
6:      number x,y,z,n,o;
7:
8:      recovery_level(0,0,0);
9:      create_cliffmem(50);
10:     declare_primal(3,pack_signature(3,lambda,-1));
11:     format_printing("%.0f");
12:
13:     /* unit pauli matrices */
14:     o=pack_pauli(1.0,0.0,0.0,0.0,0.0,0.0,0.0,0.0);
15:     x=pack_pauli(0.0,0.0,1.0,0.0,0.0,0.0,0.0,0.0);
16:     y=pack_pauli(0.0,0.0,0.0,0.0,1.0,0.0,0.0,0.0);
17:     z=pack_pauli(0.0,0.0,0.0,0.0,0.0,0.0,1.0,0.0);
18:
19:     printf(" [1] = \n");  print_pauli(o);
20:     printf(" [X] = \n");  print_pauli(x);
21:     printf(" [Y] = \n");  print_pauli(y);
22:     printf(" [Z] = \n");  print_pauli(z);
23:
24:     /* pauli central multiplication*/
25:     p[0]=central_mul(x,x,0);
26:     p[1]=central_mul(y,y,0);
27:     p[2]=central_mul(z,z,0);
28:     p[3]=central_mul(x,y,0);
29:     p[4]=central_mul(y,z,0);
30:     p[5]=central_mul(z,x,0);
31:     p[6]=central_mul(central_mul(x,y,0),z,1);
32:
33:     printf("\n[X]*[X] = \n");    print_pauli(p[0]);
34:     printf(" [Y]*[Y] = \n");     print_pauli(p[1]);
35:     printf(" [Z]*[Z] = \n");     print_pauli(p[2]);
36:     printf(" [X]*[Y] = \n");     print_pauli(p[3]);
37:     printf(" [Y]*[Z] = \n");     print_pauli(p[4]);
38:     printf(" [Z]*[X] = \n");     print_pauli(p[5]);
39:     printf(" [X]*[Y]*[Z] = \n"); print_pauli(p[6]);
40:     recycle_array(7,p);
41:
42:     /* unit extraction */
43:     p[0]=pack_pauli(1.0,2.0,3.0,4.0,5.0,6.0,7.0,8.0);
44:     p[1]=central_mul(inner_mul(p[0],o,0),inverseof_pauli
                              (inner_mul(o,o,0),1),3);
```

```
45:      p[2]=central_mul(inner_mul(p[0],x,0),inverseof_pauli
                               (inner_mul(x,x,0),1),3);
46:      p[3]=central_mul(inner_mul(p[0],y,0),inverseof_pauli
                               (inner_mul(y,y,0),1),3);
47:      p[4]=central_mul(inner_mul(p[0],z,0),inverseof_pauli
                               (inner_mul(z,z,0),1),3);
48:      recycle_number(o);  recycle_number(x);
49:      recycle_number(y);  recycle_number(z);
50:
51:      printf("\n[P] = \n");                   print_pauli(p[0]);
52:      printf("([P],[I])/([I],[I]) = \n");  print_pauli(p[1]);
53:      printf("([P],[X])/([X],[X]) = \n");  print_pauli(p[2]);
54:      printf("([P],[Y])/([Y],[Y]) = \n");  print_pauli(p[3]);
55:      printf("([P],[Z])/([Z],[Z]) = \n");  print_pauli(p[4]);
56:      recycle_array(4,p+1);
57:
58:      /* magnitude and inverse */
59:      x=magnitudeof_pauli(p[0],0);
60:      p[1]=inverseof_pauli(p[0],0);
61:      p[2]=central_mul(p[0],p[1],0);
62:      y=magnitudeof_pauli(p[2],0);
63:      o=scalar_number(1.0,0.0);
64:      z=magnitudeof_number(sub_number(p[2],o,2),1);
65:
66:      format_printing("%.3f");
67:      printf("\n[P] = \n");          print_pauli(p[0]);
68:      printf("|[P]| = ");            print_number(x,8);
69:      printf("1/[P] = \n");          print_pauli(p[1]);
70:      printf("[P]*1/[P] = \n");      printf_pauli("%.1e",p[2]);
71:      printf("|[P]*1/[P]| = ");      print_number(y,8);
72:      printf("|[P]*1/[P]-1| = ");    printf_number("%e",z,8);
73:      recycle_number(x);            recycle_number(y);
74:      recycle_number(z);
75:      recycle_array(3,p);
76:
77:      /* rotation */
78:      n=pack_R3vector(pack_R3array(1.0,0.0,0.0,a));
79:      x=normalise_vector(pack_R3vector(pack_R3array(1.0,1.0,0.0,a)),1);
80:      y=pack_pauli(0.0,0.0,1.0,0.0,0.0,0.0,0.0,0.0);
81:      z=rotate_number(y,n,x,0);
82:
83:      printf("\nunit vector n1 = "); print_vector(n);
84:      printf("unit vector n2 = ");   print_vector(x);
85:      printf("[P] =\n");             print_pauli(y);
86:      printf("n2*n1*[P]*n1*n2 = rotation of [P] in plane n1,n2;
                               2*angle between n1,n2\n");
87:      printf(" =\n");                print_pauli(z);
88:      recycle_number(x);            recycle_number(y);
89:      recycle_number(z);            recycle_number(n);
90:
91:      free_cliffmem(0);
92:  }
```

8.2.2 Example Output

The results from running the code listed in Example 8.1a are given in Example 8.1b. The unit Pauli matrices in lines 2–13 are the same as those in Eq. (8.1). Comparison of the products in lines 15–32 with the expected values in Eq. (8.2) confirms the veracity of the results.

Lines 40–51 show the coefficients correctly recovered from the Pauli matrix in Eq. (8.5). The results for the calculations involving magnitude and inverse behave as expected in lines 56–64 with accuracy of the underlying double precision data type.

Lastly, lines 66–74 show that the result produced by rotating the Pauli unit matrix X by $90°$ in the xy plane is the unit matrix Y.

Example 8.1(b)

```
 1:    demo17()
 2:    [1] =
 3:    [(1+i0)  (0+i0)]
 4:    [(0+i0)  (1+i0)]
 5:    [X] =
 6:    [(0+i0)  (1+i0)]
 7:    [(1+i0)  (0+i0)]
 8:    [Y] =
 9:    [(0+i0)  (0-i1)]
10:    [(0+i1)  (0+i0)]
11:    [Z] =
12:    [(1+i0)  (0+i0)]
13:    [(0+i0)  (-1+i0)]
14:
15:    [X]*[X] =
16:    [(1+i0)  (0+i0)]
17:    [(0+i0)  (1+i0)
18:    [Y]*[Y] =
19:    [(1+i0)  (0+i0)]
20:    [(0+i0)  (1+i0)]
21:    [Z]*[Z] =
22:    [(1+i0)  (0+i0)]
23:    [(0+i0)  (1+i0)]
24:    [X]*[Y] =
25:    [(0+i1)  (0+i0)]
26:    [(0+i0)  (0-i1)]
27:    [Y]*[Z] =
28:    [ (0+i0)  (0+i1)]
29:    [(-0+i1)  (0+i0)]
30:    [Z]*[X] =
31:    [ (0+i0)  (1+i0)]
32:    [(-1+i0)  (0+i0)]
33:    [X]*[Y]*[Z] =
34:    [(0+i1)  (0+i0)]
35:    [(0+i0)  (0+i1)]
36:
37:    [P] =
38:    [(8+i10)  (9-i1)]
```

```
39:    [(-3+i9)  (-6-i6)]
40:    ([P],[I])/([I],[I]) =
41:    [(1+i2)  (0+i0)]
42:    [(0+i0)  (1+i2)
43:    ([P],[X])/([X],[X]) =
44:    [(3+i4)  (0+i0)]
45:    [(0+i0)  (3+i4)]
46:    ([P],[Y])/([Y],[Y]) =
47:    [(5+i6)  (0+i0)]
48:    [(0+i0)  (5+i6)]
49:    ([P],[Z])/([Z],[Z]) =
50:    [(7+i8)  (0+i0)]
51:    [(0+i0)  (7+i8)]
52:
53:    [P] =
54:    [(8.000+i10.000)  (9.000-i1.000)]
55:    [(-3.000+i9.000)  (-6.000-i6.000)]
56:    |[P]| = (14.283+i0.000)e{}
57:    1/[P] =
58:    [(0.026-i0.035)  (-0.012-i0.045)]
59:    [(0.048+i0.008)  (-0.044+i0.049)]
60:    [P]*1/[P] =
61:    [ (1.0e+00-i2.8e-17)  (9.0e-17-i2.4e-17)]
62:    [(-2.1e-17+i3.1e-17)  (1.0e+00-i2.8e-17)]
63:    |[P]*1/[P]| = (1.000+i0.000)e{}
64:    |[P]*1/[P]-1| = (2.364575e-16+i0.000000e+00)e{}
65:
66:    unit vector n1 = [1.000+i0.000, 0.000+i0.000, 0.000+i0.000]
67:    unit vector n2 = [0.707+i0.000, 0.707+i0.000, 0.000+i0.000]
68:    [P] =
69:    [(0.000+i0.000)  (1.000+i0.000)]
70:    [(1.000+i0.000)  (0.000+i0.000)]
71:    n2*n1*[P]*n1*n2 = rotation of [P] in plane n1,n2;
                                2*angle between n1,n2
72:      =
73:    [(0.000+i0.000)  (0.000-i1.000)]
74:    [(0.000+i1.000)  (0.000+i0.000)]
```

Reference

Pauli, W. (1927), 'Zur quantenmechanik des magnetischen elektrons', *Zeitschrift für Physik* **43**(9–10), 601–623.

9

Bicomplex Numbers

Bicomplex numbers $C = z_1 + hz_2$ were first constructed by Segre (1892) as complex numbers composed using a new imaginary unit h. The 'real' z_1 and 'imaginary' z_2 parts of the bicomplex numbers are themselves both conventional complex numbers $z_1 = x_1 + iy_1$ and $z_2 = x_2 + iy_2$, based on the usual imaginary unit i. In isolation, the new unit h behaves exactly like the original unit i. Together, the two units are independent and do not combine under addition or multiplication, but do commute without change of sign. If either of the imaginary units is wholly absent due to the values of the corresponding coefficients being zero, then the whole bicomplex number behaves like an ordinary complex number.

9.1 Theory

Bicomplex numbers $C = z_1 \mathfrak{R} + z_2 \mathfrak{J}$ are represented in Clifford algebra by identifying new real and imaginary units $\mathfrak{R}$ and $\mathfrak{J}$ with particular Clifford units:

$$\begin{cases} \mathfrak{R} \leftrightarrow e_\emptyset \\ \mathfrak{J} \leftrightarrow e_0 e_1 \leftrightarrow h e_\emptyset \end{cases} \tag{9.1}$$

in a context where the signature λ is all -1.

Bicomplex numbers form a particular subset of two-dimensional Clifford numbers. Although two-dimensional Clifford numbers have three grades and four components, these bicomplex numbers retain only two of the grades (zero and two) and two of the components.

The new real and imaginary units obey the normal rules for multiplication of ordinary complex units:

$$\begin{cases} \mathfrak{R}^2 = \mathfrak{R} \\ \mathfrak{J}^2 = -\mathfrak{R} \\ \mathfrak{J}\mathfrak{R} = \mathfrak{R}\mathfrak{J} \end{cases} \tag{9.2}$$

These rules are honoured by the Clifford algebra when Grassmann's central multiplication is used.

A comparison of Eq. (9.1) to the equivalent identification for quaternions in Eq. (7.1) indicates that bicomplex numbers are equivalent to two of the unit quaternions, 1 and K.

Numerical Calculations in Clifford Algebra: A Practical Guide for Engineers and Scientists,
First Edition. Andrew Seagar.
© 2023 John Wiley & Sons Ltd. Published 2023 by John Wiley & Sons Ltd.

Table 9.1 Geometric operations on bicomplex numbers.

Magnitude	$\|C\| = \sqrt{(C, C^*)}$
Inversion (reciprocal)	$1/C = \check{C}/(C\check{C})$
Reflection	$\mathbb{Q}\,C = -nCn$
Inner projection	$\mathbb{P}^-\,C = \frac{1}{2}(\mathbb{I} - \mathbb{Q})\,C$
Outer projection	$\mathbb{P}^+\,C = \frac{1}{2}(\mathbb{I} + \mathbb{Q})\,C$
Rotation	$\mathbb{R}\,C = n_2 n_1 C n_1 n_2$

The other two unit quaternions are missing. This is consistent with the notion that quaternions are a generalisation of both complex and bicomplex numbers.

Table 9.1 lists the operations provided within the Clifford numerical suite for bicomplex numbers. All multiplications are central (Clifford). Vectors n, n_1, and n_2 have real coefficients and unit length.

9.1.1 Conjugate

The complex conjugate '$*$' operates on the imaginary parts of the complex coefficients z_1, z_2. The role of the complex conjugate for the pseudo-scalar unit $\mathfrak{J}$ is taken by the reversion operator:

$$(z_1\widetilde{\mathfrak{R} + z_2\mathfrak{J}}) = \widetilde{z_1 e_\emptyset} + \widetilde{z_2 e_0 e_1} = z_1 e_\emptyset + z_2 e_1 e_0 = z_1 e_\emptyset - z_2 e_0 e_1 = z_1\mathfrak{R} - z_2\mathfrak{J} \qquad (9.3)$$

so that the product of a bicomplex number with its conjugate is

$$(z_1\mathfrak{R} + z_2\mathfrak{J})(z_1\widetilde{\mathfrak{R} + z_2\mathfrak{J}}) = (z_1\mathfrak{R} + z_2\mathfrak{J})(z_1\mathfrak{R} - z_2\mathfrak{J}) = z_1^2\mathfrak{R}^2 - z_2^2\mathfrak{J}^2 = (z_1^2 + z_2^2)\mathfrak{R} \qquad (9.4)$$

9.1.2 Magnitude

The magnitude of a real valued bicomplex number is the square root of the sum of the squares of all of its coefficients. If the coefficients are complex, the product from multiplying the coefficient with its complex conjugate is taken instead of the square.

For bicomplex numbers, the square of the magnitude is the same as the inner product of the bicomplex number with its complex conjugate.

9.1.3 Inverse

The inverse of a bicomplex number is another bicomplex number. Multiplying a bicomplex number by its inverse from either side produces the unit bicomplex number $\mathfrak{R}$.

In Clifford algebra, many bicomplex numbers do have an inverse, but that is not guaranteed. Some bicomplex numbers which have non-zero magnitude do not have any inverse.

Take as example the bicomplex number $C = 1\mathfrak{R} + i\mathfrak{I}$. The square of the magnitude is obtained from inner multiplication as the non-zero value:

$$(\mathfrak{R} + i\mathfrak{I}, \mathfrak{R} + i^*\mathfrak{I}) = (\mathfrak{R} + i\mathfrak{I}, \mathfrak{R} - i\mathfrak{I}) = \mathfrak{R}\tilde{\mathfrak{R}} + \mathfrak{I}\tilde{\mathfrak{I}} = \mathfrak{R}^2 + \mathfrak{I}(-\mathfrak{I})$$

$$= \mathfrak{R} + \mathfrak{R} = 2\mathfrak{R} \neq 0 \tag{9.5}$$

However, the denominator, when attempting to calculate the inverse by putting $z_1 = 1$ and $z_2 = i$ into Eq. (9.4), produces the value:

$$(\mathfrak{R} + i\mathfrak{I})(\mathfrak{R} - i\mathfrak{I}) = (1^2 + i^2)\mathfrak{R} = 0 \tag{9.6}$$

The denominator is zero so that no inverse can be found.

9.1.4 Reflection, Projection, and Rotation

The bicomplex numbers are effectively a subset of quaternions, with the coefficients for two of the quaternion units set to zero. Therefore, the operations of reflection, projection, and rotation for bicomplex numbers behave essentially in the same way as for quaternions. However, the bicomplex numbers span only two Clifford dimensions rather than the three for quaternions, so are somewhat more limited.

For the rotation operator $\mathbb{R}$, the only plane available for rotation is the plane of the bivector $e_0 e_1$ identified with the imaginary bicomplex unit $\mathfrak{I}$. Rotation of a bivector in its own plane is an identity operation, giving as result the original bivector. Rotation of bicomplex numbers using the rotation operator in Table 9.1 is not particularly useful.

Rotation in the bicomplex plane formed by an interpretation of the scalar unit $\mathfrak{R}$ and the pseudo-scalar unit $\mathfrak{I}$ as orthogonal vectors[1] can be achieved as for ordinary complex numbers in the ordinary complex plane, multiplying one bicomplex number by a unit bicomplex number which represents the angle of rotation.

$$(z(\cos\theta)\mathfrak{R} + z(\sin\theta)\mathfrak{I})((\cos\phi)\mathfrak{R} + (\sin\phi)\mathfrak{I})$$

$$= z(\cos\theta\cos\phi - \sin\theta\sin\phi)\mathfrak{R} + z(\cos\theta\sin\phi + \sin\theta\cos\phi)\mathfrak{I}$$

$$= z\cos(\theta + \phi)\mathfrak{R} + z\sin(\theta + \phi)\mathfrak{I} \tag{9.7}$$

However, it would probably make more sense to use a two-dimensional Clifford vector basis instead, as in Chapter 6.

9.2 Practice

9.2.1 Example Code

The code shown in Example 9.1a demonstrates the use of operations on bicomplex numbers as selected from Table 9.1. A two-dimensional context with negative signature is established in lines 7–10. The bicomplex units $\mathfrak{R}$ and $\mathfrak{I}$ are created on lines 12 and 14 and printed on

---

1 Which they are not.

lines 13 and 15. The square of the imaginary bicomplex unit $\Im$ is calculated on line 17 and printed on line 18.

Example 9.1(a)

```
 1:   void demo18() /* bicomplex numbers */
 2:   {
 3:       double a[2][2];
 4:       int lambda[2];
 5:       number x,y,z,c,n;
 6:
 7:       recovery_level(0,0,0);
 8:       create_cliffmem(50);
 9:       declare_primal(2,pack_signature(2,lambda,-1));
10:       format_printing("%0.f");
11:
12:       x=pack_Rbicomplex(1.0,0.0);
13:       printf("1 = ");              print_bicomplex(x);  recycle_number(x);
14:       x=pack_Rbicomplex(0.0,1.0);
15:       printf("h = ");              print_bicomplex(x);
16:
17:       c=central_mul(x,x,1); printf("h*h = ");
18:                                    print_bicomplex(c);  recycle_number(c);
19:
20:       x=pack_Rbicomplex(1.0,2.0);     printf("C = ");
                                          print_bicomplex(x);
21:       z=magnitudeof_bicomplex(x,0);   printf("|C| = ");
22:                                       printf_number("%e",z,8);
                                          recycle_number(z);
23:       y=inverseof_bicomplex(x,0);     printf("1/C = ");
                                          printf_bicomplex("%.5f",y);
24:       z=central_mul(x,y,3);           printf("C*1/C = ");
                                          printf_bicomplex("%e",z);
25:       x=magnitudeof_number(z,0);      printf("|C*1/C| = ");
26:                                       printf_number("%e",x,8);
                                          recycle_number(x);
27:       y=scalar_number(1.0,0.0);
28:       z=subfrom_number(z,y);
29:       z=magnitudeof_number(z,1);   printf("|C*1/C-1| = ");
                                       printf_number("%e",z,8);
30:                                    recycle_number(y);
                                       recycle_number(z);
31:       a[0][0]=1.0;  a[0][1]=0.0;
32:       a[1][0]=0.0;  a[1][1]=0.0;
33:       n=pack_Nvector(2,a);
34:
35:       a[0][0]=1.0;  a[0][1]=0.0;
36:       a[1][0]=1.0;  a[1][1]=0.0;
37:       x=normalise_vector(pack_Nvector(2,a),1);
38:
39:       printf("unit vector n1 = ");       print_vector(n);
40:       printf("unit vector n2 = ");       printf_vector("%.3f",x);
41:
```

```
42:      y=pack_Rbicomplex(0.0,1.0);
43:      printf("C = ");                        printf_bicomplex("%0.f",y);
44:      printf("C = ");                        printf_quaternion("%0.f",y);
45:      printf("C = ");                        printf_number("%0.f",y,4);
46:
47:      x=rotate_number(y,n,x,7);
48:      printf("n2*n1*C*n1*n2  = rotation of C in plane n1,n2; 2*angle
                                 between n1,n2\n");
49:      printf("            = ");              printf_bicomplex("%0.f",x);
50:      printf("            = ");              printf_quaternion("%0.f",x);
51:      printf("            = ");              printf_number("%0.f",x,4);
52:      recycle_number(x);
53:
54:      free_cliffmem(0);
55:    }
```

A Clifford number is initialised to a simple bicomplex number $1\Re + 2\mathfrak{I}$ on line 20.

On line 21, the magnitude is calculated. The inverse is taken on line 23 and multiplied by the number itself on line 24. The magnitude of that number, which should be unity, is calculated on line 25, and the magnitude of the difference between unity and the product of bicomplex number and inverse, which should be zero, is calculated on line 29.

Results of the calculations of magnitude and inverse are printed throughout lines 22–29.

Examples of calculations involving rotation are given from lines 31–47. The statements, which print the results, are given from lines 49–51 in three formats (bicomplex, quaternion, and number). These should be compared the value of the bicomplex number before rotation on lines 43–45.

9.2.2 Example Output

The results of running the code listed in Example 9.1a are shown in Example 9.1b. Lines 2 and 3 demonstrate the manner in which the bicomplex units are printed, and line 4 confirms the square of the bicomplex imaginary unit.

Lines 6–10 give the results for various mixed inverse and magnitude operations for the number printed on line 5.

Example 9.1(b)

```
1:    demo18()
2:    1 = (1+i0)Re
3:    h = (1+i0)Im
4:    h*h = (-1+i0)Re
5:    C = (1+i0)Re+(2+i0)Im
6:    |C| = (2.236068e+00+i0.000000e+00)e{}
7:    1/C = (0.20000+i0.00000)Re+(-0.40000+i0.00000)Im
8:    C*1/C = (1.000000e+00+i0.000000e+00)Re
9:    |C*1/C| = (1.000000e+00+i0.000000e+00)e{}
```

(Continued)

Example 9.1(b) (Continued)

```
10:   |C*1/C-1|  =  (0.000000e+00+i0.000000e+00)e{}
11:   unit vector n1 = [1+i0, 0+i0
12:   unit vector n2 = [0.707+i0.000, 0.707+i0.000
13:   C = (1+i0)Im
14:   C = (1+i0)K
15:   C = (1+i0)e{0,1}
16:   n2*n1*C*n1*n2  = rotation of C in plane n1,n2;  2*angle between n1,n2
17:                  = (1+i0)Im
18:                  = (1+i0)K
19:                  = (1+i0)e{0,1}
```

The product from multiplying the number on line 5 with its inverse on line 7 is the real bicomplex unit on line 8, which has magnitude one on line 9. The accuracy of the inverse is given in line 10 by calculating the magnitude of value obtained by subtracting one from the product of vector and its inverse. The underlying machine precision is of type 'double' with around 15–16 decimal places. In this case that accuracy has been exceeded, by chance rather than by design.

Lines 11–19 demonstrate the 90° rotation in the plane of the basis vectors for the bivector imaginary bicomplex unit. The result shows no change in the value of the number, as expected.

Reference

Segre, C. (1892), 'Le rappresentazioni reali delle forme complesse e gli enti iperalgebrici', *Mathematische Annalen* **40**(3), 413–467.

10

Electromagnetic Fields

10.1 Theory

In Clifford algebra, electromagnetic fields are represented[1] by identifying the temporal unit vector $\hat{\mathbf{t}}$ and the spatial unit vectors $\hat{\mathbf{x}}, \hat{\mathbf{y}}, \hat{\mathbf{z}}$ with particular Clifford units:

$$\begin{cases} \hat{\mathbf{t}} \leftrightarrow ie_0 \\ \hat{\mathbf{x}} \leftrightarrow e_1 \\ \hat{\mathbf{y}} \leftrightarrow e_2 \\ \hat{\mathbf{z}} \leftrightarrow e_3 \end{cases} \tag{10.1}$$

in a context, where the signature λ is all -1.

Four dimensions are required, with the extra dimension used for time or frequency. The Clifford unit for time is imaginary rather than real so that the Clifford geometry with uniform signatures behaves like the Minkowskian geometry, which has mixed signatures.[2]

Electromagnetic fields in four dimensions form a particular subset of four-dimensional Clifford numbers. Although four-dimensional Clifford numbers have 5 grades and 16 components, electromagnetic fields retain only one of the grades (grade 2) and six of the components. Three of these components are used to accommodate the magnetic field, and the other three are used for the electric field.

10.1.1 Time and Frequency

Electromagnetic fields are in general functions of both three-dimensional space and one-dimensional time. When a problem is formulated for fields in that form, it is said to be cast within the time domain.

In cases where the field is periodic in time, it is often convenient to decompose the temporal aspect of the field into a sum of discrete sinusoidal components at frequencies separated equally by a constant value $\Delta\omega$ and examine each component of frequency in isolation.

---

1 This choice is appropriate for the flat (Minkowskian) spacetime of special relativity, to which Maxwell's classical equations of electromagnetism conform.
2 If preferred, it is also possible to represent electromagnetic fields in the Clifford numerical suite using mixed signatures.

Numerical Calculations in Clifford Algebra: A Practical Guide for Engineers and Scientists,
First Edition. Andrew Seagar.
© 2023 John Wiley & Sons Ltd. Published 2023 by John Wiley & Sons Ltd.

When a problem is formulated for fields in terms of these isolated frequencies, it is said to be cast within the frequency domain.

Time and frequency domains are related to one another by Fourier transformation, as in Chapter 21. Chapter 23 gives an example of an electromagnetic problem solved in the time domain. Chapter 24 gives an example of an electromagnetic problem solved in the frequency domain.

10.1.2 Electromagnetic Entities

The entities used to represent the electromagnetic field mathematically are listed for both time and frequency domains in Table 10.1, where $\sigma = -e_1 e_2 e_3$ is the negatively signed unit of volume for the three spatial dimensions, and where $c = 1/(\sqrt{\mu}\sqrt{\epsilon})$ is[3] the velocity of propagation for electromagnetic waves through a region where the magnetic permeability is μ and the electric permittivity is ϵ.

The bivector field F has magnetic H and electric E components, both of which are vectors in the three spatial dimensions. The source of the field consists of one or both of current J and charge ρ sources. The current source is a vector, and the charge source is a scalar. A potential P is sometimes used with vector A and scalar ϕ components. The differential operator consists of a three-dimensional vectorial spatial component:

$$\nabla = e_1 \frac{\partial}{\partial x} + e_2 \frac{\partial}{\partial y} + e_3 \frac{\partial}{\partial z} \tag{10.2}$$

and a one-dimensional component αe_0 for which the coefficient α is either a temporal derivative $(-\frac{i}{c}\frac{\partial}{\partial t})$ in the time domain or a simple scalar variable (k_m) in the frequency domain. The scalar k_m is called the wavenumber.

If a field $F(\mathbf{r}, t)$ is periodic in time, then it can be composed of a Fourier series (a sum) of individual monochromatic sinusoidal components:

$$F(\mathbf{r}, t) = \sum_m F_{k_m}(\mathbf{r}) e^{i\omega_m t} \tag{10.3}$$

where $\omega_m = m\Delta\omega$ is the angular frequency for each sinusoidal component (or *mode m*) and $k_m = \omega_m/c$ is the corresponding wavenumber. The coefficients of the individual bivector components within the single terms $F_{k_m}(\mathbf{r})$ are complex scalars, often called phasors. Here the whole term $F_{k_m}(\mathbf{r})$ is interpreted as a generalised multi-dimensional Clifford valued multi-phasor.

Table 10.1 Electromagnetic entities in time and frequency domain.

	Time domain, $F(\mathbf{r}, t)$		Frequency domain, $F_{k_m}(\mathbf{r})$	
Operator	$D = \nabla$	$- \frac{i}{c}e_0\frac{\partial}{\partial t}$	$D_{k_m} = \nabla$	$+ k_m e_0$
Field	$F = \sqrt{\mu}H\sigma - i\sqrt{\epsilon}Ee_0$		$F_{k_m} = \sqrt{\mu}H_{k_m}\sigma - i\sqrt{\epsilon}E_{k_m}e_0$	
Source	$S = \sqrt{\mu}J$	$+ \frac{i}{\sqrt{\epsilon}}\rho e_0$	$S_{k_m} = \sqrt{\mu}J_{k_m}$	$+ \frac{i}{\sqrt{\epsilon}}\rho_{k_m}e_0$
Potential	$P = \frac{1}{\sqrt{\mu}}A$	$+ i\sqrt{\epsilon}\phi e_0$	$P_{k_m} = \frac{1}{\sqrt{\mu}}A_{k_m}$	$+ i\sqrt{\epsilon}\phi_{k_m}e_0$

Source: Seagar (2012, 2016).

_____

3 For complex numbers $\sqrt{ab}$ does not always give the same value as $\sqrt{a}\sqrt{b}$.

Applying the differential operator D to a monochromatic field in the time domain is equivalent to applying the differential operator D_{k_m} to the corresponding multi-phasor in the frequency domain and then multiplying by the temporal factor $e^{i\omega_m t}$:

$$DF(\mathbf{r}, t) = D \left[F_{k_m}(\mathbf{r}) e^{i\omega_m t} \right] = (\nabla - \tfrac{i}{c} e_0 \tfrac{\partial}{\partial t}) \ [F_{k_m}(\mathbf{r}) \ e^{i\omega_m t}]$$
$$= [(\nabla + k_m e_0) \ F_{k_m}(\mathbf{r})] \ e^{i\omega_m t} = \left[D_{k_m} F_{k_m}(\mathbf{r}) \right] e^{i\omega_m t}$$

(10.4)

In the frequency domain, the temporal factor appears on both sides of all equations. For simplicity it may be omitted, leaving only the multi-phasors, and restored at any later time if required.

10.1.3 Dirac Operators

In the time domain, the differential operator D is a four-dimensional spatiotemporal Dirac operator. In the frequency domain, the differential operator D_{k_m} can be viewed in two ways: either as the temporal Fourier transform of the operator D, or as a three-dimensional spatial Dirac operator with a constant term added in a fourth (frequency) dimension. With the second interpretation, the constant term is known as a perturbation and the operator is known as a perturbed Dirac operator. Here the first interpretation is preferred, and the operator D_{k_m} is referred to as the k-Dirac operator (Axelsson et al. 2001).

10.1.4 Maxwell's Equations

In the four-dimensional Clifford context with negative signature, Maxwell's (1873a, 1873b) equations of electromagnetism in time and frequency domain relate the derivative of the field to the sources:

$$\begin{cases} DF = S \\ D_{k_m} F_{k_m} = S_{k_m} \end{cases}$$

(10.5)

The differential operators are first-order ordinary (not partial) differential operators, so these equations are first-order ordinary differential equations. The temporal factor $e^{i\omega_m t}$ which could be written on both sides of the second line of Eq. (10.5), has been omitted.

The potentials and fields are related by the differential operators in the same way as the fields and sources:

$$\begin{cases} DP = F \\ D_{k_m} P_{k_m} = F_{k_m} \end{cases}$$

(10.6)

For applications using electromagnetic fields and the solution of Maxwell's equations, see Chapters 22 to 24.

10.1.5 Simplified Notation

It is convenient when dealing with only one phasor at a time, to leave the mode m unwritten in order to simplify the notation. It is possible to do that in all equations of Sections 10.1.2

and 10.1.4 except for Eq. (10.3), where the index of summation remains strictly an integer. The simpler notation with the mode unwritten is adopted in Chapter 22.

At times it is also convenient go even further and to drop the distinction between the time and frequency domain, with both the mode m and the wavenumber k left unwritten. This approach is used in Sections 10.1.6–10.1.9 and in Chapters 24 and 25.

10.1.6 Magnitude

The magnitude of the bivector electromagnetic field is calculated by taking the square root of the sum of the products from multiplying all six coefficients by their complex conjugates:

$$|\mathcal{F}| = \sqrt{|\sqrt{\mu}H|^2 + |\sqrt{\epsilon}E|^2} \tag{10.7}$$

It is not possible to bring the material properties outside the magnitude squared as μ or ϵ unless they are purely real and positive with no imaginary part.

10.1.7 Inverse

A Clifford bivector u in a four-dimensional context with all negative signature written in the form:

$$u = a\sigma + be_0 \tag{10.8}$$

where a and b are Clifford vectors in the three spatial dimensions, may or may not have an inverse. If there is an inverse, one of the forms it can take is the following:

$$\frac{1}{a\sigma + be_0} = \frac{1}{u} = u^{-1} = \frac{(\sigma u\sigma)^2 u}{(a-b)^2(a+b)^2} = \frac{u(\sigma u\sigma)^2}{(a-b)^2(a+b)^2} \tag{10.9}$$

With $a = \sqrt{\mu}H$ and $b = -i\sqrt{\epsilon}E$, the number u plays the role of an electromagnetic field $\mathcal{F} = u = \sqrt{\mu}H\sigma - i\sqrt{\epsilon}Ee_0$, and the inverse $\mathcal{F}'$ of the field $\mathcal{F}$ can in principle be calculated as a whole using Eq. (10.9). However, not all electromagnetic fields have an inverse. Electromagnetic plane waves have no inverse.

Rather than calculating the inverse of the field as a whole, an alternative is to make the calculation in terms of magnetic and electric components of the field:

$$\mathcal{F}' = \frac{1}{\mathcal{F}} = \frac{1}{\sqrt{\mu}H\sigma - i\sqrt{\epsilon}Ee_0} = \sqrt{\mu}H'\sigma - i\sqrt{\epsilon}E'e_0 \tag{10.10}$$

where

$$\sqrt{\mu}H' = \frac{1}{\sqrt{\epsilon}E} \bigg/ \left(\frac{\sqrt{\mu}H}{\sqrt{\epsilon}E} + \frac{\sqrt{\epsilon}E}{\sqrt{\mu}H} \right) \quad \text{and} \quad \sqrt{\epsilon}E' = \frac{-1}{\sqrt{\mu}H} \bigg/ \left(\frac{\sqrt{\mu}H}{\sqrt{\epsilon}E} + \frac{\sqrt{\epsilon}E}{\sqrt{\mu}H} \right) \tag{10.11}$$

The calculation in magnetic and electric components uses the vector inverse from Table 6.1.

The inverse of an electromagnetic field is utilised in the context of monogenic boundary field extension using the Cauchy integral (see Chapter 24). It is not difficult to calculate the extension accurately for points distant from the boundary, but the accuracy deteriorates close to the boundary unless some kind of regularisation for a near-singular expression is applied. One of the methods of regularisation used in Chapter 24 requires the ratio of two electromagnetic fields F_n/F_d, which can be calculated as the product of the numerator with the inverse of the denominator, $F_n(1/F_d)$.

10.1.8 Reflection

When a planar electromagnetic wave encounters a flat perfectly conducting surface, the wave is reflected with an angle on the other side of the normal equal to the angle of incidence. Application of the reflection operator for vectors (Table 6.1) to the electromagnetic fields in Table 10.1 produces the same effect (Seagar 2016):

$$\mathbb{Q}F = \sqrt{\mu}(\mathbb{Q}H)\sigma + i\sqrt{\epsilon}(\mathbb{Q}E)e_0 \tag{10.12}$$

As well as being reflected in a geometrical sense, the electric field takes on a negative sign (changing from minus to plus) to account for the change of polarity required by the electric boundary conditions at a short circuit.

10.1.9 Projection

Applying the projection operators listed in Table 6.1 to the electromagnetic field splits the magnetic and electric fields into components normal (n) and tangential (t) to the surface (Seagar 2016):

$$\begin{cases} \mathbb{P}^+F = \sqrt{\mu}(\mathbb{P}^+H)\sigma - i\sqrt{\epsilon}(\mathbb{P}^-E)e_0 = \sqrt{\mu}H_{(t)}\sigma - i\sqrt{\epsilon}E_{(n)}e_0 \\ \mathbb{P}^-F = \sqrt{\mu}(\mathbb{P}^-H)\sigma - i\sqrt{\epsilon}(\mathbb{P}^+E)e_0 = \sqrt{\mu}H_{(n)}\sigma - i\sqrt{\epsilon}E_{(t)}e_0 \end{cases} \tag{10.13}$$

10.1.10 Rotation

Application of the reflection operator for vectors (Table 6.1) to the electromagnetic fields in Table 10.1 produces the same effect as first applying the rotation to the individual parts of the field and then combining the results as an electromagnetic bivector (Seagar 2016):

$$\mathbb{R}F = \sqrt{\mu}(\mathbb{R}H)\sigma - i\sqrt{\epsilon}(\mathbb{R}E)e_0 \tag{10.14}$$

The change in sign for the electric field in the case of reflection does not appear in the case of rotation.

10.2 Practice

10.2.1 Example Code

The code listed in Example 10.1a demonstrates simple calculations using electromagnetic fields. A four-dimensional Clifford context with all negative signature is established in lines 10–13. For simplicity, in lines 16–17, the values of the material properties are set to unity. Arrays of three complex numbers are initialised on lines 20 and 21, one for the magnetic field and one for the electric field. These are combined on line 22 into a Clifford number in the form of the field F in Table 10.1. The field is then printed in three different formats in lines 25–27. Various expressions involving magnitude and inverse of the field are calculated from line 30 to 34, and then printed on lines 36–42.

Example 10.1(a)

```
 1:  void demo19() /* electomagnetic fields */
 2:  {
 3:      double Harray[3][2],Earray[3][2];
 4:      double mu[2],epsilon[2],a[3];
 5:      int lambda[4];
 6:      number x,y,z,r,p,m,d;
 7:      int n;
 8:
 9:      /* use 4 dimensions for electromagnetic fields */
10:      recovery_level(0,0,0);
11:      create_cliffmem(50);
12:      declare_primal(4,pack_signature(4,lambda,-1));
13:      format_printing("%.0f");
14:
15:      /* complex material properties, just = 1 here */
16:      mu[0]=1.0;          mu[1]=0.0;
17:      epsilon[0]=1.0;   epsilon[1]=0.0;
18:
19:      /* arrays of complex field components */
20:      pack_3array(1.0,2.0,3.0,4.0,5.0,6.0,Harray);
21:      pack_3array(7.0,8.0,9.0,10.0,11.0,12.0,Earray);
22:      x=pack_emfield(mu,epsilon,Harray,Earray);
23:
24:      /* field as four dimensional Clifford number */
25:      printf("F = ");      print_emfield(x);
26:      printf("F = ");      print_number(x,3);
27:      printf("F = ");      print_tree(x);
28:
29:      /* magnitude and inverse */
30:      z=magnitudeof_emfield(x,0);
31:      y=inverseof_emfield(x,0);
32:      p=central_mul(x,y,1);
33:      m=magnitudeof_number(p,0);
34:      d=magnitudeof_number(sub_number(p,scalar_number(1.0,0.0),2),1);
35:      format_printing("%.3e");
36:      printf("|F| = ");            print_number(z,8);    recycle_number(z);
37:      printf("1/F = ");            printfs_emfield("%.3e","\n    ",y);
38:
39:      format_printing("%e");                             recycle_number(y);
40:      printf("F*1/F = ");          print_number(p,2);    recycle_number(p);
41:      printf("|F*1/F| = ");        print_number(m,8);    recycle_number(m);
42:      printf("|F*1/F-1| = ");      print_number(d,8);    recycle_number(d);
43:
44:      /* arrays of complex field components */
45:      pack_3array(0.0,0.0,-1.0,0.0,0.0,0.0,Harray); /* -Hy */
46:      pack_3array(0.0,0.0, 0.0,0.0,1.0,0.0,Earray); /*  Ez */
47:      y=pack_emfield(mu,epsilon,Harray,Earray);
48:
49:      /* field as four dimensional Clifford number */
50:      printf("\nF = EM field travelling in direction of x\n");
```

```
51:     printf (" = ");                    printf_emfield("%.1f",y);
52:
53:     /* reflect field opposite to direction of travel */
54:     n=pack_R03vector(pack_R3array(-1.0,0.0,0.0,a));
55:     r=reflect_vector(y,n,0);
56:     format_printing("%.3f");
57:     printf("\nunit vector n = "); print_vector(n);
                                      recycle_number(n);
58:     printf("n*F*n = reflection of wave F backwards\n");
59:     printf(" = ");                     printf_emfield("%.1f",r);
60:                                        recycle_number(r);
61:
62:     /* reflect to vertical with mirror at 45 degrees */
63:     n=normalise_vector(pack_R03vector(pack_R3array(-1.0,0.0,1.0,a)),1);
64:     r=reflect_vector(y,n,0);
65:     printf("\nunit vector n = "); print_vector(n);
66:     printf("n*F*n = reflection of EM field to vertical in
               45 degree mirror\n");
67:     printf(" = ");                     printf_emfield("%.1f",r);
68:                                        recycle_number(r);
69:
70:     /* rotate field 90 degrees to vertical */
71:     x=pack_R03vector(pack_R3array(0.0,0.0,1.0,a));
72:     r=rotate_number(y,x,n,0);
73:     printf("\nunit vector n1 = "); print_vector(x);   recycle_number(x);
74:     printf("unit vector n2 = ");    print_vector(n);  recycle_number(n);
75:     printf("n2*n1*F*n1*n2 = EM field to vertical using
               90 degrees rotation\n");
76:     printf(" = ");       printf_emfield("%.0f",r);
77:                                        recycle_number(r);
78:
79:     /* rotate field around direction of travel */
80:     n=pack_R03vector(pack_R3array(0.0,1.0,0.0,a));
81:     x=normalise_vector(pack_R03vector(pack_R3array(0.0,1.0,1.0,a)),1);
82:     r=rotate_number(y,n,x,1);
83:     printf("\nunit vector n1 = "); print_vector(n);   recycle_number(n);
84:     printf("unit vector n2 = ");    print_vector(x);  recycle_number(x);
85:     printf("n2*n1*F*n1*n2 = rotation of EM field 90 degrees in
               plane of yz\n");
86:     printf(" = ");       printf_emfield("%.0f",r);
87:                                        recycle_number(r);
88:     free_cliffmem(0);
89:   }
```

A new electromagnetic field is established on line 47, with magnetic component $-H\hat{y}$ on line 45 in the negative y direction and electric component $E\hat{z}$ on line 46 in the positive z direction. Taking the vector cross product of electric with the magnetic field $E\hat{z} \times (-H\hat{y}) = EH\hat{z} \times (-\hat{y}) = EH\hat{y} \times \hat{z} = EH\hat{x}$ indicates[4] that the electromagnetic field is propagating in the positive x direction.

---

4 According to Poynting's (1884) theorem.

The field is reflected in a mirror perpendicular to the *x* axis on line 55, and then on line 64 reflected in a mirror tilted from horizontal up 45° about the *y* axis. The results of these operations are printed on lines 59 and 67, respectively.

Rotations are applied to the field on lines 72 and 82. The first of these rotates 90° in the *xz* plane directing the field vertically, and the second rotates 90 degrees in the *yz* plane, twisting the field but not changing its direction of propagation. The results of the rotations are printed on lines 76 and 86, respectively.

10.2.2 Example Output

The results of running the code in Example 10.1a are listed in Example 10.1b. The field printed as magnetic and electric components in line 2 is contained internally in grade two of a four-dimensional Clifford number, as shown by printing as a number in lines 3 and 4 and as a tree in lines 5–22.

Example 10.1(b)

```
 1:   demo19()
 2:   F = [1+i2, 3+i4, 5+i6].H + [7+i8, 9+i10, 11+i12].E
 3:   F = (-8+i7)e{0,1}+(-10+i9)e{0,2}+(5+i6)e{1,2}
 4:     +(-12+i11)e{0,3}+(-3-i4)e{1,3}+(1+i2)e{2,3}
 5:   F = (
 6:   |     [0-15] (
 7:   |     |    [0-7] (
 8:   |     |    |    [3] (-8+i7)e{0,1}
 9:   |     |    |    [4-7] (
10:   |     |    |    |    [5] (-10+i9)e{0,2}
11:   |     |    |    |    [6] (5+i6)e{1,2}
12:   |     |    |  )
13:   |     |  )
14:   |     |    [8-15] (
15:   |     |    |    [8-11] (
16:   |     |    |    |    [9] (-12+i11)e{0,3}
17:   |     |    |    |    [10] (-3-i4)e{1,3}
18:   |     |    |  )
19:   |     |    |    [12] (1+i2)e{2,3}
20:   |     |  )
21:   |  )
22:   )
23:   |F| = (2.550e+01+i0.000e+00)e{}
24:   1/F = [-5.935e-03+i6.515e-03, -6.044e-03+i6.127e-03, -6.153e-03
              +i5.740e-03].H
25:      + [9.754e-03-i1.074e-02, 1.361e-02-i1.568e-02,
              1.747e-02-i2.062e-02].E
26:   F*1/F = (1.000000e+00+i2.775558e-17)e{}+(6.938894e-18
              +i0.000000e+00)e{0,1}
27:      +(-6.938894e-18+i2.081668e-17)e{0,2}+(-3.469447e-18
              +i1.214306e-17)e{1,2}
28:      +(-1.040834e-17-i6.938894e-18)e{0,3}+(6.938894e-18+i1.214306e-17)
              e{1,3}
```

```
29:    +(2.775558e-17+i1.040834e-17)e{2,3}+(-1.734723e-17-i5.551115e-17)
       e{0,1,2,3}
30:    |F*1/F|  = (1.000000e+00+i0.000000e+00)e{}
31:    |F*1/F-1| = (1.356416e-16+i0.000000e+00)e{}
32:
33:    F = EM field travelling in direction of x
34:      = [0.0+i0.0, -1.0+i0.0, 0.0+i0.0].H + [0.0+i0.0, 0.0+i0.0,
         1.0+i0.0].E
35:
36:    unit vector n = [0.000+i0.000, -1.000+i0.000, 0.000+i0.000,
                        0.000+i0.000
37:    n*F*n = reflection of wave F backwards
38:          = [0.0+i0.0, -1.0+i0.0, 0.0+i0.0].H + [0.0+i0.0, 0.0+i0.0,
            -1.0+i0.0].E
39:
40:    unit vector n = [0.000+i0.000, -0.707+i0.000, 0.000+i0.000,
                        0.707+i0.000
41:    n*F*n = reflection of EM field to vertical in 45 degree mirror
42:          = [0.0+i0.0, -1.0+i0.0, 0.0+i0.0].H + [-1.0+i0.0, 0.0+i0.0,
            0.0+i0.0].E
43:
44:    unit vector n1 = [0.000+i0.000, 0.000+i0.000, 0.000+i0.000,
                         1.000+i0.000
45:    unit vector n2 = [0.000+i0.000, -0.707+i0.000, 0.000+i0.000,
                         0.707+i0.000
46:    n2*n1*F*n1*n2  = EM field to vertical using 90 degrees rotation
47:                   = [0+i0, -1+i0, 0+i0].H + [-1+i0, 0+i0, 0+i0].E
48:
49:    unit vector n1 = [0.000+i0.000, 0.000+i0.000, 1.000+i0.000,
                         0.000+i0.000
50:    unit vector n2 = [0.000+i0.000, 0.000+i0.000, 0.707+i0.000,
                         0.707+i0.000
51:    n2*n1*F*n1*n2  = rotation of EM field 90 degrees in plane of yz
52:                   = [0+i0, 0+i0, -1+i0].H + [0+i0, -1+i0, 0+i0].E
```

The magnitude and inverse of the field are listed on lines 23–25. As a check, the product of the field with its inverse is unity to within the underlying double precision accuracy on lines 26–29. The magnitude of the product is unity on line 30, and the magnitude of the error is at double precision accuracy on line 31.

Comparing the electric and magnetic fields after reflection $(-H\hat{y} - E\hat{z})$ on line 38 with the originals $(-H\hat{y} + E\hat{z})$ on line 34 shows an inversion of sign for the electric field but no change to the magnetic field. That change of sign means the cross product of electric field with magnetic field also changes sign $(-\hat{z} \times (-)\hat{y} = -\hat{x})$, indicating a reversal in the direction of propagation.

Comparing the electric and magnetic fields after reflection $(-H\hat{y} - E\hat{x})$ on line 42 with the originals $(-H\hat{y} + E\hat{z})$ on line 34 shows that the electric field has changed from the z direction to the negative x direction, while the magnetic field has remained the same. The cross product of electric field with magnetic field $(-\hat{x} \times (-)\hat{y} = \hat{z})$ now indicates propagation in the z (vertical) direction.

The result of the rotation on line 47 is the same as the reflection on line 42, with the new direction of propagation in the vertical direction.

For the rotation on line 52 in the yz plane, both the magnetic and electric fields are modified with electric field $-E\hat{y}$ taking the former direction of magnetic field, and magnetic field $-H\hat{z}$ taking the opposite of the former direction of electric field. The cross product $(-\hat{y} \times (-)\hat{z} = \hat{x})$ remains unchanged and the direction of propagation remains the same. The outcome here has been to change the polarisation of the field.

The values produced by the invocation of `printf_emfield()` and `printfs _emfield()` treat the internal data as if the values for the electric permittivity and permeability are both unity. In this particular example, it is those values which are used, so the values printed are the true field values.

In cases where the material properties take other values, the routines for printing fields produce each field scaled by the square root of the value of its material property, as for the dipole fields in Eq. (22.3). That is useful when plotting the magnetic and electric fields on the same scale, to avoid one dominating the other, as in Figure 22.2. If the actual (unscaled) values are required, the sub-routine `unpack_emfield()` may be used to extract them before printing.

10.3 Field Arithmetic

The arithmetic operations of addition and multiplication maintain closure when both terms involved are either quaternions, Pauli matrices, or bicomplex numbers, producing entities of the original type. The same is not true for vectors and electromagnetic fields. In those cases, entities accommodating a wider set of units are required to support the result of multiplication.

An electromagnetic field which takes the form $p\sigma + qe_0$, where p and q are vectors in three spatial dimensions identified with Clifford units e_1, e_2, and e_3, and $\sigma = -e_1 e_2 e_3$ is the negatively signed unit of volume, contains six components with units $e_0 e_1$, $e_0 e_2$, $e_0 e_3$, $e_1 e_2$, $e_1 e_3$, and $e_2 e_3$, all in grade 2. When multiplied by another electromagnetic field $a\sigma + be_0$ containing all of the same units, the products of the individual terms fall either in grade zero (as for $e_0 e_1 e_0 e_1 = -e_\varnothing$), grade 2 ($e_0 e_1 e_1 e_2 = -e_0 e_2$), or grade 4 ($e_0 e_1 e_2 e_3$). Any further multiplication gives results which remain confined to grades 0, 2, and 4.

10.3.1 Extensions Based on Quaternions

In order to support arithmetic involving arbitrary addition and multiplication of electromagnetic fields, it is appropriate to introduce a new entity which allows in four dimensions any units which fall into grades 0, 2, and 4. This involves extending the electromagnetic field (in grade 2) by adding components supporting scalars (grade 0) and pseudo-scalars (grade 4). This leads to an entity composed of eight components in the form:

$$u = (s + p\sigma) + (q + r\sigma) e_0 = (s + p\sigma) + (q\sigma + r) \sigma e_0$$

$$= P + Q\sigma e_0 \tag{10.15}$$

for two scalars s and r, and two vectors p and q in the three spatial dimensions, or for two quaternions $P = s + p\sigma$ and $Q = r + q\sigma$ with a basis of spatial dimensions. In this case,[5] the quaternion units are $I = e_2 e_3$, $J = e_3 e_1$, and $K = e_1 e_2$.

The basis of the extended electromagnetic field can be obtained as the pair-wise product of all elements in the set of unit quaternions, $1, I, J, K$, with the set containing scalar 1 and pseudo-scalar $\tau = \sigma e_0 = e_0 e_1 e_2 e_3$:

$$\{1, I, J, K\} * \{1, \tau\} = \{1, I, J, K, I\tau, J\tau, K\tau, \tau\} \tag{10.16}$$

This follows the method by which Clifford extended quaternions to form a four-dimensional Clifford algebra. In that case, the second set in the pair-wise product was another set of quaternions (Clifford 1878, Seagar 2016) and 16 basis elements were formed. Other constructions have also been used to extend quaternions to octonians (Cayley 1845, Graves 1845, Hamilton 1848) and to extend Pauli matrices (almost quaternions) to Dirac (1928) matrices.

10.3.2 Inverses

Electromagnetic scattering problems formulated using the Cauchy integral lead to a linear systems of equations in which the coefficients of the system (the elements of a matrix) are extended electromagnetic fields (see Chapter 25). When some kind of matrix inversion is used to solve these systems, it is necessary to utilise an inverse for the individual elements in the matrix, i.e. for the extended electromagnetic fields. The inverse can be developed in at least two ways.

One possibility is to seek an inverse in the form:

$$1 = u\frac{1}{u} = u\frac{v}{uv} \tag{10.17}$$

for some number v which renders the denominator uv to a (complex) scalar. The division then reduces to a scalar division for which there is little difficulty. As an example, for a complex number u, v is the complex conjugate. When u is a vector, v is the same vector. For quaternions and Pauli matrices, v is the reversion of u. For the extended electromagnetic field, one suitable form of v is $v = \tilde{u}\sigma u\tilde{u}\sigma$, so that the inverse is the following:

$$\frac{1}{u} = \frac{\tilde{u}\sigma u\tilde{u}\sigma}{u(\tilde{u}\sigma u\tilde{u}\sigma)} = \frac{(\tilde{u}\sigma)u(\tilde{u}\sigma)}{(u\tilde{u}\sigma)^2} \tag{10.18}$$

Algebraically, the denominator $(u\tilde{u}\sigma)^2$ is guaranteed to be a (complex) scalar. However, that relies on the additive cancellation of equal and opposite values, which may not happen when imprecise numerical arithmetic leaves some non-zero residue in other grades. The issue can be circumvented either by taking only the scalar part of the result or by calculating the denominator in terms of scalars and vectors:

$$(u\tilde{u}\sigma)^2 = \left[(s - r)^2 - (p - q)^2\right]\left[(s + r)^2 - (p + q)^2\right] \tag{10.19}$$

Again, imprecise zeros may be generated by the squares of the sums and differences of the vector components, but this is easily avoided by taking the inner products rather than the squares.

---

5 Unlike in Chapter 7.

Another possibility is to seek an inverse in the form:

$$1 = u\frac{1}{u} = (P + Q\sigma e_0)(S + T\sigma e_0) \tag{10.20}$$

for two quaternions S and T. Expanding and equating components gives the 2×2 matrix equation with quaternion elements:

$$\begin{pmatrix} P & Q \\ Q & P \end{pmatrix}\begin{pmatrix} S \\ T \end{pmatrix} = \begin{pmatrix} 1 \\ 0 \end{pmatrix} \tag{10.21}$$

The inverses of quaternions are already defined so that simple matrix elimination by row operations gives the values of S and T, provided care is taken to honour the non-commutativity of quaternion algebra.

Pre-multiplying the first and second rows by $1/Q$ and $1/P$, respectively, and then subtracting the second gives

$$\begin{pmatrix} (1/Q)P - (1/P)Q & 0 \\ 0 & 0 \end{pmatrix}\begin{pmatrix} S \\ T \end{pmatrix} = \begin{pmatrix} 1/Q \\ 0 \end{pmatrix} \tag{10.22}$$

from which

$$S = \left(\tfrac{1}{Q}P - \tfrac{1}{P}Q\right)^{-1}\tfrac{1}{Q} = \tfrac{1}{Q}\left[Q\left(\tfrac{1}{Q}P - \tfrac{1}{P}Q\right)^{-1}\tfrac{1}{Q}\right] = \tfrac{1}{Q}\left[Q\left(\tfrac{1}{Q}P - \tfrac{1}{P}Q\right)\tfrac{1}{Q}\right]^{-1}$$
$$= \tfrac{1}{Q}\left(P\tfrac{1}{Q} - Q\tfrac{1}{P}\right)^{-1} = \frac{1/Q}{P/Q - Q/P} \tag{10.23}$$

or:

$$\tfrac{1}{S} = Q\left(\tfrac{1}{Q}P - \tfrac{1}{P}Q\right) = P - Q\tfrac{1}{P}Q \tag{10.24}$$

Similarly, pre-multiplying the first and second rows by $1/P$ and $1/Q$, respectively, and then subtracting the second gives:

$$\begin{pmatrix} 0 & (1/P)Q - (1/Q)P \\ 0 & 0 \end{pmatrix}\begin{pmatrix} S \\ T \end{pmatrix} = \begin{pmatrix} 1/P \\ 0 \end{pmatrix} \tag{10.25}$$

from which

$$T = \left(\tfrac{1}{P}Q - \tfrac{1}{Q}P\right)^{-1}\tfrac{1}{P} = \tfrac{1}{P}\left(Q\tfrac{1}{P} - P\tfrac{1}{Q}\right)^{-1} = \frac{1/P}{Q/P - P/Q} \tag{10.26}$$

or

$$\tfrac{1}{T} = P\left(\tfrac{1}{P}Q - \tfrac{1}{Q}P\right) = Q - P\tfrac{1}{Q}P \tag{10.27}$$

The inverses for the electromagnetic field in Section 10.1.7 are special cases of the equations in this section with particular values of s, p, q, and r. For example, with $s = r = 0, p = a$, and $q = b$ Eqs. (10.18) and (10.19) together reduce to Eq. (10.9), and the inverse

$$\frac{1}{P + Q\sigma e_0} = S + T\sigma e_0 \tag{10.28}$$

from Eq. (10.20) reduces with Eqs. (10.23), (10.24), (10.26), and (10.27) to

$$\frac{1}{a\sigma + be_0} = (a - ba^{-1}b)^{-1}\sigma + (b - ab^{-1}a)^{-1}e_0$$
$$= c^{-1}\sigma + d^{-1}e_0 = x\sigma + ye_0 \tag{10.29}$$

where $x = \sqrt{\mu}H'$ and $y = -i\sqrt{\epsilon}E'$ are vectors, as in Eqs. (10.10) and (10.11).

10.3.3 Example Code

The code listed in Example 10.2a demonstrates calculations involving the extended electromagnetic field. There are three routines of particular interest: `denom_xemfield()`, `reciprocalof_xemfield()`, and `inverseof_xemfield()`. The first of these calculates the denominator in Eq. (10.18) for the inverse, using the expression in Eq. (10.19). This can be used to check if a finite inverse exists without the need for calculating anything else. The second routine calculates both the numerator and the denominator, without attempting the division. If the denominator is non-zero, the ratio of numerator to denominator can then be calculated to provide the inverse. The third routine calculates the numerator and denominator, checks the denominator, and invokes `brexit(0)` if it is zero, or takes the ratio and returns the inverse.

A four-dimensional numerical context with all negative signature is established in lines 9–12. A test number is created and initialised with non-zero components in grades 0, 2, and 4 in lines 15–24.

Example 10.2(a)

```
1:   void demo20() /* extended EM field */
2:   {
3:     int lambda[4];
4:     cliff_comp x_comp;
5:     number r[2];
6:     number x,y,z,p,m,d,num,den;
7:
8:     /* use 4 dimensions for electromagnetic fields */
9:     recovery_level(0,0,0);
10:    create_cliffmem(50);
11:    declare_primal(4,pack_signature(4,lambda,-1));
12:    format_printing("%.0f");
13:
14:    /* EM field extended with scalar and pseudoscalar */
15:    x=fresh_number();
16:    x=merge_comp(x,pack_comp(&x_comp,1.0,2.0,spack_unit("e{}")),
                   addition_to,identity);
17:    x=merge_comp(x,pack_comp(&x_comp,3.0,4.0,spack_unit("e{0,1}")),
                   addition_to,identity);
18:    x=merge_comp(x,pack_comp(&x_comp,5.0,6.0,spack_unit("e{0,2}")),
                   addition_to,identity);
19:    x=merge_comp(x,pack_comp(&x_comp,7.0,8.0,spack_unit("e{0,3}")),
                   addition_to,identity);
20:    x=merge_comp(x,pack_comp(&x_comp,9.0,10.0,spack_unit("e{1,2}")),
                   addition_to,identity);
21:    x=merge_comp(x,pack_comp(&x_comp,11.0,12.0,spack_unit("e{1,3}")),
                   addition_to,identity);
22:    x=merge_comp(x,pack_comp(&x_comp,13.0,14.0,spack_unit("e{2,3}")),
                   addition_to,identity);
23:    x=merge_comp(x,pack_comp(&x_comp,15.0,16.0,spack_unit
                   ("e{0,1,2,3}")),
24:                                      addition_to,identity);
```

(Continued)

Example 10.2(a) (Continued)

```
25:    /* field as four dimensional Clifford number */
26:    printf("X = ");      print_number(x,5);
27:
28:    den=denom_xemfield(x);            /* get denomintor of inverse */
29:    if(is_zero(den))                  /* test for singular inverse */
30:       {
31:         printf("reciprocal of extended EM field has zero as
                      denominator\n");
32:          brexit(0);
33:       }
34:    recycle_number(den);
35:
36:    z=magnitudeof_number(x,0);        /* magnitude  */
37:    y=inverseof_xemfield(x,0);        /* get inverse as single number */
38:    p=central_mul(x,y,0);
39:    m=magnitudeof_number(p,0);
40:    d=magnitudeof_number(sub_number(p,scalar_number(1.0,0.0),2),1);
41:    format_printing("%.3e");
42:    printf("|X| = ");          print_number(z,3);     recycle_number(z);
43:    printf("1/X = ");          print_number(y,2);     recycle_number(y);
44:    format_printing("%e");
45:    printf("X*1/X = ");        print_number(p,2);     recycle_number(p);
46:    printf("|X*1/X| = ");      print_number(m,2);     recycle_number(m);
47:    printf("|X*1/X-1| = ");    print_number(d,2);     recycle_number(d);
48:
49:    reciprocalof_xemfield(x,r,0);   /* get inverse as ratio of
                                           numbers */
50:    num=r[0];                        /* numerator of inverse       */
51:    den=r[1];                        /* denominator of inverse      */
52:    y=central_mul(num,inverseof_component(den,0),2); /* y = num/den */
53:    p=central_mul(x,y,1);
54:    m=magnitudeof_number(p,0);
55:    d=magnitudeof_number(sub_number(p,scalar_number(1.0,0.0),2),1);
56:    format_printing("%.3e");
57:    printf("numerator = ");       printf_number("%.1f",num,2);
                recycle_number(num);
58:    printf("denominator = ");     printf_number("%.1f",den,2);
                recycle_number(den);
59:    printf("1/X = num/den = "); print_number(y,2);
                recycle_number(y);
60:    format_printing("%e");
61:    printf("X*1/X = ");        print_number(p,2);     recycle_number(p);
62:    printf("|X*1/X| = ");      print_number(m,2);     recycle_number(m);
63:    printf("|X*1/X-1| = ");    print_number(d,2);     recycle_number(d);
64:
65:    free_cliffmem(0);
66: }
```

The denominator for the inverse is calculated on line 28 and checked on line 29. If it is zero, the user's recovery strategy is invoked. Various expressions involving the magnitude and inverse of the field are calculated from line 36 to 40, and then printed on lines 42–47.

The same calculations with the inverse of the field are repeated on lines 52–55, this time using the numerator and denominator extracted on lines 49 through 51 to calculate the inverse on line 52. Results are printed on lines 57–63.

10.3.4 Example Output

The results of running the code in Example 10.2a are listed in Example 10.2b. The extended electromagnetic field is printed as a number lines 2–3, showing non-zero components in grades 0, 2, and 4.

Example 10.2(b)

```
 1:  demo20()
 2:  X = (1+i2)e{}+(3+i4)e{0,1}+(5+i6)e{0,2}+(9+i10)e{1,2}+(7+i8)e{0,3}
 3:    +(11+i12)e{1,3}+(13+i14)e{2,3}+(15+i16)e{0,1,2,3}
 4:  |X| = (3.868e+01+i0.000e+00)e{}
 5:  1/X = (2.030e-03-i1.816e-03)e{}+(-1.840e-03+i1.377e-03)e{0,1}
 6:    +(-4.281e-03+i4.481e-03)e{0,2}+(-5.772e-03+i6.182e-03)e{1,2}
 7:    +(-4.461e-03+i4.580e-03)e{0,3}+(-7.658e-03+i8.478e-03)e{1,3}
 8:    +(-8.393e-03+i9.385e-03)e{2,3}+(9.909e-03-i1.114e-02)e{0,1,2,3}
 9:  X*1/X = (1.000000e+00-i1.561251e-17)e{}+(1.110223e-16-i3.295975e-17)
            e{0,1}
10:    +(0.000000e+00+i1.474515e-17)e{0,2}+(0.000000e+00+i1.214306e-17)
            e{1,2}
11:    +(-5.551115e-17-i2.081668e-17)e{0,3}+(-5.551115e-17-i1.040834e-17)
            e{1,3}
12:    +(6.938894e-18+i6.938894e-18)e{2,3}+(-6.938894e-18+i1.040834e-17)
            e{0,1,2,3}
13:  |X*1/X| = (1.000000e+00+i0.000000e+00)e{}
14:  |X*1/X-1| = (1.448390e-16+i0.000000e+00)e{}
15:  numerator = (-5160.0+i3136.0)e{}+(4568.0-i2256.0)e{0,1}
16:    +(11144.0-i8032.0)e{0,2}+(15080.0-i11136.0)e{1,2}
17:    +(11576.0-i8176.0)e{0,3}+(20120.0-i15376.0)e{1,3}
18:    +(22088.0-i17056.0)e{2,3}+(-26104.0+i20272.0)e{0,1,2,3}
19:  denominator = (-2179264.0-i404736.0)e{}
20:  1/X = num/den = (2.030e-03-i1.816e-03)e{}+(-1.840e-03+i1.377e-03)
                    e{0,1}
21:    +(-4.281e-03+i4.481e-03)e{0,2}+(-5.772e-03+i6.182e-03)e{1,2}
22:    +(-4.461e-03+i4.580e-03)e{0,3}+(-7.658e-03+i8.478e-03)e{1,3}
23:    +(-8.393e-03+i9.385e-03)e{2,3}+(9.909e-03-i1.114e-02)e{0,1,2,3}
24:  X*1/X = (1.000000e+00-i1.561251e-17)e{}+(1.110223e-16-i3.295975e-17)
              e{0,1}
25:    +(0.000000e+00+i1.474515e-17)e{0,2}+(0.000000e+00+i1.214306e-17)
            e{1,2}
26:    +(-5.551115e-17-i2.081668e-17)e{0,3}+(-5.551115e-17-i1.040834e-17)
            e{1,3}
27:    +(6.938894e-18+i6.938894e-18)e{2,3}+(-6.938894e-18+i1.040834e-17)
            e{0,1,2,3}
28:  |X*1/X| = (1.000000e+00+i0.000000e+00)e{}
29:  |X*1/X-1| = (1.448390e-16+i0.000000e+00)e{}
```

The magnitude and the inverse as calculated by `inverseof_xemfield()` are listed on lines 4–8. As a check, the product of the field with its inverse is unity to within double precision accuracy on lines 9–12. The magnitude of the product is unity on line 13, and the magnitude of the error is zero at double precision accuracy on line 14.

The numerator and denominator as provided by `reciprocalof_xemfield()` are listed on lines 15 through 19. The inverse calculated from numerator and denominator is subject to all of the same checks on lines 20–29 as previously on lines 5–14, with exactly the same results.

References

Axelsson, A., Grognard, R., Hogan, J. and McIntosh, A. (2001), Harmonic analysis of Dirac operators on Lipschitz domains, *in* F. Brackx, ed., 'Clifford Analysis and its Applications', Vol. 25 of *NATO Science Series II, Mathematics Physics and Chemistry*, Kluwer, Dordrecht, Netherlands, pp. 231–246.

Cayley, A. (1845), 'On Jacobi's elliptic functions, in reply to the Rev. Brice Bronwin; and on quaternions', *Philosophical Magazine and Journal of Science* **26**(172), 208–211.

Clifford, W. K. (1878), 'Applications of Grassmann's extensive algebra', *American Journal of Mathematics* **1**(4), 350–358.

Dirac, P. A. M. (1928), 'The quantum theory of the electron', *Proceedings of the Royal Society. Series A, Containing Papers of a Mathematical and Physical Character* **117**(778), 610–624.

Graves, J. T. (1845), 'On a connection between the general theory of normal couples and the theory of complete quadratic functions of two variables', *Philosophical Magazine and Journal of Science* **26**(173), 315–320.

Hamilton, Sir W. R. (1848), 'Note by Professor Sir W. R. Hamilton, respecting the researches of John T. Graves, Esq.', *Transaction of the Royal Irish Academy* **21**, 338–341.

Maxwell, J. C. (1873a), *A Treatise on Electricity and Magnetism*, Vol. 1, Clarendon Press, Oxford, UK.

Maxwell, J. C. (1873b), *A Treatise on Electricity and Magnetism*, Vol. 2, Clarendon Press, Oxford, UK.

Poynting, J. H. (1884), 'On the transfer of energy in the electromagnetic field', *Philosophical Transactions of the Royal Society of London* **175**, 343–361.

Seagar, A. (2012), 'Calculation of electromagnetic fields in three dimensions using the Cauchy integral', *IEEE Transactions on Magnetics* **48**(2), 175–178.

Seagar, A. (2016), *Application of Geometric Algebra to Electromagnetic Scattering – the Clifford-Cauchy-Dirac Technique*, Springer, Singapore.

11

Arrays of Clifford Numbers

11.1 Theory

Arrays of Clifford numbers are created using any of the standard mechanisms provided within the 'c' programming language (King 2008) for arrays of other numbers. Arrays may be placed explicitly on the stack or in static storage with pre-determined size either within the scope of a sub-routine or globally. They may also be allocated dynamically by invoking standard library routines such as `malloc()`.

In any such case, when initialised, the array itself contains only the tokens which link to the internal Clifford memory structures. Copying the tokens from one array of datatype 'number' to another does not duplicate the underlying memory, but only creates two points of access. If something is modified and the token for one of the points of access is updated, the token for the other point of access may be no longer valid. Whether the memory accessed through the second token remains valid or not depends on the particular operations which have occurred. Some operations occur *in situ*, constructing the results by overwriting memory. Some operations construct the results from new memory and leave the original memory undamaged. Yet other operations use fresh memory for results but damage the contents of the original memory during the process and then recycle it at the end.

Copying tokens from one array to another without duplicating the underlying storage can be achieved by invoking the sub-routine `copy_array()`. To duplicate the underlying memory, the sub-routine `clone_array()` should be used instead.

Most of the routines which operate on arrays consist of a simple loop which sequentially applies an operation on the individual elements to which the arrays refer. As simple examples, the program code for the sub-routines `copy_array()` and `clone_array()` are listed in Examples 11.1 and 11.2.

Example 11.1

```
1:   number *copy_array(n,a,b)
2:        int n;
3:        number *a,*b; /* arrays a[n], b[n] */
4:   {
5:     int i;
```

(Continued)

Numerical Calculations in Clifford Algebra: A Practical Guide for Engineers and Scientists,
First Edition. Andrew Seagar.
© 2023 John Wiley & Sons Ltd. Published 2023 by John Wiley & Sons Ltd.

Example 11.1 (Continued)

```
6:
7:    for(i=0;i<n;i++)
8:        {
9:            b[i]=a[i]; /* the underlying storage is shared */
10:       }
11:    return(b);
12:  }
```

Example 11.2

```
1:    number *clone_array(n,a,b)
2:          int n;
3:          number *a,*b; /* arrays a[n], b[n] */
4:    {
5:      int i;
6:
7:      for(i=0;i<n;i++)
8:          {
9:              b[i]=cloneof_number(a[i]); /* the underlying storage is
                                             separate */
10:         }
11:    return(b);
12:  }
```

It is also possible to construct much more sophisticated procedures operating on arrays. For some non-trivial examples, see the solution of linear systems in Chapter 20, the fast Fourier transform in Chapter 21, and the finite difference time domain (FDTD) method of wave propagation in Chapter 23.

11.2 Practice

11.2.1 Example Code

The program code listed in Example 11.3a operates on an array of numbers 'a [n]' passed along with other parameters in the argument list. The individual elements of the array are modified one at a time on lines 23 and 32 within the loops from line 17 to 25 and from line 26 to 34. The purpose of these loops is to construct the linear rising and falling edges of a triangular-shaped electromagnetic pulse, on lines 19–22 and 28–31, respectively.

Example 11.3(a)

```
1:    void linear_pulse(n,a,w,mu,epsilon) /* add triangular shape EM
                                             pulse to field array a[n] */
2:          int n,w;                       /* size, width */
3:          cliff_comp *mu,*epsilon; /* material properties */
```

```
 4:        number *a;                    /* array of size n, already ini-
tialised externally */
 5:    {
 6:        double Harray[3][2],Earray[3][2];
 7:        double x,y;
 8:        number F;
 9:        int i,i1,i2,ic;
10:
11:        printf("# triangle\n");
12:        if(w>n) w=n;
13:        y=0.0;                                    /* imaginary part */
14:        i1=floor(0.5+(n-w)/2.0);
15:        i2=floor(0.5+(n+w)/2.0);
16:        ic=floor(0.5+n/2.0);
17:        for(i=i1;i<ic;i++)
18:          {
19:            x=2.0*(i-i1)/w;                             /* real part */
20:            pack_3array(0.0,0.0,0.0,0.0,x,y,Harray);   /* Hz */
21:            pack_3array(0.0,0.0,x,y,0.0,0.0,Earray);   /* Ey */
22:            F=pack_emfield(mu->coef,epsilon->coef,Harray,Earray);
                                                     /* field value */
23:            a[i]=addto_number(a[i],F);
24:            recycle_number(F);
25:          }
26:        for(i=ic;i<i2;i++)
27:          {
28:            x=(2.0*(i2-i))/w;                          /* real part */
29:            pack_3array(0.0,0.0,0.0,0.0,x,y,Harray);   /* Hz */
30:            pack_3array(0.0,0.0,x,y,0.0,0.0,Earray);   /* Ey */
31:            F=pack_emfield(mu->coef,epsilon->coef,Harray,Earray);
                                                     /* field value */
32:            a[i]=addto_number(a[i],F);
33:            recycle_number(F);
34:          }
35:    }
```

The array is not initialised within the sub-routine so that must be done externally. The array may be initialised to either zero or non-zero values, onto which the shape of the triangular pulse is superimposed. Apart from the need to recycle the variable 'F' on lines 24 and 33, there is little here to indicate the underlying workings of the Clifford numerical mechanisms.

Example 11.3b lists a main routine which invokes the sub-routine of Example 11.3a to construct a pulse of triangular (or other) shape and calculate its Fourier transform. Details of the Fourier transform itself are found in Chapter 21.

The overall size of the pulse is determined on lines 21 and 22. A four-dimensional Clifford context suitable for electromagnetic fields is established in lines 25–28. The extended printing format is used in line 28 so that only the real parts of the results are printed.

Two arrays of numbers are created dynamically by invoking the standard library routine `malloc()` on lines 31 and 32. Two arrays are required because the particular Fourier transform used does not perform the operation *in situ* on the input array and requires a second array for the output.

The order of creating the Clifford context and creating the arrays of numbers is not of importance and could equally well be the other way around. However, both must have been created before the arrays can be used to contain Clifford numbers.

The array 'a[n]' is taken to be the input array and is initialised with all elements zero on line 41. If the appropriate argument (1) is supplied to the main sub-routine `applica-tion0()`, then the triangular pulse is added into the array on line 45. The array of numbers is passed in the argument list to the sub-routine `linear_pulse()` in the same manner as with any array of any other type of data. After the sub-routine returns, the values of the tokens in the array not overwritten by the triangular pulse remain the same as before, and those in places where the pulse was added now have new values.

Example 11.3(b)

```
 1:   void application0(argc,argv) /* calculation with arrays */
 2:         int argc; char **argv;
 3:   {
 4:      int lambda[4];
 5:      cliff_comp mu,epsilon;
 6:      number *a,*b;
 7:      int n,width,pulse;
 8:
 9:      pulse=0;
10:      if(argc==2)
11:        {
12:          printf("%s %s\n",argv[0],argv[1]);
13:          sscanf(argv[1],"%d",&pulse);
14:        }
15:      if(pulse<0||pulse>6||argc!=2)
16:        {
17:          printf("useage: application 0 pulse\n pulse=0(square),
                       1(triangle)");
18:          printf(" 2(quadratic), 3(cubic), 4(sine), 5(gaussian),
                       6(bell)\n");
19:          exit(0);
20:        }
21:      n=256;        /* samples in spatial dimension */
22:      width=n/4;    /* width of pulse in samples */
23:
24:      /* establish Clifford maths context */
25:      recovery_level(0,0,0);
26:      create_cliffmem(4*n);
27:      declare_primal(4,pack_signature(4,lambda,-1));
                     /* four dimensions for EM fields */
28:      format_printing("%.3e#r");
29:
30:      /* allocate arrays dynamically */
31:      a=(number *)malloc(n*sizeof(number));
32:      b=(number *)malloc(n*sizeof(number));
33:
34:      /* material properties */
35:      pack_comp(&mu,1.0,0.0,0);        /* permeability=1.0 */
36:      pack_comp(&epsilon,1.0,0.0,0);  /* permittivity=1.0 */
```

```
37:     printf("# permittivity = ");   print_comp(&epsilon);
                                        printf("\n");
38:     printf("# permeability = ");   print_comp(&mu);
                                        printf("\n");
39:
40:     /* set up pulse shape */
41:     fresh_array(n,a);              /* initialise all values to zero */
42:     switch(pulse)
43:         {
44:         case 0: constant_pulse(n,a,width,&mu,&epsilon);    break;
45:         case 1: linear_pulse(n,a,width,&mu,&epsilon);      break;
46:         case 2: quadratic_pulse(n,a,width,&mu,&epsilon);   break;
47:         case 3: cubic_pulse(n,a,width,&mu,&epsilon);       break;
48:         case 4: cosine_pulse(n,a,width,&mu,&epsilon);      break;
49:         case 5: gaussian_pulse(n,a,width,&mu,&epsilon);    break;
50:         case 6: bell_pulse(n,a,width,&mu,&epsilon);        break;
51:         default:
52:             printf("pulse shape (%d) outside range (0-6) ",pulse);
53:             printf("in routine: application0()\n");
54:             brexit(0);
55:             break;
56:         }
57:     printf("# starting pulse\n");
58:     print_magnitudeof_array(n,a); /* show magnitude of field */
59:
60:     /* do transform, destroy, and recycle array a[n] in process */
61:     transform(n,a,b);
62:
63:     /* show spectrum */
64:     printf("\n# spectral values\n");
65:     print_magnitudeof_array(n,b);
66:
67:     /* dispose of numbers in array b[n] */
68:     recycle_array(n,b);
69:
70:     /* dispose of arrays */
71:     free((void *)a);
72:     free((void *)b);
73:     free_cliffmem(1);
74: }
```

The magnitude of all elements in the array is printed on line 58. The Fourier transform is taken on line 61, destroying and recycling array 'a [n]' and filling array 'b [n]' with the results. The magnitude of the results are printed on line 65.

All of the numbers in the array 'b [n]' are recycled on line 68. This must be done before the array is freed on line 72. However, it does not matter whether the freeing of the arrays and the Clifford memory are the order of lines 71–73 or in any other order at all.

11.2.2 Example Output

Figure 11.1 shows the triangular pulse produced by running the code listed in Example 11.3b. The amplitude is close to 1.4 rather than unity because the magnitude is the

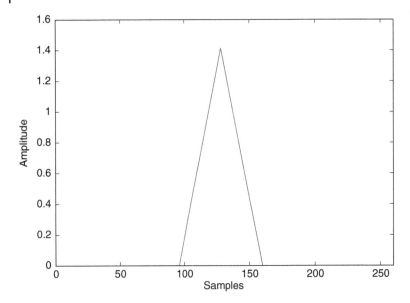

Figure 11.1 Triangular pulse.

square root of the sums of the squares of both magnetic and electric fields. Each of those takes a maximum value of unity from the variable 'x' on lines 19 and 28 of Example 11.3a.

Figure 11.2 shows the magnitude of the spectrum of the triangular pulse produced by the Fourier transform. The low positive frequency components start on the left. The frequency increases towards the middle of the figure. The right half of the figure contains the negative

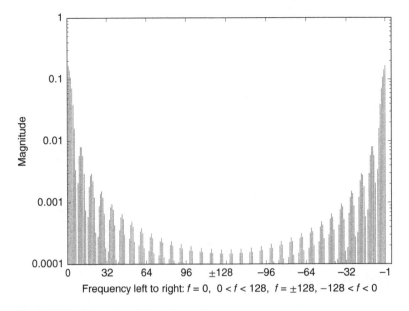

Figure 11.2 Spectrum of triangular pulse.

frequency components, starting with the most negative ones (in the middle of the figure) and increasing to the least negative ones on the right hand end.

Reference

King, K. N. (2008), *C Programming: A Modern Approach*, 2nd edn, W. W. Norton, New York, US-NY.

12

Power Series

Power series can be used to implement polynomials and other functions. Within the Clifford numerical suite, pre-defined power series are provided for calculating trigonometric and hyperbolic functions, logarithmic and exponential functions, reciprocals, and roots. Being implemented as a summation of series, these functions have finite ranges of convergence and have accuracy limited by loss of precision due to imprecise cancellation of alternating terms and by truncation of an infinite number of terms.

12.1 Theory

Power series for a function f are implemented in the general form of Eq. (12.1), where the user can specify the values of the coefficients a_k:

$$f(x) = \sum_{k=0}^{K-1} a_k(x - c)^k \tag{12.1}$$

The value c is the centre of convergence for the expansion. Usually, there are some limits on how far away from c the series will converge. Convergence is covered in more detail in Section 12.1.3.

Within the Clifford numerical suite, methods are provided for the user to construct functions by defining their own series, as in Section 12.1.1. Pre-defined series for some of the most common functions are described in Section 12.1.2.

12.1.1 User Defined

Table 12.1 lists the first few terms of the series for the Clifford–Bessel function (Greenhill 1919) of order n; $f(x) = C_n(x)$, which is a variation of the more common (regular) Bessel functions of the first kind $J_n(x)$. The series converges in principle for all real values x provided enough terms are used.

The series for the Clifford–Bessel function is not pre-defined within the Clifford numerical suite. In order to construct the series and evaluate the function, the user must first fill an array 'a [K]' with the coefficients a_k for the first K terms. It is only necessary to initialise the array one time.

Numerical Calculations in Clifford Algebra: A Practical Guide for Engineers and Scientists, First Edition. Andrew Seagar.
© 2023 John Wiley & Sons Ltd. Published 2023 by John Wiley & Sons Ltd.

Table 12.1 Power series for Bessel–Clifford function $f(x) = C_n(x)$ of order n.

Series	Range
$f(x) = \frac{1}{0!n!} - \frac{1}{1!(1+n)!}x + \frac{1}{2!(2+n)!}x^2 - \frac{1}{3!(3+n)!}x^3 + \cdots$	$-\infty < x < \infty$

Once the array is initialised with the coefficients, evaluation of the corresponding function is achieved by passing the array 'a[K]' and the argument 'x' for the function to the sub-routine poly_number(n,a,x,c,d). When evaluating the function, it is possible to use fewer terms than coefficients by setting the first argument n ≤ K. The fourth argument 'c' is the centre of convergence, which for the Clifford–Bessel power series is zero. The fifth argument 'd' is the disposal flag, which can be used to recycle the numbers associated with 'a', 'x', and 'c'. The sub-routine poly_number() is implemented using Horner's (1819) method with terms accumulated in the order from highest power to lowest.

The function can be re-evaluated as many times as required by passing different values for the argument 'x'. Different functions can be evaluated by initialising other arrays 'a2[K2]', 'a3[K3]', 'a4[K4]' to the coefficients for the corresponding series, and then passing those arrays to the sub-routine poly_number() in place of the second argument 'a'.

The method of initialising an array with the coefficients of the Clifford–Bessel series and evaluating the corresponding (Clifford–Bessel) function, is demonstrated in Section 12.2.1.

12.1.2 Predefined

The coefficients for the power series listed in Table 12.2 are pre-defined within the Clifford numerical suite. These series (Spiegel 1968) have centres of convergence at either $c = 0$ (in the upper half of the table) or $c = 1$ (in the lower half of the table).

Table 12.2 Pre-defined power series.

Label	Function	Series	Range
COS	Cosine	$f(x) = 1 - \frac{1}{2!}x^2 + \frac{1}{4!}x^4 - \frac{1}{6!}x^6 + \cdots$	$-\infty < x < \infty$
SIN	Sine	$f(x) = x - \frac{1}{3!}x^3 + \frac{1}{5!}x^5 - \frac{1}{7!}x^7 + \cdots$	$-\infty < x < \infty$
COSH	Hyperbolic cosine	$f(x) = 1 + \frac{1}{2!}x^2 + \frac{1}{4!}x^4 + \frac{1}{6!}x^6 + \cdots$	$-\infty < x < \infty$
SINH	Hyperbolic sine	$f(x) = x + \frac{1}{3!}x^3 + \frac{1}{5!}x^5 + \frac{1}{7!}x^7 + \cdots$	$-\infty < x < \infty$
EXP	Exponential	$f(x) = 1 + x + \frac{1}{2!}x^2 + \frac{1}{3!}x^3 + \cdots$	$-\infty < x < \infty$
LOG	Logarithm	$f(x) =$ $(x-1) - \frac{1}{2}(x-1)^2 + \frac{1}{3}(x-1)^3 - \frac{1}{4}(x-1)^4 + \cdots$	$0 < x \leq 2$
CM_INV	Inverse under central multiplication	$f(x) = 1 - (x-1) + (x-1)^2 - (x-1)^3 + \cdots$	$0 < x < 2$
SQRT	Square root	$f(x) = 1 + \frac{1}{2}(x-1) - \frac{1}{2\times4}(x-1)^2 + \frac{1\times3}{2\times4\times6}(x-1)^3 - \cdots$	$0 < x \leq 2$
ROOT	nth root	$f(x) = 1 + \frac{1}{n}(x-1) + \frac{1}{2!\,n}\left(\frac{1}{n}-1\right)(x-1)^2$ $+ \frac{1}{3!\,n}\left(\frac{1}{n}-1\right)\left(\frac{1}{n}-2\right)(x-1)^3 + \cdots$	$0 < x \leq 2$

The series are properly defined for scalars, i.e. real or complex numbers. They can be applied formally to Clifford numbers, because all of the operations used are properly defined for Clifford numbers. When the Clifford numbers contain only scalar components, the values calculated are as expected. When the Clifford numbers contain non-scalar components, the values calculated are sometimes not as one might naïvely expect. Care is needed in interpreting the results.

Familiar identities such as $(ab)^k = a^k b^k$ do not apply unless a and b commute, $ab = ba$. That is usually not the case for Clifford numbers, but is true if one of the factors is a scalar.

With Clifford numbers, exponential and trigonometric functions are related through their series as follows:

$$e^{ax} = \cos x + a \sin x \tag{12.2}$$

under the conditions that x must be a scalar, to guarantee commutativity, and $a^2 = -1$. The latter condition is commonly enforced with $a = i$, the imaginary unit as the square root of -1. However, within Clifford algebra (taking a negative signature for all primal units), there are multiple entities which square to -1. Examples are e_0, e_1, $e_0 e_1$, and $\frac{1}{\sqrt{2}}(e_0 + e_1)$. As a consequence, there are many relationships such as

$$e^{e_0 x} = \cos x + e_0 \sin x \tag{12.3}$$

If positive signatures are taken for the primal units, then the trigonometric functions in Eqs. (12.2) and (12.3) are replaced by their hyperbolic counterparts.

Putting $x = \pi$ in Eq. (12.2) gives $e^{a\pi} = -1$, so that the logarithm of the real number -1 within the realm of Clifford numbers is

$$\ln(-1) = a\pi \tag{12.4}$$

where a is any number which squares to -1. Since there are more than one of those, the logarithm of -1 is multivalued and therefore, the logarithm is not a function at that point on the real line. The usual result for complex numbers, $\ln(-1) = i\pi$, can no longer be used with any assurance.

12.1.3 Convergence

The ranges of convergence listed in the right hand column of Table 12.2 are for real values of x. The ranges are purely formal and should not be taken to indicate that numerical accuracy can be obtained over the whole range. Convergence and accuracy are most likely when the terms subject to exponentiation have magnitude less than unity, $|x - c| < 1$, because then the series are sure to be monotonically decreasing (although that by itself is no guarantee of convergence). If the series are not monotonically decreasing, loss of precision can occur from the near cancellation of large valued adjacent terms.

Within the Clifford numerical suite, three implementations of the pre-defined power series are provided: `series()`, `x_series()` and `xx_series()`. The version `series()` offers all of the functions in Table 12.2 implemented directly in that form. The second version `x_series()` provides an extended range of convergence, but only for the functions `LOG`, `CM_INV`, `SQRT`, and `ROOT`. The cost of the extended range is in accommodating only values of the argument x which have a non-zero scalar component. The third version `xx_series()` extends the range even further, but only for the function `CM_INV`. In this case, it is also possible to accommodate arguments with zero scalar component.

12.1.4 Factorisation

A Clifford number written in the form of a scalar component s plus a non-scalar component u can be factorised as follows:

$$(s + u) = (s + t + u - t) = (s + t)\left(1 + \frac{u - t}{s + t}\right) \tag{12.5}$$

provided that the number t is also a scalar. This factorisation offers a means of evaluating some functions in two parts: a scalar first part and a non-scalar second part, as shown in Table 12.3. The first part is evaluated using the standard mathematical library functions (King 2008) for scalars (complex numbers), and the second part is evaluated using the series in the lower half of Table 12.2, which have a centre of convergence $c = 1$.

In order to optimise convergence, the value of t is chosen:

$$t = \frac{|u|^2}{s^*} \tag{12.6}$$

minimising the magnitude of the fractional term $(u - t)/(s + t)$. This keeps the part evaluated using the series as close as possible to the centre of convergence. For example, in the case where the scalar part is very much larger than the non-scalar part:

$$s + u = 10 + e_1 = \left(10 + \frac{1}{10}\right)\left(1 + \frac{e_1 - \frac{1}{10}}{10 + \frac{1}{10}}\right) = \left(10 + \frac{1}{10}\right)\left(1 + \frac{10e_1 - 1}{101}\right) \tag{12.7}$$

the magnitude of the quotient in the second factor is significantly less than unity:

$$|x - 1| = \sqrt{\frac{10^2 + 1^2}{101^2}} = 0.0995 \tag{12.8}$$

In the case where the scalar part is very much smaller than the non-scalar part:

$$s + u = 0.1 + e_1 = \left(\frac{1}{10} + 10\right)\left(1 + \frac{e_1 - 10}{\frac{1}{10} + 10}\right) = \left(\frac{1}{10} + 10\right)\left(1 + \frac{10e_1 - 100}{101}\right) \tag{12.9}$$

the magnitude is marginally less than unity:

$$|x - 1| = \sqrt{\frac{10^2 + 100^2}{101^2}} = 0.995 \tag{12.10}$$

The powers of $(x - 1)^k = (0.0995)^k$ in the series in the lower half of Table 12.2 should diminish quickly in the first case leading to rapid convergence. In the second case, the powers $(0.995)^k$ do diminish, albeit slowly, so that convergence will be slow and accuracy may be poor due to errors accumulating during the summation of many terms.

Table 12.3 Evaluation of logarithm and binomial functions as a product of scalar and non-scalar parts.

$\ln(s + u)$ $\rightarrow$	$\ln(s + t + u - t)$ $\rightarrow$	$\ln\left[(s + t)\left(1 + \frac{u-t}{s+t}\right)\right]$ $\rightarrow$	$\ln(s + t) + \ln\left(1 + \frac{u-t}{s+t}\right)$
$(s + u)^k$ $\rightarrow$	$(s + t + u - t)^k$ $\rightarrow$	$\left[(s + t)\left(1 + \frac{u-t}{s+t}\right)\right]^k$ $\rightarrow$	$(s + t)^k\left(1 + \frac{u-t}{s+t}\right)^k$

This method is used to extend the range of convergence within the Clifford numerical suite for the functions LOG, CM_INV, SQRT, and ROOT. The latter three of these are based on the binomial series. The extended range of convergence is accessed using the sub-routine x_series() instead of series(). Note that this is only possible if the number, $x = s + u$, for which the function is calculated has a non-zero scalar part s.

12.1.5 Squaring

A second method of factorisation can be used to extend the range of convergence further if the argument of the function has no scalar part. Squaring numbers often introduces a scalar part. For example, for negative signatures, $(e_0 + e_1)^2 = -2$. Unfortunately in some cases the result is zero, as for $(ie_0 + e_1)^2 = 0$. This problem is resolved if the complex conjugate is used, yielding $(ie_0 + e_1)(ie_0 + e_1)^* = (ie_0 + e_1)(-ie_0 + e_1) = 2(-1 + ie_0e_1)$. However, the problem persists for other numbers, such as $(e_0 + e_1e_2e_3)(e_0 + e_1e_2e_3)^* = 0$.

In those cases where the result has a non-zero scalar part, squaring can be used to extend the reciprocal function to numbers without scalar parts using the factorisation:

$$\frac{1}{x} = x^* \left(\frac{1}{xx^*}\right) \tag{12.11}$$

The doubly extended range of convergence for the reciprocal function is accessed within the Clifford numerical suite using the sub-routine xx_series(), instead of series() or x_series().

It is tempting to try the same method for the other functions in the lower half of Table 12.2 using the relationships $\ln(x) \rightarrow \frac{1}{2}\ln(x^2)$, $x^k \rightarrow (x^2)^{k/2}$ and $\sqrt[k]{x} \rightarrow x^{1/k} \rightarrow (x^2)^{1/2k}$. Unfortunately, these don't apply for Clifford numbers with non-scalar components. The squaring operation destroys too much information from the original Clifford number. Unless that information is re-introduced elsewhere, as the leading factor x^* does in the product of Eq. (12.11), the value calculated is usually not as intended.

12.2 Practice

12.2.1 User Defined

The code listed in Examples 12.1a,b demonstrates the method of a user defining their own series and evaluating the corresponding function. Example 12.1a shows a small sub-routine bessel_clifford_series(n,a,order) which fills on line 13 an array of numbers 'a' of length 'n=K' with the coefficients 'ak' taken from Table 12.1 for the series of order specified by the variable 'order'. Here the coefficients are real valued scalars. If necessary, the coefficients can take values which are Clifford numbers.

Example 12.1(a)

```
1:   number *bessel_clifford_series(n,a,order)
2:        int n,order;
3:        number *a;
```

(Continued)

Example 12.1(a) (Continued)

```
4:   {
5:       double ak;
6:       int k;
7:
8:       ak=1.0/factorial(order);
9:       a[0]=scalar_number(ak,0.0);
10:      for(k=1;k<n;k++)
11:         {
12:            ak=-ak/(double)(k*(k+order));
13:            a[k]=scalar_number(ak,0.0);
14:         }
15:      return(a);
16:  }
```

The code listed in Example 12.1b applies the sub-routine bessel_clifford
_series() on line 22 to fill the array 'a' with the coefficients of the series. This is done
for values of 'order' equal to zero and one as set by the loop on line 19. The length of
the array is chosen so as to contain 16 coefficients, on lines 3 and 14. The coefficients are
printed on line 23.

Within the loop from line 19 to 46, the function corresponding to the series is evaluated
at lines 30 and 40 using the sub-routine poly_number(). On line 30, the argument 'x' is
a real valued scalar, and on line 40, the argument is a non-scalar value with a single primal
unit, as established on the lines immediately above. In each case, the argument is evaluated
for 101 values (as set on line 16) within loops from line 27 to 31 and line 37 to 41. The
values are stored in an array 'fx'. The disposal flag in the last argument of the sub-routine
poly_number() is set to the value 4 so as to recycle the argument 'x' once it has been
used. The arrays of values 'fx' are printed on lines 32 and 42. Only the real parts are printed
as specified on line 12.

Example 12.1(b)

```
1:   void demo21() /* users own series function */
2:   {
3:       number a[16]; /* array for coefficients of series */
4:       number *fx;
5:       number x,centre;
6:       int lambda[3];
7:       int k,n,p,order;
8:
9:       recovery_level(0,0,0);
10:      create_cliffmem(100);
11:      declare_primal(3,pack_signature(3,lambda,-1));
12:      format_printing("%.4e#r");
13:
14:      n=16;                          /* number of coefficients */
15:      centre=fresh_number();   /* zero is centre of convergence */
```

```
16:      p=101;                          /* number of points calculated */
17:      fx=(number *)malloc(p*sizeof(number));
18:
19:      for(order=0;order<2;order++)
20:        {
21:            /* calculate coefficients for series */
22:            bessel_clifford_series(n,a,order);
23:            print_array(n,a,8); /* coefficients */
24:            printf("\n");
25:
26:            /* scalar (grade 0) argument */
27:            for(k=0;k<p;k++)
28:              {
29:                x=scalar_number(25.0*k/(double)(p-1),0.0);
30:                fx[k]=poly_number(n,a,x,centre,4);
31:              }
32:            print_array(p,fx,8); /* function values */
33:            printf("\n");
34:            recycle_array(p,fx);
35:
36:            /* grade 1 argument */
37:            for(k=0;k<p;k++)
38:              {
39:                x=simple_number(25.0*k/(double)(p-1),0.0,
                                    spack_unit("e{1}"));
40:                fx[k]=poly_number(n,a,x,centre,4);
41:              }
42:            print_array(p,fx,8); /* function values */
43:            printf("\n");
44:            recycle_array(p,fx);
45:            recycle_array(n,a);
46:        }
47:      recycle_number(centre);
48:      free((void *)fx);
49:
50:      free_cliffmem(1);
51:    }
```

Figure 12.1 shows plotted as a graph the 101 values calculated for the Clifford–Bessel function with the scalar argument. Both orders zero and one are shown.

Figure 12.2 shows the values calculated for the Clifford–Bessel function with the non-scalar argument. In this case, the value of the function has both scalar and non-scalar components. Orders zero and one are shown for each.

12.2.2 Predefined

The code listed in Example 12.2 provides a framework to run tests on the pre-defined series in Table 12.2. One of three tests is selected from the list on line 4 according to the value 1, 2, or 3 of the integer 'test_num' provided on line 11 in the argument 'argv[1]'. The three test routines are listed in Examples 12.3a, 12.4a, and 12.5a. The first test applies the series

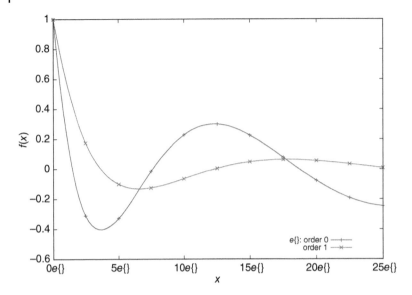

Figure 12.1 Clifford–Bessel function $f(x) = C_n(x)$ with scalar argument for orders $n = 0$ and 1.

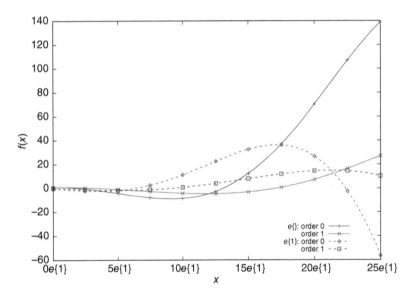

Figure 12.2 Clifford–Bessel function $f(x) = C_n(x)$ with non-scalar argument for orders $n = 0$ and 1.

directly as shown in Table 12.2. The second test applies the factorisations in Table 12.3 for extended convergence of the series in the lower half of Table 12.2. The third test applies in addition the second factorisation in Eq. (12.11) to further extend the range of the reciprocal (central multiplicative inverse) function.

Example 12.2

```
1:   void application1(argc,argv) /* series functions */
2:       int argc; char **argv;
3:   {
4:     static void (*tests[])(number x, int p)={test_series,
                             test_x_series,test_xx_series};
5:     int lambda[3];
6:     number x1,x2,x3,x4,temp;
7:     double x;
8:     int n,test_num;
9:
10:    test_num=0;
11:    if(argc==2) sscanf(argv[1],"%d",&test_num);
12:    if(test_num<1||test_num>3||argc!=2)
13:       {
14:        printf("useage: application 1 test\n test=1(series),
                   2(x_series)");
15:        printf(" 3(xx_series)\n");
16:        exit(0);
17:       }
18:    recovery_level(0,0,0);
19:    create_cliffmem(220);
20:    declare_primal(3,pack_signature(3,lambda,-1));
21:    format_printing("%.5f");
22:
23:    n=23;    /* terms beyond this power not used */
24:    make_series(n,3); /* 3 for ROOT = cube root */
25:
26:    x=M_PI/6.0;                          /* 30 degrees */
27:    x1=scalar_number(x,0.0);             /* normal range of
                                               convergence */
28:    x2=simple_number(x,0.0,spack_unit("e{0}"));
29:    x3=simple_number(x,0.0,spack_unit("e{0,1}"));
30:    x4=add_number(add_number(x1,x2,0),x3,1);
31:    if(test_num>1)                       /* extended range of
                                               convergence */
32:       {
33:        temp=scalar_number(5.0,0.0);     x1=addto_number(x1,temp);
34:        x2=addto_number(x2,temp);        x3=addto_number(x3,temp);
35:        x4=addto_number(x4,temp);        recycle_number(temp);
36:       }
37:    if(test_num>2)                       /* doubly extended range */
38:       {
39:        temp=simple_number(1.0,0.0,spack_unit("e{2}"));
40:        x1=central_mul(x1,temp,1);       x2=central_mul(x2,temp,1);
41:        x3=central_mul(x3,temp,1);       x4=central_mul(x4,temp,3);
42:       }
43:    tests[test_num-1](x1,n);   recycle_number(x1);   printf("\n");
44:    tests[test_num-1](x2,n);   recycle_number(x2);   printf("\n");
45:    tests[test_num-1](x3,n);   recycle_number(x3);   printf("\n");
```

(Continued)

Example 12.2 (Continued)

```
46:      tests[test_num-1](x4,n);    recycle_number(x4);
47:
48:      free_cliffmem(0);
49:  }
```


Each of the tests evaluates its functions on lines 43–46 of Example 12.2 using the values "x1" to "x4" as arguments to the functions. The accuracy of the functions is estimated by calculating the error in whichever identities that apply to each of them. The accuracy is governed by the highest power utilised in the series, which is set to the value "n=23" on line 23. When generating the coefficients for the series on line 24, the cube root is chosen for the ROOT function.

For the first test, with standard range of convergence, the values "x1" to "x4" are established on lines 27–30 with a real scalar component at a distance less than 1 away from the centre of convergence. The value $x = \pi/6$ on line 26 serves as the angle $\theta = 30°$ for the trigonometric functions and helps to recognise parts of the results in those components, where $\sin\theta = 1/2$ and $\cos\theta = \sqrt{3}/2$ appear.

For the second test, the values "x1" to "x4" are placed a distance further than 1 away from the centre of convergence by adding the number 5 in lines 33–35. For the third test, those values are further modified on lines 39–41 by multiplying all values by $1e_2$, leaving no scalar parts.

The context in which the tests are executed is established in lines 18–21. In this case the signatures for the primal units are all negative on line 20. If the signatures chosen are all positive, different numerical results are obtained. Some of the differences are obvious, with some components of the trigonometric functions trading places with corresponding components of the hyperbolic functions.

12.2.2.1 Standard Convergence

The code listed in Example 12.3a tests the functions in the upper half of Table 12.2 using the series without any enhancements. Lines 14–27 cover the trigonometric functions, lines 29–42 deal with the hyperbolic functions, and lines 44–58 test the exponential function. In order to check the accuracy (or at least consistency), the cosine and sine functions are squared, added, and compared to 1 from line 18 to 25. Similarly, the hyperbolic sine and cosine are squared, and their difference is compared to 1 from line 33 to 40. The check for the exponential function is a comparison of the square of the exponential with the exponential of twice the value, from line 49 to 55. The maximum error of all of the tests is captured on lines 26, 41, and 56 to be printed on line 60.

Example 12.3(a)

```
1:  void test_series(x,p)  /* series() standard convergence */
2:         number x;
3:         int p;
4:  {
```

```
 5:     number y,z,c,s,e1,e2;
 6:     double error,maxerror;
 7:     int n;
 8:
 9:     n=p+1;
10:     printf("number x is ");
11:     print_number(x,8);
12:     maxerror=0.0;
13:
14:     /* cosine(x) and sine(x)*/
15:     printf("\npolynomial cosine and sine series (%d) terms\n",n);
16:     c=series(COS,x,p);  printf("cos(x)=> ");  print_number(c,8);
17:     s=series(SIN,x,p);  printf("sin(x)=> ");  print_number(s,8);
18:     /* some checks */
19:     printf("cos(x)*cos(x)+sin(x)*sin(x)=> ");
20:     z=add_number(central_mul(c,c,0),central_mul(s,s,0),3);
21:     printf_number("%.3e",z,2);
22:     printf("cos(x)*cos(x)+sin(x)*sin(x)-1=> ");
23:     printf_number("%.3e",z=sub_number(z,scalar_number(1.0,0.0),3),2);
24:     printf("|cos(x)*cos(x)+sin(x)*sin(x)-1|=> ");
25:     printf_number("%.3e",z=magnitudeof_number(z,1),8);
26:     maxerror=get_real(z,0,0);
27:     recycle_number(s);        recycle_number(c);   recycle_number(z);
28:
29:     /* hyperbolic cosine(x) and hyperbolic sine(x) */
30:     printf("\npolynomial hyperbolic cosine and sine series
                (%d) terms\n",n);
31:     c=series(COSH,x,p);        printf("cosh(x)=> ");  print_number(c,8);
32:     s=series(SINH,x,p);        printf("sinh(x)=> ");  print_number(s,8);
33:     /* some checks */
34:     printf("cosh(x)*cosh(x)-sinh(x)*sinh(x)=> ");
35:     z=sub_number(central_mul(c,c,0),central_mul(s,s,0),3);
36:     printf_number("%.3e",z,8);
37:     printf("cosh(x)*cosh(x)-sinh(x)*sinh(x)-1=> ");
38:     printf_number("%.3e",z=sub_number(z,scalar_number(1.0,0.0),3),8);
39:     printf("|cosh(x)*cosh(x)-sinh(x)*sinh(x)-1|=> ");
40:     printf_number("%.3e",z=magnitudeof_number(z,1),8);
41:     error=get_real(z,0,0);  if(error>maxerror) maxerror=error;
42:     recycle_number(s);        recycle_number(c);   recycle_number(z);
43:
44:     /* exponential(x) and exponential(2x) */
45:     printf("\npolynomial exponential series (%d) terms\n",n);
46:     y=clonefuncof_number(x,twice);
47:     e1=series(EXP,x,p);     printf("exp(x)=> ");  print_number(e1,8);
48:     e2=series(EXP,y,p);     printf("exp(2x)=> ");  print_number(e2,8);
49:     /* some checks */
50:     printf("exp(x)*exp(x)=> ");
51:     printf_number("%.3e",z=central_mul(e1,e1,0),2);
52:     printf("exp(x)*exp(x)-exp(2x)=> ");
53:     printf_number("%.3e",z=sub_number(z,e2,1),8);
54:     printf("|exp(x)*exp(x)-exp(2x)|=> ");
55:     printf_number("%.3e",z=magnitudeof_number(z,1),8);
56:     error=get_real(z,0,0);  if(error>maxerror) maxerror=error;
```

(Continued)

Example 12.3(a) (Continued)

```
57:     recycle_number(e1);      recycle_number(e2);   recycle_number(y);
58:     recycle_number(z);
59:
60:     printf("\nmaximum error %e\n",maxerror);
61: }
```

Example 12.3b lists the results of running the test code in Example 12.3a. There are four sections in the results corresponding to the four different places where the functions are evaluated (lines 43–46 in Example 12.2). Within each section, there are three subsections: the first for the trigonometric functions, the second for the hyperbolic functions, and the third for the exponential function.

In the first section on lines 4–25, the functions are evaluated at the point $x = \pi/6$ on the real line. The cosine and sine functions return the expected results on lines 5 and 6. On lines 7–9, the results are checked using trigonometric identities. In this case, no measurable error is detected, as reported on line 9.

The hyperbolic functions produce the results on lines 12 and 13. The checks using hyperbolic identities on lines 14–16 indicate an error (on line 16) close to the double precision accuracy of the underlying data type. The situation is much the same for the exponential function on lines 18–23. The overall maximum error for all three sets of functions is reported on line 25.

Example 12.3(b)

```
 1:    application1()
 2:    number x is (0.52360+i0.00000)e{}
 3:
 4:    polynomial cosine and sine series (24) terms
 5:    cos(x)=> (0.86603+i0.00000)e{}
 6:    sin(x)=> (0.50000+i0.00000)e{}
 7:    cos(x)*cos(x)+sin(x)*sin(x)=> (1.000e+00+i0.000e+00)e{}
 8:    cos(x)*cos(x)+sin(x)*sin(x)-1=> (0.000e+00+i0.000e+00)e{}
 9:    |cos(x)*cos(x)+sin(x)*sin(x)-1|=> (0.000e+00+i0.000e+00)e{}
10:
11:    polynomial hyperbolic cosine and sine series (24) terms
12:    cosh(x)=> (1.14024+i0.00000)e{}
13:    sinh(x)=> (0.54785+i0.00000)e{}
14:    cosh(x)*cosh(x)-sinh(x)*sinh(x)=> (1.000e+00+i0.000e+00)e{}
15:    cosh(x)*cosh(x)-sinh(x)*sinh(x)-1=> (-2.220e-16+i0.000e+00)e{}
16:    |cosh(x)*cosh(x)-sinh(x)*sinh(x)-1|=> (2.220e-16+i0.000e+00)e{}
17:
18:    polynomial exponential series (24) terms
19:    exp(x)=> (1.68809+i0.00000)e{}
20:    exp(2x)=> (2.84965+i0.00000)e{}
21:    exp(x)*exp(x)=> (2.850e+00+i0.000e+00)e{}
22:    exp(x)*exp(x)-exp(2x)=> (4.441e-16+i0.000e+00)e{}
23:    |exp(x)*exp(x)-exp(2x)|=> (4.441e-16+i0.000e+00)e{}
```

```
24:
25:    maximum error 4.440892e-16
26:
27:    number x is (0.52360+i0.00000)e{0}
28:
29:    polynomial cosine and sine series (24) terms
30:    cos(x)=> (1.14024+i0.00000)e{}
31:    sin(x)=> (0.54785+i0.00000)e{0}
32:    cos(x)*cos(x)+sin(x)*sin(x)=> (1.000e+00+i0.000e+00)e{}
33:    cos(x)*cos(x)+sin(x)*sin(x)-1=> (-2.220e-16+i0.000e+00)e{}
34:    |cos(x)*cos(x)+sin(x)*sin(x)-1|=> (2.220e-16+i0.000e+00)e{}
35:
36:    polynomial hyperbolic cosine and sine series (24) terms
37:    cosh(x)=> (0.86603+i0.00000)e{}
38:    sinh(x)=> (0.50000+i0.00000)e{0}
39:    cosh(x)*cosh(x)-sinh(x)*sinh(x)=> (1.000e+00+i0.000e+00)e{}
40:    cosh(x)*cosh(x)-sinh(x)*sinh(x)-1=> (0.000e+00+i0.000e+00)e{}
41:    |cosh(x)*cosh(x)-sinh(x)*sinh(x)-1|=> (0.000e+00+i0.000e+00)e{}
42:
43:    polynomial exponential series (24) terms
44:    exp(x)=> (0.86603+i0.00000)e{}+(0.50000+i0.00000)e{0}
45:    exp(2x)=> (0.50000+i0.00000)e{}+(0.86603+i0.00000)e{0}
46:    exp(x)*exp(x)=> (5.000e-01+i0.000e+00)e{}
                        +(8.660e-01+i0.000e+00)e{0}
47:    exp(x)*exp(x)-exp(2x)=> (1.110e-16+i0.000e+00)e{}
48:    |exp(x)*exp(x)-exp(2x)|=> (1.110e-16+i0.000e+00)e{}
49:
50:    maximum error 2.220446e-16
51:
52:    number x is (0.52360+i0.00000)e{0,1}
53:
54:    polynomial cosine and sine series (24) terms
55:    cos(x)=> (1.14024+i0.00000)e{}
56:    sin(x)=> (0.54785+i0.00000)e{0,1}
57:    cos(x)*cos(x)+sin(x)*sin(x)=> (1.000e+00+i0.000e+00)e{}
58:    cos(x)*cos(x)+sin(x)*sin(x)-1=> (-2.220e-16+i0.000e+00)e{}
59:    |cos(x)*cos(x)+sin(x)*sin(x)-1|=> (2.220e-16+i0.000e+00)e{}
60:
61:    polynomial hyperbolic cosine and sine series (24) terms
62:    cosh(x)=> (0.86603+i0.00000)e{}
63:    sinh(x)=> (0.50000+i0.00000)e{0,1}
64:    cosh(x)*cosh(x)-sinh(x)*sinh(x)=> (1.000e+00+i0.000e+00)e{}
65:    cosh(x)*cosh(x)-sinh(x)*sinh(x)-1=> (0.000e+00+i0.000e+00)e{}
66:    |cosh(x)*cosh(x)-sinh(x)*sinh(x)-1|=> (0.000e+00+i0.000e+00)e{}
67:
68:    polynomial exponential series (24) terms
69:    exp(x)=> (0.86603+i0.00000)e{}+(0.50000+i0.00000)e{0,1}
70:    exp(2x)=> (0.50000+i0.00000)e{}+(0.86603+i0.00000)e{0,1}
71:    exp(x)*exp(x)=> (5.000e-01+i0.000e+00)e{}
                        +(8.660e-01+i0.000e+00)e{0,1}
72:    exp(x)*exp(x)-exp(2x)=> (1.110e-16+i0.000e+00)e{}
73:    |exp(x)*exp(x)-exp(2x)|=> (1.110e-16+i0.000e+00)e{}
```

(Continued)

Example 12.3(b) (Continued)

```
74:
75:   maximum error 2.220446e-16
76:
77:   number x is  (0.52360+i0.00000)e{}+(0.52360+i0.00000)e{0}
                        +(0.52360+i0.00000)e{0,1}
78:
79:   polynomial cosine and sine series (24) terms
80:   cos(x)=>  (1.11450+i0.00000)e{}+(-0.28639+i0.00000)e{0}
                    +(-0.28639+i0.00000)e{0,1}
81:   sin(x)=>  (0.64346+i0.00000)e{}+(0.49604+i0.00000)e{0}
                    +(0.49604+i0.00000)e{0,1}
82:   cos(x)*cos(x)+sin(x)*sin(x)=>  (1.000e+00+i0.000e+00)e{}
                                     +(-3.331e-16+i0.000e+00)e{0}
83:    +(-3.331e-16+i0.000e+00)e{0,1}
84:   cos(x)*cos(x)+sin(x)*sin(x)-1=>  (-2.220e-16+i0.000e+00)e{}
                                       +(-3.331e-16+i0.000e+00)e{0}
85:    +(-3.331e-16+i0.000e+00)e{0,1}
86:   |cos(x)*cos(x)+sin(x)*sin(x)-1|=>  (5.207e-16+i0.000e+00)e{}
87:
88:   polynomial hyperbolic cosine and sine series (24) terms
89:   cosh(x)=>  (0.84166+i0.00000)e{}+(0.26135+i0.00000)e{0}
                     +(0.26135+i0.00000)e{0,1}
90:   sinh(x)=>  (0.40440+i0.00000)e{}+(0.54394+i0.00000)e{0}
                     +(0.54394+i0.00000)e{0,1}
91:   cosh(x)*cosh(x)-sinh(x)*sinh(x)=>  (1.000e+00+i0.000e+00)e{}
92:   cosh(x)*cosh(x)-sinh(x)*sinh(x)-1=>  (-2.220e-16+i0.000e+00)e{}
93:   |cosh(x)*cosh(x)-sinh(x)*sinh(x)-1|=>  (2.220e-16+i0.000e+00)e{}
94:
95:   polynomial exponential series (24) terms
96:   exp(x)=>  (1.24606+i0.00000)e{}+(0.80529+i0.00000)e{0}
                    +(0.80529+i0.00000)e{0,1}
97:   exp(2x)=>  (0.25566+i0.00000)e{}+(2.00688+i0.00000)e{0}
                     +(2.00688+i0.00000)e{0,1}
98:   exp(x)*exp(x)=>  (2.557e-01+i0.000e+00)e{}
                        +(2.007e+00+i0.000e+00)e{0}
99:    +(2.007e+00+i0.000e+00)e{0,1}
100:  exp(x)*exp(x)-exp(2x)=>  (0.000e+00+i0.000e+00)e{}
101:  |exp(x)*exp(x)-exp(2x)|=>  (0.000e+00+i0.000e+00)e{}
102:
103:  maximum error 5.207408e-16
```

The same tests are repeated on lines 29–50 for the non-scalar number $x = \pi e_0/6$. Detailed results of the trigonometric, hyperbolic, and exponential tests are listed on lines 29–34, lines 36–41, and lines 43–48, respectively. The overall maximum error on line 50 is again close to the double precision accuracy of the underlying data type.

In this case, it is worth noting that the coefficients of the values produced in the cosine and sine functions on lines 30 and 31 are the same as those for the corresponding hyperbolic functions on lines 12 and 13. This happens when the argument to the function squares to a negative value. Normally, this is seen with an imaginary argument, but equally well here with a Clifford unit of negative signature.

Exactly the reverse is noted for the results of the hyperbolic functions on lines 37 and 38, where the coefficients take the same values as the trigonometric functions on lines 5 and 6.

The exponential series delivers on line 44 the cosine value in the scalar component and the sine value in the non-scalar component, in accordance with Eq. (12.3). When doubling the value of the argument, the components on line 45 are reversed. This is not general but a special case because here (in degrees) the two angles are $\theta = 30°$ and $2\theta = 60° = 90° - \theta$.

The maximum error for all identity checks for the second section of code is reported on line 50, similar to before.

The third section of tests from line 52 to 75 behaves almost exactly the same as the second section because (for negative signatures) both Clifford units $e_0 e_1$ and e_0 square to give -1.

The fourth section of tests from line 77 to 103 evaluates the function at an argument with components in grades 0, 1, and 2. In this case, the consistency checks using the identities again give an error on line 103 close to machine precision. However, it is much less easy to make any observations relating the values produced to those for the earlier tests.

12.2.2.2 Extended Convergence

The code listed in Example 12.4a tests the functions in the lower half of Table 12.2 using the series with the factorisation in Eq. (12.5) to provide an extended range of convergence.

Lines 14–28 cover the logarithmic function, lines 30–42 deal with the reciprocal function, lines 44–56 test the square root function, and lines 58–70 cover the case $n = 3$ for the nth root function.

Consistency checks for the logarithm involve comparing the logarithm of a number squared to the logarithm of the number added to itself in lines 19–25. To check the reciprocal, it is multiplied by the original number in lines 34–40, and the square root is squared as a check in lines 48–54. Finally, to check for the cube root in lines 62–68, the value is cubed. The maximum error of all of the tests is captured on lines 26, 41, 55, and 69 to be printed on line 72.

Example 12.4(a)

```
1:    void test_x_series(x,p) /* x_series() can use number outside
                                   normal range of convergence */
2:            number x;
3:            int p;
4:    {
5:       number y,z,c,s,l1,l2;
6:       double error,maxerror;
7:       int n;
8:
9:       n=p+1;
10:      printf("number x is ");
11:      print_number(x,8);
12:      maxerror=0.0;
13:
14:      /* log(x) and log(x*x) */
15:      printf("\npolynomial logarithm series (%d) terms\n",n);
```

(Continued)

Example 12.4(a) (Continued)

```
16:    y=central_mul(x,x,0);
17:    l1=x_series(LOG,x,p);    printf("log(x)=> ");
                                print_number(l1,8);
18:    l2=x_series(LOG,y,p);    printf("log(x*x)=> ");
                                print_number(l2,8);
19:    /* some checks */
20:    printf("log(x)+log(x)=> ");
21:    printf_number("%.3e",z=add_number(l1,l1,0),2);
22:    printf("log(x)+log(x)-log(x*x)=> ");
23:    printf_number("%.3e",z=sub_number(z,l2,1),2);
24:    printf("|log(x)+log(x)-log(x*x)|=> ");
25:    printf_number("%.3e",z=magnitudeof_number(z,1),8);
26:    maxerror=get_real(z,0,0);
27:    recycle_number(l1);     recycle_number(l2);   recycle_number(y);
28:    recycle_number(z);
29:
30:    /* central multiplicative inverse 1/x */
31:    printf("\npolynomial reciprocal series (%d) terms\n",n);
32:    y=x_series(CM_INV,x,p);
33:    printf("(1/x)=> ");       print_number(y,8);
34:    /* some checks */
35:    printf("x*(1/x)=> ");
36:    printf_number("%.3e",z=central_mul(x,y,0),8);
37:    printf("x*(1/x)-1=> ");
38:    printf_number("%.3e",z=sub_number(z,scalar_number(1.0,0.0),3),8);
39:    printf("|x*(1/x)-1|=> ");
40:    printf_number("%.3e",z=magnitudeof_number(z,1),8);
41:    error=get_real(z,0,0);  if(error>maxerror) maxerror=error;
42:    recycle_number(y);       recycle_number(z);
43:
44:    /* square root */
45:    printf("\npolynomial square root series (%d) terms\n",n);
46:    s=x_series(SQRT,x,p);
47:    printf("sqrt(x)=> ");   print_number(s,8);
48:    /* some checks */
49:    printf("sqrt(x)*sqrt(x)=> ");
50:    printf_number("%.3e",z=central_mul(s,s,0),2);
51:    printf("sqrt(x)*sqrt(x)-x=> ");
52:    printf_number("%.3e",z=subfrom_number(z,x),2);
53:    printf("|sqrt(x)*sqrt(x)-x|=> ");
54:    printf_number("%.3e",z=magnitudeof_number(z,1),8);
55:    error=get_real(z,0,0);  if(error>maxerror) maxerror=error;
56:    recycle_number(s);       recycle_number(z);
57:
58:    /* nth root, with n=3 */
59:    printf("\npolynomial cube root series (%d) terms\n",n);
60:    c=x_series(ROOT,x,p);
61:    printf("3root(x)=> ");  print_number(c,8);
62:    /* some checks */
63:    printf("3root(x)*3root(x)*3root(x)=> ");
64:    printf_number("%.3e",z=central_mul(central_mul(c,c,0),c,1),2);
65:    printf("3root(x)*3root(x)*3root(x)-x=> ");
66:    printf_number("%.3e",z=subfrom_number(z,x),2);
```

```
67:     printf("|3root(x)*3root(x)*3root(x)-x|=> ");
68:     printf_number("%.3e",z=magnitudeof_number(z,1),8);
69:     error=get_real(z,0,0);    if(error>maxerror) maxerror=error;
70:     recycle_number(c);        recycle_number(z);
71:
72:     printf("\nmaximum error %e\n",maxerror);
73: }
```

Example 12.4b lists the results of running the test code in Example 12.4a. There are four sections in the results corresponding to the four different places, where the functions are evaluated. Within each section, there are four subsections, the first for the logarithmic function, then the reciprocal, the square root, and the cube root.

Example 12.4(b)

```
 1:  application2()
 2:  number x is (5.52360+i0.00000)e{}
 3:
 4:  polynomial logarithm series (24) terms
 5:  log(x)=> (1.70903+i0.00000)e{}
 6:  log(x*x)=> (3.41806+i0.00000)e{}
 7:  log(x)+log(x)=> (3.418e+00+i0.000e+00)e{}
 8:  log(x)+log(x)-log(x*x)=> (0.000e+00+i0.000e+00)e{}
 9:  |log(x)+log(x)-log(x*x)|=> (0.000e+00+i0.000e+00)e{}
10:
11:  polynomial reciprocal series (24) terms
12:  (1/x)=> (0.18104+i0.00000)e{}
13:  x*(1/x)=> (1.000e+00+i0.000e+00)e{}
14:  x*(1/x)-1=> (-1.110e-16+i0.000e+00)e{}
15:  |x*(1/x)-1|=> (1.110e-16+i0.000e+00)e{}
16:
17:  polynomial square root series (24) terms
18:  sqrt(x)=> (2.35023+i0.00000)e{}
19:  sqrt(x)*sqrt(x)=> (5.524e+00+i0.000e+00)e{}
20:  sqrt(x)*sqrt(x)-x=> (0.000e+00+i0.000e+00)e{}
21:  |sqrt(x)*sqrt(x)-x|=> (0.000e+00+i0.000e+00)e{}
22:
23:  polynomial cube root series (24) terms
24:  3root(x)=> (1.76770+i0.00000)e{}
25:  3root(x)*3root(x)*3root(x)=> (5.524e+00+i0.000e+00)e{}
26:  3root(x)*3root(x)*3root(x)-x=> (0.000e+00+i0.000e+00)e{}
27:  |3root(x)*3root(x)*3root(x)-x|=> (0.000e+00+i0.000e+00)e{}
28:
29:  maximum error 1.110223e-16
30:
31:  number x is (5.00000+i0.00000)e{}+(0.52360+i0.00000)e{0}
32:
33:  polynomial logarithm series (24) terms
34:  log(x)=> (1.61489+i0.00000)e{}+(0.10434+i0.00000)e{0}
35:  log(x*x)=> (3.22978+i0.00000)e{}+(0.20868+i0.00000)e{0}
```

(Continued)

Example 12.4(b) (Continued)

```
36:   log(x)+log(x)=> (3.230e+00+i0.000e+00)e{}
                      +(2.087e-01+i0.000e+00)e{0}
37:   log(x)+log(x)-log(x*x)=> (2.776e-17+i0.000e+00)e{0}
38:   |log(x)+log(x)-log(x*x)|=> (2.776e-17+i0.000e+00)e{}
39:
40:   polynomial reciprocal series (24) terms
41:   (1/x)=> (0.19783+i0.00000)e{}+(-0.02072+i0.00000)e{0}
42:   x*(1/x)=> (1.000e+00+i0.000e+00)e{}+(-1.388e-17+i0.000e+00)e{0}
43:   x*(1/x)-1=> (-1.388e-17+i0.000e+00)e{0}
44:   |x*(1/x)-1|=> (1.388e-17+i0.000e+00)e{}
45:
46:   polynomial square root series (24) terms
47:   sqrt(x)=> (2.23912+i0.00000)e{}+(0.11692+i0.00000)e{0}
48:   sqrt(x)*sqrt(x)=> (5.000e+00+i0.000e+00)e{}
                        +(5.236e-01+i0.000e+00)e{0}
49:   sqrt(x)*sqrt(x)-x=> (-1.110e-16+i0.000e+00)e{0}
50:   |sqrt(x)*sqrt(x)-x|=> (1.110e-16+i0.000e+00)e{}
51:
52:   polynomial cube root series (24) terms
53:   3root(x)=> (1.71205+i0.00000)e{}+(0.05957+i0.00000)e{0}
54:   3root(x)*3root(x)*3root(x)=> (5.000e+00+i0.000e+00)e{}
                                   +(5.236e-01+i0.000e+00)e{0}
55:   3root(x)*3root(x)*3root(x)-x=> (8.882e-16+i0.000e+00)e{}
56:   |3root(x)*3root(x)*3root(x)-x|=> (8.882e-16+i0.000e+00)e{}
57:
58:   maximum error 8.881784e-16
59:
60:   number x is (5.00000+i0.00000)e{}+(0.52360+i0.00000)e{0,1}
61:
62:   polynomial logarithm series (24) terms
63:   log(x)=> (1.61489+i0.00000)e{}+(0.10434+i0.00000)e{0,1}
64:   log(x*x)=> (3.22978+i0.00000)e{}+(0.20868+i0.00000)e{0,1}
65:   log(x)+log(x)=> (3.230e+00+i0.000e+00)e{}
                      +(2.087e-01+i0.000e+00)e{0,1}
66:   log(x)+log(x)-log(x*x)=> (2.776e-17+i0.000e+00)e{0,1}
67:   |log(x)+log(x)-log(x*x)|=> (2.776e-17+i0.000e+00)e{}
68:
69:   polynomial reciprocal series (24) terms
70:   (1/x)=> (0.19783+i0.00000)e{}+(-0.02072+i0.00000)e{0,1}
71:   x*(1/x)=> (1.000e+00+i0.000e+00)e{}+(-1.388e-17+i0.000e+00)e{0,1}
72:   x*(1/x)-1=> (-1.388e-17+i0.000e+00)e{0,1}
73:   |x*(1/x)-1|=> (1.388e-17+i0.000e+00)e{}
74:
75:   polynomial square root series (24) terms
76:   sqrt(x)=> (2.23912+i0.00000)e{}+(0.11692+i0.00000)e{0,1}
77:   sqrt(x)*sqrt(x)=> (5.000e+00+i0.000e+00)e{}
                        +(5.236e-01+i0.000e+00)e{0,1}
78:   sqrt(x)*sqrt(x)-x=> (-1.110e-16+i0.000e+00)e{0,1}
79:   |sqrt(x)*sqrt(x)-x|=> (1.110e-16+i0.000e+00)e{}
80:
81:   polynomial cube root series (24) terms
82:   3root(x)=> (1.71205+i0.00000)e{}+(0.05957+i0.00000)e{0,1}
```

```
 83:   3root(x)*3root(x)*3root(x)=>  (5.000e+00+i0.000e+00)e{}
                                        +(5.236e-01+i0.000e+00)e{0,1}
 84:   3root(x)*3root(x)*3root(x)-x=>  (8.882e-16+i0.000e+00)e{}
 85:   |3root(x)*3root(x)*3root(x)-x|=>  (8.882e-16+i0.000e+00)e{}
 86:
 87:   maximum error 8.881784e-16
 88:
 89:   number x is (5.52360+i0.00000)e{}+(0.52360+i0.00000)e{0}
                                        +(0.52360+i0.00000)e{0,1}
 90:
 91:   polynomial logarithm series (24) terms
 92:   log(x)=> (1.71794+i0.00000)e{}+(0.09423+i0.00000)e{0}
                                        +(0.09423+i0.00000)e{0,1}
 93:   log(x*x)=> (3.43587+i0.00000)e{}+(0.18846+i0.00000)e{0}
                                        +(0.18846+i0.00000)e{0,1}
 94:   log(x)+log(x)=> (3.436e+00+i0.000e+00)e{}
                           +(1.885e-01+i0.000e+00)e{0}
 95:    +(1.885e-01+i0.000e+00)e{0,1}
 96:   log(x)+log(x)-log(x*x)=> (-4.441e-16+i0.000e+00)e{}
                                  +(-2.776e-17+i0.000e+00)e{0}
 97:    +(-2.776e-17+i0.000e+00)e{0,1}
 98:   |log(x)+log(x)-log(x*x)|=> (4.458e-16+i0.000e+00)e{}
 99:
100:   polynomial reciprocal series (24) terms
101:   (1/x)=> (0.17785+i0.00000)e{}+(-0.01686+i0.00000)e{0}
                                        +(-0.01686+i0.00000)e{0,1}
102:   x*(1/x)=> (1.000e+00+i0.000e+00)e{}
103:   x*(1/x)-1=> (0.000e+00+i0.000e+00)e{}
104:   |x*(1/x)-1|=> (0.000e+00+i0.000e+00)e{}
105:
106:   polynomial square root series (24) terms
107:   sqrt(x)=> (2.35548+i0.00000)e{}+(0.11114+i0.00000)e{0}
                                        +(0.11114+i0.00000)e{0,1}
108:   sqrt(x)*sqrt(x)=> (5.524e+00+i0.000e+00)e{}
                           +(5.236e-01+i0.000e+00)e{0}
109:    +(5.236e-01+i0.000e+00)e{0,1}
110:   sqrt(x)*sqrt(x)-x=> (-8.882e-16+i0.000e+00)e{}
                              +(-1.110e-16+i0.000e+00)e{0}
111:    +(-1.110e-16+i0.000e+00)e{0,1}
112:   |sqrt(x)*sqrt(x)-x|=> (9.019e-16+i0.000e+00)e{}
113:
114:   polynomial cube root series (24) terms
115:   3root(x)=> (1.77120+i0.00000)e{}+(0.05567+i0.00000)e{0}
                                        +(0.05567+i0.00000)e{0,1}
116:   3root(x)*3root(x)*3root(x)=> (5.524e+00+i0.000e+00)e{}
                                     +(5.236e-01+i0.000e+00)e{0}
117:    +(5.236e-01+i0.000e+00)e{0,1}
118:   3root(x)*3root(x)*3root(x)-x=> (-8.882e-16+i0.000e+00)e{}
                                        +(-2.220e-16+i0.000e+00)e{0}
119:    +(-2.220e-16+i0.000e+00)e{0,1}
120:   |3root(x)*3root(x)*3root(x)-x|=> (9.421e-16+i0.000e+00)e{}
121:
122:   maximum error 9.420555e-16
```

All of the values for which the functions are evaluated fall outside the normal range of convergence for the series in Table 12.2. Nevertheless, the consistency checks show that the identities are honoured to an accuracy approaching double precision, on lines 29, 58, 87, and 122.

12.2.2.3 Doubly Extended Convergence

The code listed in Example 12.5a tests the central multiplicative inverse function (reciprocal) using both factorisations in Eqs. (12.5) and (12.11) to provide a doubly extended range of convergence. The power series is evaluated on line 15, and consistency checks performed from line 17 to 23. The error from the final check is captured on line 24 and is printed on line 28.

Example 12.5(a)

```
 1:   void test_xx_series(x,p)  /* xx_series() can use number with zero
                                      scalar part */
 2:        number x;
 3:        int p;
 4:   {
 5:      number y,z;
 6:      double maxerror;
 7:      int n;
 8:
 9:      n=p+1;
10:      printf("number x is ");
11:      print_number(x,8);
12:
13:      /* central multiplicative inverse 1/x */
14:      printf("\npolynomial reciprocal series (%d) terms\n",n);
15:      y=xx_series(CM_INV,x,p);
16:      printf(" (1/x)=> ");       print_number(y,2);
17:      /* some checks */
18:      printf("x*(1/x)=> ");
19:      printf_number("%.3e",z=central_mul(x,y,0),2);
20:      printf("x*(1/x)-1=> ");
21:      printf_number("%.3e",z=sub_number(z,scalar_number(1.0,0.0),3),2);
22:      printf("|x*(1/x)-1|=> ");
23:      printf_number("%.3e",z=magnitudeof_number(z,1),2);
24:      maxerror=get_real(z,0,0);
25:      recycle_number(y);
26:      recycle_number(z);
27:
28:      printf("\nmaximum error %e\n",maxerror);
29:   }
```

Results from running the code in Example 12.5a are listed in Example 12.5b. The checks are honoured to an accuracy approaching double precision, on lines 10, 20, 30, and 43.

Example 12.5(b)

```
 1:   application3()
 2:   number x is (5.52360+i0.00000)e{2}
 3:
 4:   polynomial reciprocal series (24) terms
 5:   (1/x)=> (-0.18104-i0.00000)e{2}
 6:   x*(1/x)=> (1.000e+00+i1.225e-16)e{}
 7:   x*(1/x)-1=> (0.000e+00+i1.225e-16)e{}
 8:   |x*(1/x)-1|=> (1.225e-16+i0.000e+00)e{}
 9:
10:   maximum error 1.224647e-16
11:
12:   number x is (5.00000+i0.00000)e{2}+(0.52360+i0.00000)e{0,2}
13:
14:   polynomial reciprocal series (24) terms
15:   (1/x)=> (-0.19783-i0.00000)e{2}+(-0.02072-i0.00000)e{0,2}
16:   x*(1/x)=> (1.000e+00+i1.225e-16)e{}
17:   x*(1/x)-1=> (0.000e+00+i1.225e-16)e{}
18:   |x*(1/x)-1|=> (1.225e-16+i0.000e+00)e{}
19:
20:   maximum error 1.224647e-16
21:
22:   number x is (5.00000+i0.00000)e{2}+(0.52360+i0.00000)e{0,1,2}
23:
24:   polynomial reciprocal series (24) terms
25:   (1/x)=> (-0.19783-i0.00000)e{2}+(0.02072+i0.00000)e{0,1,2}
26:   x*(1/x)=> (1.000e+00+i1.225e-16)e{}+(1.110e-16+i1.233e-32)e{0,1}
27:   x*(1/x)-1=> (-1.110e-16+i1.225e-16)e{}+(1.110e-16+i1.233e-32)e{0,1}
28:   |x*(1/x)-1|=> (1.991e-16+i0.000e+00)e{}
29:
30:   maximum error 1.991218e-16
31:
32:   number x is (5.52360+i0.00000)e{2}+(0.52360+i0.00000)e{0,2}
                                        +(0.52360+i0.00000)e{0,1,2}
33:
34:   polynomial reciprocal series (24) terms
35:   (1/x)=> (-0.17785-i0.00000)e{2}+(-0.01686-i0.00000)e{0,2}
36:   +(-0.00000-i0.00000)e{1,2}+(0.01686+i0.00000)e{0,1,2}
37:   x*(1/x)=> (1.000e+00+i1.225e-16)e{}+(3.761e-21+i1.541e-33)e{0}
38:   +(3.469e-18+i7.704e-34)e{1}+(-4.163e-17-i6.163e-33)e{0,1}
39:   x*(1/x)-1=> (2.220e-16+i1.225e-16)e{}+(3.761e-21+i1.541e-33)e{0}
40:   +(3.469e-18+i7.704e-34)e{1}+(-4.163e-17-i6.163e-33)e{0,1}
41:   |x*(1/x)-1|=> (2.570e-16+i0.000e+00)e{}
42:
43:   maximum error 2.569957e-16
```

References

Greenhill, S. (1919), 'The Bessel-Clifford function and its applications', *Philosophical Magazine* **38**(227), 501–528.

Horner, W. G. (1819), 'A new method of solving numerical equations of all orders, by continuous approximation', *Philosophical Transactions of the Royal Society of London* **109**, 308–335.

King, K. N. (2008), *C Programming: A Modern Approach*, 2nd edn, W. W. Norton, New York, US-NY.

Spiegel, M. R. (1968), *Mathematical Handbook of Formulas and Tables*, Schaum's outline series, McGraw–Hill, Cambridge, UK.

13

Matrices of Clifford Numbers

Systems of linear equations in which the coefficients[1] are Clifford entities of a particular type, can be solved by matrix inversion when the inverses of the coefficients (elements of the matrix) can be calculated, and when the inverses and products of those coefficients yield entities of the same type. This is the case when the Clifford entities are scalars, bicomplex numbers, quaternions, Pauli matrices, and extended electromagnetic fields.

Any inversion technique which applies to scalar coefficients (these being commutative) can be extended to coefficients of higher grades, as long as care is taken to honour non-commutative multiplication. As a practical example, the inversion of systems with extended electromagnetic fields as coefficients in the context of electromagnetic scattering is examined in Chapter 25. Here the fundamentals are demonstrated by applying, for simplicity, Gaussian elimination (Grcar 2011) on smaller systems involving bicomplex numbers.

13.1 Background

Matrices are accessed within the Clifford numerical suite using the special data type 'matrix'. The data type is defined as the structure shown in Example 13.1.

Example 13.1

```
1:  /* matrix type definition */
2:  typedef struct{
3:      int nrows;          /* number of rows */
4:      int ncols;          /* number of columns */
5:      int stride;         /* offset between 1st elements in adjacent
                               columns */
6:      number *data;       /* first element of data */
7:  } matrix;
```

1 The coefficients here are the known constant terms in the linear equations, which may or may not be Clifford numbers. They should not be confused with the (other) coefficients within the components of the Clifford numbers themselves.

Numerical Calculations in Clifford Algebra: A Practical Guide for Engineers and Scientists,
First Edition. Andrew Seagar.
© 2023 John Wiley & Sons Ltd. Published 2023 by John Wiley & Sons Ltd.

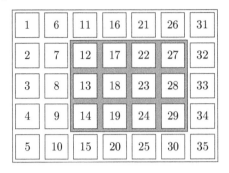

Figure 13.1 Matrix with five rows and seven columns containing a sub-matrix of three rows and four columns.

Routines are provided for simple matrix manipulation, such as the addition, subtraction, multiplication, and inverse of whole matrices. In addition, lower-level routines for scaling and adding or subtracting rows from one another offer a convenient mechanism allowing the user to develop additional functionality. A facility for the definition of sub-matrices within existing matrices provides a means to implement new algorithms in a recursive fashion where that suits.

Matrices are stored in memory in column major format, with consecutive elements in each column occupying consecutive locations in memory. When a new matrix is initialised whole columns follow in memory one from the previous, with the first element in each column separated from the first element in the previous column by the number of rows. The distance between the first element in one column and the next is called the 'stride'.

A sub-matrix is defined from an existing matrix *in situ* without copying or cloning the underlying data. This is achieved by setting the values in the 'matrix' structure for the sub-matrix to identify the corresponding data in the original matrix. Operations on the sub-matrix modify parts of the original matrix.

The value of the stride for the memory associated with a sub-matrix is the same as for the original matrix, but the number of rows, columns, and the pointer to the first element of data probably differ. Consecutive elements in each column occupy consecutive locations in memory, but the first element in each column is separated from the first element in the previous column by the value of the stride, which is no longer necessarily the same as the number of rows. As a consequence the adjacent columns in the sub-matrix can no longer be treated as contiguous.

As an example, Figure 13.1 shows a matrix with five rows and seven columns in 35 contiguous memory locations, containing a sub-matrix of three rows and four columns. For both matrices, the value of the stride is 5. For the full matrix, the columns run contiguously in memory, from 5 to 6, 10 to 11, and so on, whereas the columns in the sub-matrix are separated by 'stride-nrows=2' elements, as between 14 and 17, 19 and 22, etc.

13.2 Inversion

Matrix inversion or the solution of a linear system, implemented here using Gaussian elimination in the form of LU decomposition, is effected by invoking the routine solve_matrix(). A crucial part of the inversion is the inverse of individual elements

Table 13.1 Function prototype of routine to declare function for inversion of matrix elements.

Inverse declaration	Page
`number (*declare_inverse(number (*new_inverse)` `(number xn))) (number xo);`	337

Table 13.2 Pre-defined functions for matrix inversion.

`new_inverse()`	Page
`number inv_scalar(number x);`	359
`number inv_bicomplex(number x);`	358
`number inv_quaternion(number x);`	358
`number inv_pauli(number x);`	358
`number inv_xemfield(number x);`	360

in the matrix. Before inverting a matrix, it is necessary to declare an inverse function which matches the nature of the elements in the matrix. This is achieved as shown in Table 13.1 by passing the appropriate inverse function, represented here by the name `new_inverse()`, but in practice defined and named by the user, as an argument to the sub-routine `declare_inverse()`. The invocation returns the previously declared inverse function `old_inverse()`, which may be saved and reinstated when required with a subsequent invocation of `declare_inverse()`.

The sub-routines listed in Table 13.2 provide a variety of pre-defined inverse functions which may be used in place of the argument `new_inverse()`.

As an example, the inverse function `inv_bicomplex()` for a matrix containing bicomplex numbers is listed in Example 13.2 and declared for use in matrix inversion on line 31 of Example 13.3a. The function checks on line 9 for division by zero, which can occur if the number has no inverse, and invokes the user's recovery strategy on line 12 if appropriate. Attempted division by zero will certainly occur if the matrix is numerically singular, but may also occur for a non-singular matrix if a pivot with no inverse is encountered during elimination. The matrix inverse as implemented here is not sufficiently sophisticated to recover by itself if this happens.

Example 13.2

```
1:   /* bicomplex inverse with check for zero denominator */
2:   number inv_bicomplex(x)
3:        number x;
4:   {
```

(Continued)

Example 13.2 (Continued)

```
 5:     number y[2];
 6:     number inv;
 7:
 8:     reciprocalof_bicomplex(x,y,0);
 9:     if(is_zero(y[1]))
10:        {
11:          printf("zero denominator in routine: inv_bicomplex()\n");
12:          brexit(0);
13:        }
14:     inv=central_mul(y[0],inverseof_component(y[1],1),3);
15:     return(inv);
```

13.3 Practice

13.3.1 Example Code

The code listed in Example 13.3a demonstrates the inversion of a matrix containing a random selection of bicomplex numbers. A numerical context of dimension 2 is established in lines 10–13. The printing format for matrices is set to print columns on line 14, and the size of the matrix is defined with four rows and columns in lines 17 and 18. Memory space is allocated for three such matrices in lines 21–23, and the structures for the matrices are initialised in lines 26–28. At this stage, the matrices are empty.

Example 13.3(a)

```
 1:  void demo22() /* matrix inverse */
 2:  {
 3:     int lambda[2];
 4:     matrix m0,m1,m2;
 5:     number *element,*data0,*data1,*data2;
 6:     double error;
 7:     int nrows,ncols,nr,nc,st,i,j;
 8:
 9:     /* keep simple with 2 Clifford dimensions and bicomplex
          numbers */
10:     recovery_level(0,0,0);
11:     create_cliffmem(200);
12:     declare_primal(2,pack_signature(2,lambda,-1));
13:     format_printing("%.2f");
14:     format_matrix(1);            /* 1=print columns, 0=print rows */
15:
16:     /* choose size of matrix */
17:     nrows=4;
18:     ncols=4;
19:
20:     /* allocate memory for three matrices */
```

```
21:     data0=malloc(nrows*ncols*sizeof(number));
22:     data1=malloc(nrows*ncols*sizeof(number));
23:     data2=malloc(nrows*ncols*sizeof(number));
24:
25:     /* define matrices: size, symbolic name, memory for data */
26:     declare_matrix(nrows,ncols,&m0,data0);
27:     declare_matrix(nrows,ncols,&m1,data1);
28:     declare_matrix(nrows,ncols,&m2,data2);
29:
30:     /* declare the function to be used when inverting matrices */
31:     declare_inverse(inv_bicomplex);
32:
33:     /* set up some random data in matrix m0 */
34:     random_seed(0xABCDEF);
35:     nr=rows(&m0);                        /* get number of rows from
                                                 matrix structure */
36:     nc=columns(&m0);                     /* get number of columns from
                                                 matrix structure */
37:     st=stride(&m0);                      /* get value of stride from
                                                 matrix structure */
38:     element=access_element(0,0,&m0);  /* get pointer to 1st
                                                 element of matrix/sub-matrix */
39:     for(j=0;j<nc;j++)                    /* loop over number of columns */
40:       {
41:         for(i=0;i<nr;i++)                /* loop over number of rows */
42:           {
43:             element[i]=random_bicomplex(); /* initialise elements in
                                                   all rows in column */
44:           }
45:         element=element+st;              /* move to next column using
                                                 value of stride */
46:       }
47:
48:     /* construct clone of matrix for checking inverse later */
49:     clone_matrix(&m0,&m1);
50:
51:     /* set up identity matrix in matrix m2 */
52:     nr=rows(&m2);
53:     nc=columns(&m2);
54:     st=stride(&m2);
55:     element=access_element(0,0,&m2);
56:     for(j=0;j<nc;j++)
57:       {
58:         for(i=0;i<nr;i++)
59:           {
60:           if(i!=j) element[i]=fresh_number();
61:           else     element[i]=scalar_number(1.0,0.0);
62:           }
63:         element=element+st;                       /* move to next
                                                         column */
64:       }
65:
66:     /* print matrix m0 and m2 (identity) */
67:     printf("matrix to invert\n");
```

(Continued)

Example 13.3(a) (Continued)

```
68:      print_matrix(&m0,4);
69:      printf("set of right hand side vectors\n");
70:      print_matrix(&m2,4);
71:
72:      /* eliminate below leading diagonal */
73:      printf("eliminate below leading diagonal\n");
74:      eliminate_below(&m0,&m2);
75:      printf("should be upper triangular matrix\n");
76:      print_matrix(&m0,4);
77:      printf("set of right hand side vectors\n");
78:      print_matrix(&m2,4);
79:
80:      /* eliminate above leading diagonal */
81:      printf("eliminate above leading diagonal\n");
82:      eliminate_above(&m0,&m2);
83:      printf("should be unity diagonal matrix\n");
84:      print_matrix(&m0,4);
85:      printf("set of right hand side vectors = inverse\n");
86:      print_matrix(&m2,4);
87:
88:      /* check accuracy of elimination for first matrix */
89:      error=check_identity(&m0);
90:      printf("error in elimination is: %.5e\n",error);
91:
92:      /* recycle all Clifford numbers, but keep memory for tokens */
93:      /* matrix ready to use just as if declared */
94:      refresh_matrix(&m0);
95:
96:      /* multiply clone of original matrix onto inverse */
97:      multiply_matrix(&m1,&m2,&m0,0);
98:
99:      /* check accuracy of product */
100:     printf("product matrix and inverse should be identity\n");
101:     print_matrix(&m0,4);
102:     error=check_identity(&m0);
103:     printf("error in product is: %.5e\n",error);
104:
105:     /* recycle all Clifford numbers and free the memory for tokens */
106:     recycle_matrix(&m0);
107:     recycle_matrix(&m1);
108:     recycle_matrix(&m2);
109:
110:     free_cliffmem(1);
111: }
```

Line 31 declares the inverse function for matrix inversion as the pre-defined function `inv_bicomplex()` as is appropriate for the bicomplex number entities intended as the elements of the matrix.

Random data values are established in matrix 'm0' in lines 34–46 to serve as the matrix A in the linear system $Ax = b$. There is a routine `random_matrix()` which has the same effect using only a single line of code, but the aim here is to demonstrate proper access to the elements of the matrix (or sub-matrix).

A clone of the matrix, 'm1', is created on line 49 for the purpose of checking the inverse because the inversion process destroys the original matrix.

Lines 52–64 establish as a matrix 'm2' a set of columns to serve as the known data b on the right-hand side of the linear system. After solution, the right-hand side b is replaced in 'm2' by the solution x. Here the set of columns are initialised to represent the identity matrix[2] so that after solution, the same set of columns contains the inverse matrix. In general, the matrix for the set of columns on the right-hand side need not be square but can contain however many columns are required.

The matrix for inversion and the matrix containing the set of columns on the right-hand side are printed on lines 67–70 prior to preforming the elimination.

Normally, the solution can be effected by a single invocation of the routine solve _matrix(); however, here the elimination is carried in two stages in order to check the intermediate results. Line 74 eliminates non-zero elements below the leading diagonal of matrix 'm0' (A) reducing it to an upper triangular form and performs the same operations on all of the columns of data in the matrix 'm2' (b). The upper triangular matrix is printed on line 76, and the intermediate solution in the matrix 'm2' is printed on line 78.

The solution is completed by eliminating the non-zero elements above the leading diagonal of matrix 'm0' on line 82 reducing it to a diagonal form, again carrying the same operations on all of the data in the matrix 'm2'. The diagonal matrix is printed on line 84, and the final solution in the matrix 'm2' is printed on line 86.

If the codes on lines 74 and 82 are exchanged, the matrix is inverted using UL decomposition rather than LU decomposition.

The accuracy of the elimination is checked against the identity matrix in line 89, and the error is printed on line 90. The numbers in each of the elements in the matrix are all recycled on line 94, while retaining the memory allocated to contain the elements so that the matrix can be reused.

The clone of the original matrix 'm1' is multiplied by the inverse 'm2' on line 97 with the product, which should in principle be the identity matrix, saved in the refreshed (empty) matrix 'm0'. The result of the multiplication is printed on line 101 and checked for accuracy against the identity matrix on line 102. The error in the product is printed on line 103.

Finally, the three matrices are recycled on lines 106–109. Recycling matrices is more thorough than refreshing matrices. When recycling a matrix, the numbers in each of the elements in the matrix are all recycled individually, and in addition, the memory allocated to contain the elements, here on lines 21–23 of Example 13.3a, is liberated by invoking the standard library call free().

13.3.2 Example Output

The results from running the code in Examples 13.2 and 13.3a are listed in Example 13.3b. The initial matrix of bicomplex numbers is printed on lines 3–23. The columns are printed

---

2 Again, there is a routine identity_matrix() which has the same effect using only a single line of code.

one below the other rather than one beside the other because the columns usually cannot all fit on one line for larger matrices with entities of higher grades. Rows are printed instead of columns if the argument of `format_matrix()` on line 14 of Example 13.3a is set to zero.

The initial columns of data on the right-hand side are printed on lines 25–45, showing the identity matrix.

Example 13.3(b)

```
 1:   demo22()
 2:   matrix to invert
 3:   matrix size (4,4)
 4:   column 0
 5:   (0,0)  (0.08-i0.13)e{}+(-0.03-i0.82)e{0,1}
 6:   (1,0)  (-0.37+i0.27)e{}+(0.40+i0.80)e{0,1}
 7:   (2,0)  (-0.33-i0.40)e{}+(-0.34-i0.50)e{0,1}
 8:   (3,0)  (-0.06-i1.00)e{}+(0.96-i0.69)e{0,1}
 9:   column 1
10:   (0,1)  (-0.51-i0.70)e{}+(-0.79+i0.56)e{0,1}
11:   (1,1)  (0.34+i0.67)e{}+(0.52+i0.99)e{0,1}
12:   (2,1)  (-0.60-i0.96)e{}+(-0.59+i0.72)e{0,1}
13:   (3,1)  (-0.81-i0.13)e{}+(-0.89+i0.27)e{0,1}
14:   column 2
15:   (0,2)  (0.74+i0.08)e{}+(0.45-i0.62)e{0,1}
16:   (1,2)  (-0.65-i0.14)e{}+(-0.83+i0.03)e{0,1}
17:   (2,2)  (0.46-i0.17)e{}+(0.52-i0.60)e{0,1}
18:   (3,2)  (-0.17+i0.49)e{}+(-0.29+i0.32)e{0,1}
19:   column 3
20:   (0,3)  (0.79-i0.08)e{}+(-0.12+i0.13)e{0,1}
21:   (1,3)  (-0.41-i0.60)e{}+(0.12-i0.01)e{0,1}
22:   (2,3)  (-0.56+i0.54)e{}+(-0.30-i0.38)e{0,1}
23:   (3,3)  (-0.59-i0.19)e{}+(0.90-i0.85)e{0,1}
24:   set of right hand side vectors
25:   matrix size (4,4)
26:   column 0
27:   (0,0)  (1.00+i0.00)e{}
28:   (1,0)  (0.00+i0.00)e{}
29:   (2,0)  (0.00+i0.00)e{}
30:   (3,0)  (0.00+i0.00)e{}
31:   column 1
32:   (0,1)  (0.00+i0.00)e{}
33:   (1,1)  (1.00+i0.00)e{}
34:   (2,1)  (0.00+i0.00)e{}
35:   (3,1)  (0.00+i0.00)e{}
36:   column 2
37:   (0,2)  (0.00+i0.00)e{}
38:   (1,2)  (0.00+i0.00)e{}
39:   (2,2)  (1.00+i0.00)e{}
40:   (3,2)  (0.00+i0.00)e{}
41:   column 3
42:   (0,3)  (0.00+i0.00)e{}
43:   (1,3)  (0.00+i0.00)e{}
44:   (2,3)  (0.00+i0.00)e{}
```

```
45:   (3,3) (1.00+i0.00)e{}
46:   eliminate below leading diagonal
47:   should be upper triangular matrix
48:   matrix size (4,4)
49:   column 0
50:   (0,0) (1.00+i0.00)e{}+(0.00+i0.00)e{0,1}
51:   (1,0) (-0.00+i0.00)e{}+(0.00+i0.00)e{0,1}
52:   (2,0) (-0.00-i0.00)e{}+(-0.00-i0.00)e{0,1}
53:   (3,0) (0.00-i0.00)e{}+(0.00-i0.00)e{0,1}
54:   column 1
55:   (0,1) (-0.48-i0.95)e{}+(-0.84+i0.40)e{0,1}
56:   (1,1) (1.00+i0.00)e{}+(0.00-i0.00)e{0,1}
57:   (2,1) (-0.00+i0.00)e{}+(0.00+i0.00)e{0,1}
58:   (3,1) (0.00+i0.00)e{}+(-0.00-i0.00)e{0,1}
59:   column 2
60:   (0,2) (0.64+i0.67)e{}+(0.16-i0.73)e{0,1}
61:   (1,2) (-0.45+i0.38)e{}+(-0.00+i0.03)e{0,1}
62:   (2,2) (1.00-i0.00)e{}+(-0.00-i0.00)e{0,1}
63:   (3,2) (0.00-i0.00)e{}
64:   column 3
65:   (0,3) (-0.25+i0.01)e{}+(-0.10-i0.99)e{0,1}
66:   (1,3) (-0.06-i0.14)e{}+(0.53+i0.04)e{0,1}
67:   (2,3) (1.40-i1.41)e{}+(1.50-i0.82)e{0,1}
68:   (3,3) (1.00-i0.00)e{}
69:   set of right hand side vectors
70:   matrix size (4,4)
71:   column 0
72:   (0,0) (-0.13+i0.18)e{}+(0.00-i1.20)e{0,1}
73:   (1,0) (-0.04-i0.00)e{}+(0.23+i0.41)e{0,1}
74:   (2,0) (0.86-i0.29)e{}+(1.23-i0.15)e{0,1}
75:   (3,0) (0.47+i0.25)e{}+(0.11+i0.13)e{0,1}
76:   column 1
77:   (0,1) (0.00+i0.00)e{}
78:   (1,1) (0.18-i0.03)e{}+(0.07+i0.53)e{0,1}
79:   (2,1) (-0.55+i0.39)e{}+(0.73+i0.12)e{0,1}
80:   (3,1) (-0.05+i0.28)e{}+(0.32+i0.11)e{0,1}
81:   column 2
82:   (0,2) (0.00+i0.00)e{}
83:   (1,2) (0.00+i0.00)e{}
84:   (2,2) (-1.35+i0.08)e{}+(-1.15+i0.34)e{0,1}
85:   (3,2) (-0.56-i0.16)e{}+(-0.05+i0.32)e{0,1}
86:   column 3
87:   (0,3) (0.00+i0.00)e{}
88:   (1,3) (0.00+i0.00)e{}
89:   (2,3) (0.00+i0.00)e{}
90:   (3,3) (0.10-i0.01)e{}+(-0.09-i0.17)e{0,1}
91:   eliminate above leading diagonal
92:   should be unity diagonal matrix
93:   matrix size (4,4)
94:   column 0
95:   (0,0) (1.00+i0.00)e{}
96:   (1,0) (0.00-i0.00)e{}+(0.00+i0.00)e{0,1}
97:   (2,0) (0.00-i0.00)e{}+(0.00+i0.00)e{0,1}
98:   (3,0) (0.00-i0.00)e{}+(0.00-i0.00)e{0,1}
```

(Continued)

Example 13.3(b) (Continued)

```
 99:  column 1
100:  (0,1)  (0.00+i0.00)e{}
101:  (1,1)  (1.00+i0.00)e{}+(0.00-i0.00)e{0,1}
102:  (2,1)  (-0.00-i0.00)e{}+(0.00+i0.00)e{0,1}
103:  (3,1)  (0.00+i0.00)e{}+(-0.00-i0.00)e{0,1}
104:  column 2
105:  (0,2)  (0.00+i0.00)e{}
106:  (1,2)  (0.00+i0.00)e{}
107:  (2,2)  (1.00-i0.00)e{}
108:  (3,2)  (0.00-i0.00)e{}
109:  column 3
110:  (0,3)  (-0.00+i0.00)e{}+(-0.00-i0.00)e{0,1}
111:  (1,3)  (-0.00-i0.00)e{}+(0.00+i0.00)e{0,1}
112:  (2,3)  (0.00-i0.00)e{}+(0.00-i0.00)e{0,1}
113:  (3,3)  (1.00-i0.00)e{}
114:  set of right hand side vectors = inverse
115:  matrix size (4,4)
116:  column 0
117:  (0,0)  (-0.14-i0.09)e{}+(-0.47-i0.40)e{0,1}
118:  (1,0)  (0.10+i0.15)e{}+(-0.08+i0.20)e{0,1}
119:  (2,0)  (0.12+i0.11)e{}+(-0.01-i0.17)e{0,1}
120:  (3,0)  (0.47+i0.25)e{}+(0.11+i0.13)e{0,1}
121:  column 1
122:  (0,1)  (-0.17-i0.12)e{}+(-0.29+i0.15)e{0,1}
123:  (1,1)  (0.10+i0.08)e{}+(0.08+i0.43)e{0,1}
124:  (2,1)  (-0.30-i0.18)e{}+(-0.02-i0.04)e{0,1}
125:  (3,1)  (-0.05+i0.28)e{}+(0.32+i0.11)e{0,1}
126:  column 2
127:  (0,2)  (-0.18+i0.19)e{}+(-0.06+i0.09)e{0,1}
128:  (1,2)  (-0.09+i0.13)e{}+(-0.15+i0.16)e{0,1}
129:  (2,2)  (-0.15+i0.03)e{}+(-0.55-i0.38)e{0,1}
130:  (3,2)  (-0.56-i0.16)e{}+(-0.05+i0.32)e{0,1}
131:  column 3
132:  (0,3)  (0.00+i0.12)e{}+(-0.02-i0.24)e{0,1}
133:  (1,3)  (-0.22+i0.06)e{}+(0.14-i0.00)e{0,1}
134:  (2,3)  (-0.39-i0.02)e{}+(0.23+i0.21)e{0,1}
135:  (3,3)  (0.10-i0.01)e{}+(-0.09-i0.17)e{0,1}
136:  error in elimination is: 1.15581e-16
137:  product matrix and inverse should be identity
138:  matrix size (4,4)
139:  column 0
140:  (0,0)  (1.00+i0.00)e{}+(-0.00-i0.00)e{0,1}
141:  (1,0)  (0.00+i0.00)e{}+(0.00-i0.00)e{0,1}
142:  (2,0)  (0.00+i0.00)e{}+(-0.00+i0.00)e{0,1}
143:  (3,0)  (-0.00+i0.00)e{}+(-0.00-i0.00)e{0,1}
144:  column 1
145:  (0,1)  (0.00-i0.00)e{}+(-0.00+i0.00)e{0,1}
146:  (1,1)  (1.00+i0.00)e{}+(0.00-i0.00)e{0,1}
147:  (2,1)  (0.00+i0.00)e{}+(0.00+i0.00)e{0,1}
148:  (3,1)  (0.00-i0.00)e{}+(0.00-i0.00)e{0,1}
149:  column 2
150:  (0,2)  (0.00+i0.00)e{}+(-0.00-i0.00)e{0,1}
151:  (1,2)  (-0.00+i0.00)e{}+(0.00+i0.00)e{0,1}
152:  (2,2)  (1.00-i0.00)e{}+(0.00-i0.00)e{0,1}
```

```
153:    (3,2)  (-0.00+i0.00)e{}+(-0.00-i0.00)e{0,1}
154:    column 3
155:    (0,3)  (0.00+i0.00)e{}+(0.00-i0.00)e{0,1}
156:    (1,3)  (-0.00-i0.00)e{}+(0.00+i0.00)e{0,1}
157:    (2,3)  (0.00-i0.00)e{}+(-0.00+i0.00)e{0,1}
158:    (3,3)  (1.00-i0.00)e{}+(0.00-i0.00)e{0,1}
159:    error in product is: 2.70043e-16
160:    statistics for structural heap:
161:    initial size (200)
162:    peak size      (49), as % of initial size (24%)
163:    full size      (200), expansion count (0)
164:    free count     (200), free start (40)
165:    statistics for data heap:
166:    initial size (200)
167:    peak size      (100), as % of initial size (50%)
168:    full size      (200), expansion count (0)
169:    free count     (200), free start (90)
```

After elimination below the leading diagonal, the partially reduced matrix is seen as upper triangular on lines 48–68, and the partially reduced right-hand side is lower triangular on lines 70–90.

Following elimination above the leading diagonal, the fully reduced matrix is diagonal on lines 93–113 (very close to the identity matrix) and the fully reduced right-hand side is full, now containing the inverse of the original matrix on lines 115–135.

The difference between the fully reduced (diagonal) matrix and the identity matrix printed on line 136 is close to machine precision.

Lines 138–158 list the product of the original matrix with the inverse, different from the identity matrix by the value close to machine precision on line 159.

Lines 160–169 verify that all memory has been properly recycled with the free count equal to the full size for both structural and data memory.

Reference

Grcar, J. F. (2011), 'How ordinary elimination became Gaussian elimination', *Historia Mathematica* **38**(2), 163–218.

Part II

Customisation

14

Memory

Apart from one initial block, there is no need to allocate memory. Memory grows as needed. For efficiency, the user can specify when particular variables or intermediate results are no longer required so that unused memory can be reclaimed. Apart from values which are zero, reclamation is not implemented automatically.

14.1 Memory Usage

Variables representing Clifford numbers are identified by tokens of integer value, which are interpreted by the various routines as indexes into a pair of tables not intended for direct access by the user. The tables contain nodes, in the form of internal structural elements (forks) and terminal data elements (leaves), which are assembled, as in Table 14.1, into an indexed extended binary tree. Structural and data nodes are utilised from the tables as needed and reclaimed when the corresponding data are zero or when flagged as no longer required by the user.

Structural nodes (forks) contain two branches, each of which leads to either another structural or a data node. The data nodes are the leaves at the extremity of the tree's branches. These contain the numerical data values as complex coefficients coupled to a Clifford unit. Nodes also contain indexes; structural information about the range of Clifford units accommodated by the tree beyond that node. This permits the location of individual units deep within the tree to be accessed via a linear path of maximum length equal to the dimension of the Clifford number, with no need to traverse irrelevant branches. It also permits nodes to be added to a partially filled tree directly at the proper location according to the canonical ordering adopted for the Clifford units, with never any need to sort or resort the data in the tree. The tree stays in an ordered state automatically. With the exception of the tree's root, any leaves with zero numerical value are never stored and always eliminated as an integral part of any operation in which they are produced.

There are usually as many nodes which are not leaves as there are nodes which are leaves, so the memory overhead is in direct proportion (roughly a factor of 2) to the amount of non-zero data stored. For example, Table 14.2 shows a Clifford number of dimension 3, with all $2^3 = 8$ complex components set to non-zero values. The binary tree is fully populated. Greater distance from the root within the tree is represented as greater levels of

Numerical Calculations in Clifford Algebra: A Practical Guide for Engineers and Scientists,
First Edition. Andrew Seagar.
© 2023 John Wiley & Sons Ltd. Published 2023 by John Wiley & Sons Ltd.

Table 14.1 Forks, branches, and leaves of binary tree.

Token	Node	Structural (Internal)	Fork	Left branch	– Token
				Right branch	– Token
		Data (Terminal)	Leaf	Unit	– Bitmap
				Coefficient [0]	– Real
				[1]	– Imaginary

Table 14.2 Binary tree for full Clifford number of three dimensions.

```
(
|   [0-7] (
|   |   [0-3] (
|   |   |   [0-1] (
|   |   |   |   [0] (1.00+i2.00) e{}
|   |   |   |   [1] (3.00+i4.00) e0{0}
|   |   |   )
|   |   |   [2-3] (
|   |   |   |   [2] (5.00+i6.00) e1{1}
|   |   |   |   [3] (9.00+i10.00) e{0,1}
|   |   |   )
|   |   )
|   |   [4-7] (
|   |   |   [4-5] (
|   |   |   |   [4] (7.00+i8.00) e{2}
|   |   |   |   [5] (11.00+i12.00) e{0,2}
|   |   |   )
|   |   |   [6-7] (
|   |   |   |   [6] (13.00+i14.00) e{1,2}
|   |   |   |   [7] (15.00+i16.00) e{0,1,2}
|   |   |   )
|   |   )
|   )
)
```

indentation to the right. The numbers in brackets '[...]' record the range of Clifford units accommodated beyond each node of the tree, permitting fast traversal and direct access to the location of every component.

There are three mechanisms which provide explicit control over memory usage. Before utilising Clifford numbers, the numerical suite must be initialised by invoking the sub-routine create_cliffmem(). The single argument passed to this routine is the number of components for which to allocate memory. If this is not enough, the memory expands automatically by about 19% every time more is needed. Four expansions $(1.19)^4 = 2.005$ give double the memory.

When all operations on Clifford numbers are complete, the numerical suite can be deactivated with all memory allocated to it recovered and available for use by other parts of the code via calls to the 'c' language (King 2008) standard library routines for memory allocation. Deactivation is achieved by calling the sub-routine `free_cliffmem()` with a single integer argument of value either 0 or 1. With a zero argument, the routine executes silently. With an argument of one, the routine prints a summary of memory usage. The summary can be used to determine how much total memory was utilised. If that number is requested as the initial allocation next time the program runs, no automatic expansion of memory is required and execution time is reduced somewhat.

In order to limit the expansion of memory during calculations, numbers which are no longer required may be recycled. The memory associated with such numbers can then be used for others. Recycling numbers is achieved either explicitly by invoking the sub-routine `recycle_number()` with a single integer argument set to the value of the token for the number, or implicitly as a flag in the argument list of some other routine (see Section 14.2.3). It is a good practice to ensure that all Clifford variables local and private to any sub-routine are recycled before returning from the sub-routine. Whole arrays of numbers can be recycled at one stroke by a single invocation of the sub-routine `recycle_array()`.

14.2 Examples

14.2.1 Memory Tree Sparsity

The code listed in Example 14.1a constructs a Clifford number one component at a time in lines 14–30. There are a variety of ways to do this. In this case, the sub-routine `merge_comp()` is used to merge new components into an existing tree 'u', which is created and initialised to value zero on line 14. The sub-routine `merge_comp()` navigates through the tree structure using the unit of the new component as a target. If a value is already associated with that unit in the tree, the new coefficient is merged with the existing one using the sub-routine `addition_to()`, the second to last argument in the invocation of `merge_comp()`. If no value is associated with the unit, the new coefficient is inserted into the structure of the tree after applying the sub-routine `identity()`.

The sub-routine `merge_comp()` is one of the lower-level worker routines which manipulates tree structures and values. It is invoked by almost every numerical operation on Clifford numbers, such as `add_number()` and `sub_number()`. Within `add_number()`, the last two arguments in the invocation of `merge_comp()` are `addition_to()` and `identity()` as here. Within `sub_number()`, the last two arguments are `subtraction_from()` and `negation()`. Other arguments can be used to achieve other effects.

Example 14.1(a)

```
1:   void demo23() /* memory */
2:   {
3:      int lambda[3];
```

(Continued)

Example 14.1(a) (Continued)

```
4:      cliff_comp u_comp;
5:      number u,u_k;
6:      int k;
7:
8:      recovery_level(0,0,0);
9:      create_cliffmem(50);
10:     declare_primal(3,pack_signature(3,lambda,-1));
11:     format_printing("%.0f");
12:
13:     print_mem_usage("1> ");
14:     u=fresh_number();
15:     print_mem_usage("2> ");
16:     u=merge_comp(u,pack_comp(&u_comp,1.0,2.0,spack_unit("e{}")),
                     addition_to,identity);
17:     print_mem_usage("3> ");
18:     u=merge_comp(u,pack_comp(&u_comp,3.0,4.0,spack_unit("e{0}")),
                     addition_to,identity);
19:     print_mem_usage("4> ");
20:     u=merge_comp(u,pack_comp(&u_comp,5.0,6.0,spack_unit("e{1}")),
                     addition_to,identity);
21:     print_mem_usage("5> ");
22:     u=merge_comp(u,pack_comp(&u_comp,7.0,8.0,spack_unit("e{2}")),
                     addition_to,identity);
23:     print_mem_usage("6> ");
24:     u=merge_comp(u,pack_comp(&u_comp,9.0,10.0,spack_unit("e{0,1}")),
                     addition_to,identity);
25:     print_mem_usage("7> ");
26:     u=merge_comp(u,pack_comp(&u_comp,11.0,12.0,spack_unit("e{0,2}")),
                     addition_to,identity);
27:     print_mem_usage("8> ");
28:     u=merge_comp(u,pack_comp(&u_comp,13.0,14.0,spack_unit("e{1,2}")),
                     addition_to,identity);
29:     print_mem_usage("9> ");
30:     u=merge_comp(u,pack_comp(&u_comp,15.0,16.0,spack_unit("e{0,1,2}")),
                     addition_to,identity);
31:     print_mem_usage("10>");
32:
33:     printf("dimension 3 Clifford number:\n");
34:     print_tree(u);
35:
36:     for(k=0;k<4;k++)
37:       {
38:         u_k=onegradeof_number(u,k,0);
39:         print_mem_usage("11>");
40:         printf("grade %d of dimension 3 Clifford number:\n",k);
41:         print_tree(u_k);
42:         recycle_number(u_k);
43:         print_mem_usage("12>");
44:       }
45:     recycle_number(u);
46:     print_mem_usage("13>");
47:
48:     free_cliffmem(1);
49:   }
```

After each addition of a new component into the tree, a short report of current total memory usage is generated by invoking the sub-routine `print_mem_usage()`. The argument to the sub-routine `print_mem_usage()` is printed as text immediately preceding the report. This can be used to identify the place in the code where the report is generated. If no message is required, the argument may be set to a null pointer (zero). After adding all components, the full number is printed in tree format on line 34.

The loop between lines 36 and 44 creates a new number (tree) 'u_k' containing a copy of grade 'k' from the original number 'u'. The tree structure for that grade is printed on line 41. The number (tree) is recycled on line 42 and the same variable 'u_k' is used for the next grade in the next cycle of the loop. The total memory usage is printed after the number 'u_k' has been constructed on line 39 and on line 43 after it has been recycled.

If the number 'u_k' is not recycled in the loop, memory grows, and the tokens for the trees of the earlier values of number 'u_k' in previous cycles of the loop, contained as integers in the integer variable 'u_k', are overwritten and lost. The memory associated with the lost tokens cannot be recycled individually, although all memory can be reclaimed by invoking the sub-routine `recycle_allnumber()`. However, this reclaims memory globally and is not limited to within the scope of any particular routine.

Reclamation of all memory used in the example is achieved by recycling 'u' on line 45 after the loop, since this is the only remaining number not yet recycled. As a check, memory usage is printed immediately afterwards on line 46. A more detailed summary of total memory usage is printed on line 48 before returning from the example's routine.

The results of running the code listed in Example 14.1a are shown in Example 14.1b. The amount of memory grows almost linearly between lines 2 and 11 (1> and 10>) as components are added to the number one at a time. The first number in the triplets '(a/b/c)' indicates the current memory usage. The second number indicates the highest usage up to this stage. The third number indicates the full size of memory allocated (not the free memory). The numbers are always in non-decreasing order.

There is no increase in memory between lines 3 and 4 (2> and 3>) because the number zero is overwritten by the complex number $1 + 2i$. When there is only one element of data, no structural nodes are required.

The tree for the full number of eight non-zero components in three dimensions is listed in lines 13–36. This example was constructed so that the order in which the coefficients were added is reflected by their increasing numeric value, not the order of the single unit numbers in square brackets. The pairs of unit numbers in square brackets are contained in the structural nodes and allow efficient traversal of the tree. It is never necessary to venture down the wrong branch.

Line 37 reports the memory used after creating a new Clifford number 'u_k' from grade 0 of the first number 'u'. The value reported for used data is now greater than before being the total memory for both numbers. The new number is printed in lines 39–41. Although this is a three-dimensional Clifford number, only the non-zero components occupy storage.

Line 42 reports the memory used after recycling the number 'u_k', and line 43 reports recreating the number 'u_k' from grade 1 of the first number 'u'. Whereas the number in grade 0 has one component and is a scalar, this new number has three components and is a three-dimensional vector. The Clifford units of the components are reported as the set of integers in the braces on the right-hand side of the complex values of the coefficients on lines 48, 49, and 51, and are also reported as bit patterns in the square brackets on the left-hand side of the coefficients.

The memory utilised after recycling the vector is reported in line 54. The recycling operation has reclaimed two structural nodes, indicated by the pairs of integers in square brackets on line 46 and 47, and also three data nodes, indicated by the single integers in square brackets on lines 48, 49, and 51.

Example 14.1(b)

```
 1:   demo23()
 2:   1>  used/peak/full nodes  (0/0/50) structural,  (0/0/50) data.
 3:   2>  used/peak/full nodes  (0/0/50) structural,  (1/1/50) data.
 4:   3>  used/peak/full nodes  (0/0/50) structural,  (1/1/50) data.
 5:   4>  used/peak/full nodes  (1/1/50) structural,  (2/2/50) data.
 6:   5>  used/peak/full nodes  (2/2/50) structural,  (3/3/50) data.
 7:   6>  used/peak/full nodes  (3/3/50) structural,  (4/4/50) data.
 8:   7>  used/peak/full nodes  (4/4/50) structural,  (5/5/50) data.
 9:   8>  used/peak/full nodes  (5/5/50) structural,  (6/6/50) data.
10:   9>  used/peak/full nodes  (6/6/50) structural,  (7/7/50) data.
11:   10> used/peak/full nodes  (7/7/50) structural,  (8/8/50) data.
12:   dimension 3 Clifford number:
13:   (
14:   |    [0-7] (
15:   |    |    [0-3] (
16:   |    |    |    [0-1] (
17:   |    |    |    |    [0] (1+i2)e{}
18:   |    |    |    |    [1] (3+i4)e{0}
19:   |    |    |    )
20:   |    |    |    [2-3] (
21:   |    |    |    |    [2] (5+i6)e{1}
22:   |    |    |    |    [3] (9+i10)e{0,1}
23:   |    |    |    )
24:   |    |    )
25:   |    |    [4-7] (
26:   |    |    |    [4-5] (
27:   |    |    |    |    [4] (7+i8)e{2}
28:   |    |    |    |    [5] (11+i12)e{0,2}
29:   |    |    |    )
30:   |    |    |    [6-7] (
31:   |    |    |    |    [6] (13+i14)e{1,2}
32:   |    |    |    |    [7] (15+i16)e{0,1,2}
33:   |    |    |    )
34:   |    |    )
35:   |    )
36:   )
37:   11> used/peak/full nodes  (7/7/50) structural,  (9/9/50) data.
38:   grade 0 of dimension 3 Clifford number:
39:   (
40:   |    [0] (1+i2)e{}
41:   )
42:   12> used/peak/full nodes  (7/7/50) structural,  (8/9/50) data.
43:   11> used/peak/full nodes  (9/9/50) structural,  (11/11/50) data.
```

```
44:   grade 1 of dimension 3 Clifford number:
45:   (
46:   |     [0-7] (
47:   |     |    [0-3] (
48:   |     |    |    [1] (3+i4) e{0}
49:   |     |    |    [2] (5+i6) e{1}
50:   |     |    )
51:   |     |    [4] (7+i8) e{2}
52:   |     )
53:   )
54:   12> used/peak/full nodes (7/9/50) structural, (8/11/50) data.
55:   11> used/peak/full nodes (9/9/50) structural, (11/11/50) data.
56:   grade 2 of dimension 3 Clifford number:
57:   (
58:   |     [0-7] (
59:   |     |    [3] (9+i10) e{0,1}
60:   |     |    [4-7] (
61:   |     |    |    [5] (11+i12) e{0,2}
62:   |     |    |    [6] (13+i14) e{1,2}
63:   |     |    )
64:   |     )
65:   )
66:   12> used/peak/full nodes (7/9/50) structural, (8/11/50) data.
67:   11> used/peak/full nodes (7/9/50) structural, (9/11/50) data.
68:   grade 3 of dimension 3 Clifford number:
69:   (
70:   |     [7] (15+i16) e{0,1,2}
71:   )
72:   12> used/peak/full nodes (7/9/50) structural, (8/11/50) data.
73:   13> used/peak/full nodes (0/9/50) structural, (0/11/50) data.
74:   statistics for structural heap:
75:   initial size (50)
76:   peak size     (9), as % of initial size (18%)
77:   full size     (50), expansion count (0)
78:   free count    (50), free start (6)
79:   statistics for data heap:
80:   initial size (50)
81:   peak size     (11), as % of initial size (22%)
82:   full size     (50), expansion count (0)
83:   free count    (50), free start (7)
```

Lines 55–72 repeat the process for grades 2 (bivectors) and 3 (trivectors, or alternatively the pseudo-scalar for dimension 3). In each case, only the non-zero components occupy storage. The memory structure is always as sparse as the data allow.

The memory usage after recycling all memory is reported in brief on line 73 as a result of invoking the sub-routine print_mem_usage(), and in much more detail on lines 74–83 as a result of invoking the sub-routine free_cliffmem(). The 'initial size' reports the amount allocated initially by invoking the sub-routine create_cliffmem(). The 'full size' reports the amount allocated at the time of invoking the sub-routine free_cliffmem(). If the memory fills up and expands automatically, the full size is greater than the initial size and the 'expansion count' is non-zero. That has not happened in this case. When the 'free count' is less than the full size, it is an indication of memory not being fully recycled. This could be due to a memory leak, which should probably be plugged by modifying the source code in an appropriate manner. The value of 'free start' is the index into the internal tables of structural and data nodes. The value should lie somewhere in the range from zero to one less than full size. If it is not in that range the internal tables have somehow become corrupted.

14.2.2 Memory Expansion

Memory expands automatically when and as often as needed. If the initial allocation using create_cliffmem() is found inadequate at run time, a further 19% is added. It that is not enough, the memory is expanded by another 19%.

The code listed in Example 14.2 allocates enough memory for 50 components on line 8, and then proceeds to construct an array containing 100 Clifford numbers, all scalars of zero value, in the loop on line 14. Since more memory is used than initially allocated, it expands. The expansion is entirely transparent to the user. All valid tokens remain valid.

Example 14.2(a)

```
1:    void demo24() /* memory expansion */
2:    {
3:        int lambda[3];
4:        number *x;
5:        int i,n;
6:
7:        recovery_level(0,0,0);
8:        create_cliffmem(50);
9:        declare_primal(3,pack_signature(3,lambda,-1));
10:       format_printing("%.0f");
11:
12:       n=100;
13:       x=(number *)malloc(n*sizeof(number));
14:       for(i=0;i<n;i++) { x[i]=fresh_number();  }
15:       for(i=0;i<n;i++) { recycle_number(x[i]); }
16:       free((void *)x);
17:
18:       free_cliffmem(1);
19:   }
```

In this example, the memory is never actually used. The loop on line 15 recycles everything. It is the memory statistics printed on line 18 by invoking the sub-routine free_cliffmem() which are of interest here.

The results of running the code in Example 14.2a are shown in Example 14.2b. The numbers created are scalars, which require no structural nodes. The initial allocation of 50 on line 3 is therefore adequate, and no expansion is required. The expansion count is zero on line 5.

However, twice as much as the original allocation is required for the 100 data nodes. This is reported in line 9 with a peak size of 100 being 200% of the initial size. The expansion count on line 10 indicates that the internal tables have been copied and expanded four times. For this example that could be avoided by allocating an initial size of 100 when invoking `create_cliffmem()` on line 8 of Example 14.2a.

Example 14.2(b)

```
 1:   demo24()
 2:   statistics for structural heap:
 3:   initial size (50)
 4:   peak size    (0), as % of intial size (0%)
 5:   full size    (50), expansion count (0)
 6:   free count   (50), free start (0)
 7:   statistics for data heap:
 8:   initial size (50)
 9:   peak size    (100), as % of initial size (200%)
10:   full size    (100), expansion count (4)
11:   free count   (100), free start (99)
```

14.2.3 Memory Recycling

Memory is recycled either by explicitly invoking certain sub-routines such as `recycle _number()`, `recycle_array()`, and `recycle_allnumber()`, or implicitly by setting a flag (in the position of the last argument) in a routine being used to perform some other operation. The flag is a integer representing a bit pattern, one bit marking each of the preceding arguments which should be recycled. For example, integer values zero to four indicate (0) no recycling, (1) recycling the first argument, (2) recycling the second argument, (3) the first and second arguments, and (4) the third argument. If an argument which cannot be recycled is marked, an error is generated.

The implicit method can be used to nest operations, recycling intermediate results which are not required as separate values.

14.2.3.1 Explicit and Implicit

The code listed in Example 14.3a demonstrates the use of both explicit and implicit memory recycling. In lines 12–15, two numbers 'a' and 'b' are multiplied using central multiplication to produce the result 'x'. The third (last) argument in the invocation to the sub-routine `central_mul()` is the disposal flag which indicates which of the arguments 'a' and 'b' should be recycled prior to returning from the sub-routine `central_mul()`. A value of zero as here, with no bits set, indicates that neither of the arguments should be recycled. Both of the numbers 'a' and 'b' are recycled explicitly in line 14.

Example 14.3(a)

```
 1:   void demo25() /* explicit and implicit memory recycling */
 2:   {
 3:     int lambda[3];
```

(Continued)

Example 14.3(a) (Continued)

```
 4:     number a,b,x;
 5:
 6:     recovery_level(0,0,0);
 7:     create_cliffmem(50);
 8:     declare_primal(3,pack_signature(3,lambda,-1));
 9:     format_printing("%.0f");
10:
11:     /* x=(a*b) */
12:     a=scalar_number(1.0,2.0);  b=scalar_number(3.0,4.0);
                                   print_mem_usage(0);
13:     x=central_mul(a,b,0);
                                   print_mem_usage(0);
14:     recycle_number(a);         recycle_number(b);
                                   print_mem_usage(0);
15:     print_number(x,2);         recycle_number(x);
16:
17:     a=scalar_number(1.0,2.0);  b=scalar_number(3.0,4.0);
                                   print_mem_usage(0);
18:     x=central_mul(a,b,1);
                                   print_mem_usage(0);
19:                                recycle_number(b);
                                   print_mem_usage(0);
20:     print_number(x,2);         recycle_number(x);
21:
22:     a=scalar_number(1.0,2.0);  b=scalar_number(3.0,4.0);
                                   print_mem_usage(0);
23:     x=central_mul(a,b,2);
                                   print_mem_usage(0);
24:     recycle_number(a);
                                   print_mem_usage(0);
25:     print_number(x,2);         recycle_number(x);
26:
27:     a=scalar_number(1.0,2.0);  b=scalar_number(3.0,4.0);
                                   print_mem_usage(0);
28:     x=central_mul(a,b,3);
                                   print_mem_usage(0);
29:     print_number(x,2);         recycle_number(x);
30:
31:     free_cliffmem(1);
32:   }
```

By way of comparison, lines 17–20 perform the same multiplication, recycling the first argument 'a' implicitly by setting the disposal flag to value 1 on line 18, and recycling the second argument 'b' with and explicit invocation of recycle_number() on line 19.

Lines 22–25 perform the same multiplication, this time recycling the second argument 'b' implicitly by setting the disposal flag to value 2 on line 23, and recycling the first argument 'a' with an explicit invocation of recycle_number() on line 24.

Finally, lines 27–29 repeat the process, this time recycling the both arguments implicitly by setting the disposal flag to value 3 (setting bit 0 for the first argument and bit 1 for the second argument) on line 28.

Running the code listed in Example 14.3a produces the results shown in Example 14.3b. For each of the four multiplications, the value of the product is the same, on lines 5, 9,

13, and 16. Of more interest is the memory usage immediately after each multiplication, on lines 3, 7, 11, and 15. On line 3, there are three data nodes reported in use, one for each number 'a', 'b', and 'x'. On line 7, there are two data nodes reported in use, one for each number 'a' and 'x'. On line 11, there are also two data nodes reported in use, one for each number 'b' and 'x'. On line 15, there is only one data node reported in use, for the result 'x'.

Example 14.3(b)

```
 1:   demo25()
 2:   used/peak/full nodes (0/0/50) structural, (2/2/50) data.
 3:   used/peak/full nodes (0/0/50) structural, (3/3/50) data.
 4:   used/peak/full nodes (0/0/50) structural, (1/3/50) data.
 5:   (-5+i10)e{}
 6:   used/peak/full nodes (0/0/50) structural, (2/3/50) data.
 7:   used/peak/full nodes (0/0/50) structural, (2/3/50) data.
 8:   used/peak/full nodes (0/0/50) structural, (1/3/50) data.
 9:   (-5+i10)e{}
10:   used/peak/full nodes (0/0/50) structural, (2/3/50) data.
11:   used/peak/full nodes (0/0/50) structural, (2/3/50) data.
12:   used/peak/full nodes (0/0/50) structural, (1/3/50) data.
13:   (-5+i10)e{}
14:   used/peak/full nodes (0/0/50) structural, (2/3/50) data.
15:   used/peak/full nodes (0/0/50) structural, (1/3/50) data.
16:   (-5+i10)e{}
17:   statistics for structural heap:
18:   initial size (50)
19:   peak size    (0), as % of initial size (0%)
20:   full size    (50), expansion count (0)
21:   free count   (50), free start (0)
22:   statistics for data heap:
23:   initial size (50)
24:   peak size    (3), as % of initial size (6%)
25:   full size    (50), expansion count (0)
26:   free count   (50), free start (1)
```

14.2.3.2 Implicit and Nested

Use of implicit recycling makes it possible to nest operations, as shown in Example 14.4a. Line 19 adds two pairs of numbers 'a+b' and 'c+d', then multiplies the two results together. Implicit recycling is used in the invocations of sub-routine add_number() to dispose of the arguments 'a, b' and 'c, d'. Implicit recycling is also used in the invocation of sub-routine central_mul() to dispose of the intermediate results generated by each invocation of add_number().

Example 14.4(a)

```
 1:   void demo26() /* implicit and nested memory recycling */
 2:   {
 3:      int lambda[3];
 4:      number p,q,difference;
 5:      number a1,b1,c1,d1,x1;
```

(Continued)

Example 14.4(a) (Continued)

```
6:      number a2,b2,c2,d2,x2;
7:
8:      recovery_level(0,0,0);
9:      create_cliffmem(50);
10:     declare_primal(3,pack_signature(3,lambda,-1));
11:     format_printing("%.5e");
12:
13:     a1=scalar_number(1.0,2.0);      a2=cloneof_number(a1);
14:     b1=scalar_number(3.0,4.0);      b2=cloneof_number(b1);
15:     c1=scalar_number(5.0,6.0);      c2=cloneof_number(c1);
16:     d1=scalar_number(7.0,8.0);      d2=cloneof_number(d1);
17:
18:     /* x=(a+b)*(c+d) */
19:     x1=central_mul(add_number(a1,b1,3),add_number(c1,d1,3),3);
20:
21:     p=add_number(a2,b2,0);  recycle_number(a2);  recycle_number(b2);
22:     q=add_number(c2,d2,0);  recycle_number(c2);  recycle_number(d2);
23:     x2=central_mul(p,q,0);  recycle_number(p);   recycle_number(q);
24:
25:     difference=magnitudeof_number(sub_number(x1,x2,3),1);
26:     print_number(difference,4);
27:     recycle_number(difference);
28:
29:     free_cliffmem(1);
30:     }
```

For comparison, the same addition and multiplication are performed in lines 21–23 with explicit recycling rather than implicit recycling. The use of implicit recycling and nested operations allows for more compact code, here one statement of code instead of nine separate statements.

To check that the two calculations give the same result, the difference is calculated on line 25. Again, nested operations and implicit recycling are used to condense the code into the fewest statements. The sub-routine print_number() does not support implicit recycling so that the final result 'difference' is recycled explicitly on line 27. As an alternative, the statement print_number() can be nested inside recycle_number() as recycle_number(print_number(...)). This is allowed because print_number() and most other printing routines pass back as return value the token of the number they have just printed.

The result of running the code listed in Example 14.4a is shown in Example 14.4b. The difference in the results calculated in the two alternative ways is reported (as zero) on line 2. The memory statistics on lines 3–12 confirm that all memory has been properly recycled.

Example 14.4(b)

```
1:   demo26()
2:   (0.00000e+00+i0.00000e+00)e{}
3:   statistics for structural heap:
4:   initial size (50)
```

```
 5:   peak size      (0), as % of intial size (0%)
 6:   full size      (50), expansion count (0)
 7:   free count     (50), free start (0)
 8:   statistics for data heap:
 9:   initial size (50)
10:   peak size      (8), as % of intial size (16%)
11:   full size      (50), expansion count (0)
12:   free count     (50), free start (7)
```

Reference

King, K. N. (2008), *C Programming: A Modern Approach*, 2nd edn, W. W. Norton, New York, US-NY.

15

Errors

There are two types of errors. The first type is system errors, which indicate something has gone badly wrong in the use of the underlying data structures. When the various routines are passed valid arguments, even if their values are inappropriate in a numerical sense, system errors should never be seen.

The second type of errors is user errors. These occur when the values passed in arguments are inappropriate, often leading some numerical conflict or indeterminacy, such as division by zero.

All errors print messages with the location of the error are identified by the name of the routine in which it was encountered. Recovery from some errors is possible, but not possible from others.

15.1   User Errors

Table 15.1 list all user errors detected. Properly constructed code may result in user errors if values generated at run time are passed to routines where they are outside the range of values for which proper results can be generated. Division by zero is a clear example.

For some routines, the parameters passed can all legally address the same instance of the same value, whereas others are designed with the assumption that each parameter passed identifies a separate tree. It is possible that even if separate variables are used for different parameters, they can be assigned the same values, with identical tokens at run time. Errors of this type can be rather subtle.

When user errors are encountered, the action taken is to print an error message and invoke the sub-routine `brexit()` with an argument as listed in Table 15.1. Invoking `brexit()` causes termination of the code if the argument is non-zero, or activation of the recovery strategy set in place by the user if the argument is zero.

Recovery strategies are established early at run time by invoking the sub-routine `recovery_level()` with arguments of appropriate value. Specific strategies allow for memory retention or recycling, attempting to restart a specified number of times with or without prompting the user, or termination. All possible recovery strategies are described on page 438 in Chapter 27. In all cases, an error message is printed.

Numerical Calculations in Clifford Algebra: A Practical Guide for Engineers and Scientists,
First Edition. Andrew Seagar.
© 2023 John Wiley & Sons Ltd. Published 2023 by John Wiley & Sons Ltd.

Table 15.1 User errors, A–Z.

error message	action
attempt to free memory not allocated in routine: xxx()	brexit(2)
attempt to restart with bad state pointer (%0lX) not possible in routine: xxx()	brexit(8)
attempt to scale by number with (%d) > (1) components in routine: xxx()	brexit(0)
attempt to take inverse of unit with zero signature in routine: xxx()	brexit(6)
attempt to take inverse of zero in routine: xxx()	brexit(0)
attempt to take log of zero in routine: xxx()	brexit(0)
attempt to take square root of non-scalar component with unit (%d) in routine: xxx()	brexit(0)
cannot use number with zero as square in routine: xxx()	brexit(0)
dimension (%d) < (0) in routine: xxx()	brexit(0)
dimension (%d) not > (0) in routine: xxx()	brexit(0)
dimension of vector (%d) > Clifford dimension (%d) in routine: xxx()	brexit(0)
function (%s) not found in routine: xxx()	brexit(0)
highest power (%d) > maximum (%d) in routine: xxx()	brexit(0)
illegal attempt to add different units (%d,%d) in routine: xxx()	brexit(6)
illegal attempt to add function of number to itself in routine: xxx()	brexit(6)
illegal attempt to add matrix of size (%d,%d) to different size (%d,%d) in routine: xxx()	brexit(0)
illegal attempt to add number to itself in routine: xxx()	brexit(6)
illegal attempt to clone from matrix size (%d,%d) to different size (%d,%d) in routine: xxx()	brexit(0)
illegal attempt to clone into sub-matrix at level (%d) > (0) in routine: xxx()	brexit(0)
illegal attempt to copy matrix of size (%d,%d) to different size (%d,%d) in routine: xxx()	brexit(0)
illegal attempt to multiply matrix of size (%d,%d) by matrix of size (%d,%d) in routine: xxx()	brexit(0)
illegal attempt to multiply row of size (%d) by column of size (%d) in routine: xxx()	brexit(0)
illegal attempt to overwrite original matrix with sub-matrix in routine: xxx()	brexit(0)
illegal attempt to recycle argument (%d...) in routine: xxx()	brexit(6)
illegal attempt to recycle sub-matrix at level (%d) > (0) in routine: xxx()	brexit(0)
illegal attempt to save matrix of size (%d,%d) to different size (%d,%d) in routine: xxx()	brexit(0)
illegal attempt to set sub-matrix as identity at level (%d) > (0) in routine: xxx()	brexit(0)
illegal attempt to set sub-matrix as random at level (%d) > (0) in routine: xxx()	brexit(0)
illegal attempt to subtract different units (%d,%d) in routine: xxx()	brexit(6)
illegal attempt to subtract function of number from itself in routine: xxx()	brexit(6)

Table 15.1 (Continued)

error message	action
illegal attempt to subtract matrix of size (%d,%d) from different size (%d,%d) in routine: xxx()	`brexit(0)`
illegal attempt to subtract number from itself in routine: xxx()	`brexit(6)`
illegal change to unit (%d→%d) in routine: xxx()	`brexit(6)`
illegal signature array [...] for Pauli matrix in routine: xxx()	`brexit(6)`
illegal signature array [...] for bicomplex number in routine: xxx()	`brexit(6)`
illegal signature array [...] for electromagnetic field in routine: xxx()	`brexit(6)`
illegal signature array [...] for quaternion in routine: xxx()	`brexit(6)`
illegal signature array [...] for vector in routine: xxx()	`brexit(6)`
length (%d) > maximum (%d) for format '%s' in routine: xxx()	`brexit(0)`
length (%d) > maximum (%d) for format in routine: xxx()	`brexit(0)`
length of array (%d) not > (0) in routine: xxx()	`brexit(0)`
length of array (%d) not a power of two in routine: xxx()	`brexit(0)`
matrix (%d,%d) not square in routine: xxx()	`brexit(0)`
matrix with (%d) rows too small to make sub-matrix in routine: xxx()	`brexit(0)`
matrix with rows and columns (%d,%d) too small to make sub-matrix in routine: xxx()	`brexit(0)`
number with no scalar component found in routine: xxx()	`brexit(0)`
primal unit (%d) not < Clifford dimension (%d) in routine: xxx()	`brexit(0)`
primal unit (%d) not zero or greater in routine: xxx()	`brexit(0)`
radius (%e) not > (0) in routine: xxx()	`brexit(0)`
rows (%d) on right hand side not the same as rows (%d) in matrix in routine: xxx()	`brexit(0)`
series for function (%s) not found in routine: xxx()	`brexit(6)`
signature (%d) not (+1, −1) in routine: xxx()	`brexit(0)`
signature (%d) not (+1, 0, −1) in routine: xxx()	`brexit(0)`
signature not currently set in routine: xxx()	`brexit(6)`
size of matrix (%d,%d) not positive in routine: xxx()	`brexit(0)`
unexpectedly found > (1) component in routine: xxx()	`brexit(0)`
unexpectedly found non-scalar in routine: xxx()	`brexit(0)`
unknown function found in routine: xxx()	`brexit(6)`
unrecognised matrix format (%d) in routine: xxx()	`brexit(0,6)`
zero denominator in routine: xxx()	`brexit(0)`

Recovery is not possible for all user errors. Recovery is usually possible when the error can be corrected by passing fresh data to the same sub-routine as that in which the error has been detected. In these cases, the sub-routine detecting the error will invoke `brexit()` with an argument of zero. Recovery is not possible when it is necessary to supply fresh data to a sub-routine which is different from the one in which the error has been detected. In these cases, the sub-routine detecting the error will invoke `brexit()` with an argument which is non-zero. Use of the sub-routine `brexit()` is described in detail on page 324 in Chapter 27.

A recovery strategy of printing a message and restarting the code at a particular point is most appropriate when the code is executed interactively, and the user has the ability to react to the error by entering fresh data, thereby avoiding the problem.

15.1.1 Syntax Errors and Messages

Routines which interpret strings of characters as representing Clifford units check the syntax of those strings. If syntax errors are found when attempting to convert such strings to internal (integer bit pattern) format, using the sub-routine `spack_unit()`, syntax error messages are generated as listed in Table 15.2, and the user's recovery strategy is activated.

In order to avoid generation of syntax errors, the sub-routine `sscan_unit()` is provided to check the syntax prior to attempting conversion. When an error is found during the syntax check, a string is returned with a numeric code and a descriptive message, as listed in Table 15.3. No error message is printed, and no recovery mechanism is activated.

Table 15.2 Syntax errors.

error message	action
character '%c' at offset (%d) not numeral or comma in string '%s' in routine: xxx()	`brexit(0)`
character '}' not found at end of string '%s' in routine: xxx()	`brexit(0)`
characters 'e{' not found at start of string '%s' in routine: xxx()	`brexit(0)`
comma then number not found at offset (%d) in string '%s' in routine: xxx()	`brexit(0)`
number not found at offset (2) in string '%s' in routine: xxx()	`brexit(0)`
primal unit (%d) not < Clifford dimension (%d) in string '%s' in routine: xxx()	`brexit(0)`
primal unit (%d) not > previous one in string '%s' in routine: xxx()	`brexit(0)`

Table 15.3 Syntax messages.

(err.1) characters 'e{' not found at start of string
(err.2) character '}' not found at end of string
(err.3) character '%c' at offset (%d) not numeral or comma
(err.4) number not found at offset (2)
(err.5) primal unit (%d) not < Clifford dimension (%d)
(err.6) comma then number not found at offset (%d)
(err.7) primal unit (%d) not > previous one

It is the responsibility of the user's code to interpret the syntax messages and provide an appropriate action.

15.2 System Errors

Table 15.4 lists all possible system errors. These are invariably fatal, with no possibility of sensible continuation. For most system errors, the action taken is to invoke the sub-routine `brexit()` with an argument of six, to recycle and free all memory and terminate execution.

System errors can arise from improperly constructed code, such as attempting some operation before the required initialisation has been effected. System errors can also occur when a token which has already been recycled but not reissued is passed as a number for the purpose of some operation. The operation addresses memory assuming that it is complete and properly structured. That is unlikely to be the case, so that either the logic underlying tree manipulation fails, or invalid data are read from the wrong places, or data are written

Table 15.4 System errors.

error message	action
attempt to access corrupt heap in routine: xxx()	brexit(6)
attempt to recycle token (%d) from empty heap in routine: xxx()	brexit(6)
attempt to recycle token (%d) out of array bounds [0,%d] in routine: xxx()	brexit(6)
bounds of two branches (%d,%d) illegally overlap in routine: xxx()	brexit(6)
branch location ambiguous (%d,%d) in routine xxx()	brexit(6)
branch type (%d) outside range (1–4) in routine: xxx()	brexit(6)
corrupt heap detected in routine: xxx()	brexit(6)
depth (%d) > dimension+1 (%d+1) in routine: xxx()	brexit(6)
failed to allocate heap in routine: xxx()	brexit(6)
failed to allocate memory array[%d] in routine: xxx()	brexit(6)
failed to expand heap to (%d)% of current size in routine: xxx()	brexit(6)
fork token (%d) invalid value < (1) in routine: xxx()	brexit(6)
illegal signature (%d) in routine: xxx()	brexit(6)
illegal value (%d) for sign of branch (−1, +1) in routine: xxx()	brexit(6)
leaf token (%d) invalid value > (−1) in routine: xxx()	brexit(6)
next free link (%d) from last free node not value expected (%d) in routine: xxx()	brexit(6)
null token (0) invalid value in routine: xxx()	brexit(6)
unexpectedly failed to find absolute unit in routine: xxx()	brexit(6)
unexpectedly found fork (%d) in routine: xxx()	brexit(6)
unexpectedly found leaf (%d) in routine: xxx()	brexit(6)
unit found (%d) not as expected (%d) in routine: xxx()	brexit(6)
unrecognised switch value (%c) in routine: xxx()	brexit(6)
unrecognised switch value (%d) in routine: xxx()	brexit(6)

to the wrong places destroying valid data in other trees. Such errors require rewriting the user's source code and recompilation.

15.3 Recovery

The code listed in Example 15.1a demonstrates a user error without recovery. A numerical context is established on lines 7–12 where recovery is deactivated with all three arguments to the sub-routine `recovery_level()` set to zero. This is the simplest possible usage being recommended in all cases for which nothing special is intended. More advanced usage is given in Examples 15.2 and 15.3. The full capabilities of the sub-routine `recovery_level()` are described on page 438 in Chapter 27.

Three data values are initialised on lines 15–17. The third number is (mistakenly) set to an incorrect value of zero. When the inverse of that number is calculated on line 21, an error occurs within the sub-routine `inverseof_component()`.

Example 15.1(a)

```
 1:   void demo27() /* error without recovery */
 2:   {
 3:       int lambda[3];
 4:       number one,two,three;
 5:       int k,n;
 6:
 7:       n=0;     /* number of errors tolerated before aborting */
 8:       k=0;     /* flag indicating to not attempt recovery    */
 9:       recovery_level(n,k,0);
10:       create_cliffmem(50);
11:       declare_primal(3,pack_signature(3,lambda,-1));
12:       format_printing("%.4f");
13:
14:       printf("will divide by zero\n");
15:       one=scalar_number(1.0,0.0);
16:       two=scalar_number(2.0,0.0);
17:       three=scalar_number(0.0,0.0); /* error in data value */
18:
19:       recycle_number(print_number(inverseof_component(one,1),4));
20:       recycle_number(print_number(inverseof_component(two,1),4));
21:       recycle_number(print_number(inverseof_component(three,1),4));
22:
23:       printf("will not divide by zero\n"); /* will never reach here */
24:       one=scalar_number(1.0,0.0);
25:       two=scalar_number(2.0,0.0);
26:       three=scalar_number(3.0,0.0); /* no error in data value */
27:
28:       recycle_number(print_number(inverseof_component(one,1),4));
29:       recycle_number(print_number(inverseof_component(two,1),4));
30:       recycle_number(print_number(inverseof_component(three,1),4));
31:
32:       free_cliffmem(1);
33:   }
```

The result of running the code listed in Example 15.1a is given in Example 15.1b. The inverse of the first two data values is reported correctly on lines 3 and 4. When the division by the next value (zero) is attempted, the message on line 5 is printed and the code terminates. The message on line 5 reports the error in a sub-routine inverse_simple(), which is a lower-level routine invoked by inverseof_component() in order to effect the operation. There is no traceback to the level of the user's code.

Example 15.1(b)

```
1:   demo27()
2:   will divide by zero
3:   (1.0000+i0.0000)e{}
4:   (0.5000+i0.0000)e{}
5:   attempt to take inverse of zero in routine: inverse_simple()
```

For comparison, the code listed in Example 15.2a slightly extends Example 15.1a, establishing a context in lines 8–13 for five attempts at recovery on error before terminating the program. The flag value 'k=4' in the second argument to the sub-routine recovery_level() selects a recovery strategy whereby no prompt is given and all memory is recycled. The third argument declares to the Clifford numerical suite the location of memory reserved by the user for storage of program state information used to restart the program when an error is detected.

Line 18 establishes the location at which the execution will be re-established when an error is encountered, with the invocation of the routine sigsetjmp(). This routine is provided within the 'c' programming language (King 2008) set of libraries. When invoked, the program counter and other state information are stored within the variable indicated by the first argument, and the routine returns a value of zero. When the companion routine siglongjmp() is invoked from within brexit() after detecting an error, the state information saved earlier is used to re-establish program control at the original invocation of sigsetjmp() on line 18 with a non-zero return value. The memory area provided when invoking the routine sigsetjmp() should be the same as the one provided on line 10 as the third argument to sub-routine recovery_level().

Example 15.2(a)

```
1:   void demo28() /* error with recovery */
2:   {
3:      static sigjmp_buf prog_stack; /* restart location saved here */
4:      int lambda[3];
5:      number one,two,three;
6:      int status,k,n;
7:
8:      n=5;     /* number of errors tolerated before aborting */
9:      k=4;     /* do not prompt on restart, recycle all memory */
10:     recovery_level(n,k,&prog_stack);
```

(Continued)

Example 15.2(a) (Continued)

```
11:     create_cliffmem(50);
12:     declare_primal(3,pack_signature(3,lambda,-1));
13:     format_printing("%.4f");
14:
15:     printf("you are permitted %d errors\n",n);
16:
17:     /* program restarts here after error */
18:     status=sigsetjmp(prog_stack,1);
19:     if(status==0) /* status=0 if returning from line above */
20:       {
21:         printf("starting %d\n",status); /* first time arrive here */
22:         printf("will divide by zero\n");
23:         one=scalar_number(1.0,0.0);
24:         two=scalar_number(2.0,0.0);
25:         three=scalar_number(0.0,0.0); /* error in data value */
26:       }
27:     else /* status not zero if jumping from other point in code */
28:       {
29:         printf("restarting %d\n",status); /* after error arrive
                                                  here */
30:         printf("will not divide by zero\n");
31:         one=scalar_number(1.0,0.0);
32:         two=scalar_number(2.0,0.0);
33:         three=scalar_number(3.0,0.0); /* no error in data value */
34:       }
35:     recycle_number(print_number(inverseof_component(one,1),4));
36:     recycle_number(print_number(inverseof_component(two,1),4));
37:     recycle_number(print_number(inverseof_component(three,1),4));
38:
39:     /* reach here only after restarting */
40:
41:     free_cliffmem(1);
42:   }
```

The first time control is returned through `sigsetjmp()` at line 18, the status is zero, and lines 21–25 are executed. A bad value of data is established in line 25, so that division by zero occurs at line 37. When that error is detected inside the Clifford numerical suite, the sub-routine `brexit()` is invoked, and control is returned through `siglongjmp()`, appearing at line 18 as a return from `sigsetjmp()` with a non-zero return value. The code on lines 29–33 is then executed, setting up the data values without error. On the second pass, lines 35–37 execute without error, and the program terminates normally.

The code listed in Example 15.2a produces the results given in Example 15.2b. The first two divisions are printed correctly on lines 5 and 6, but the third one generates the error message on line 7. After restarting as indicated on line 8, the correct data values are properly divided and produce results on lines 10–12 without error. The program continues to normal termination. The statistics of memory usage on lines 13–22 show that all memory has been properly recycled.

Example 15.2(b)

```
 1:   demo28()
 2:   you are permitted 5 errors
 3:   starting 0
 4:   will divide by zero
 5:   (1.0000+i0.0000)e{}
 6:   (0.5000+i0.0000)e{}
 7:   attempt to take inverse of zero in routine: inverse_simple()
 8:   restarting 1
 9:   will not divide by zero
10:   (1.0000+i0.0000)e{}
11:   (0.5000+i0.0000)e{}
12:   (0.3333+i0.0000)e{}
13:   statistics for structural heap:
14:   initial size (50)
15:   peak size      (0), as % of initial size (0%)
16:   full size      (50), expansion count (0)
17:   free count     (50), free start (0)
18:   statistics for data heap:
19:   initial size (50)
20:   peak size      (6), as % of initial size (12%)
21:   full size      (50), expansion count (0)
22:   free count     (50), free start (5)
```

15.4 Beneficial Usage

The recovery mechanism can be used to good advantage for correcting errors made when entering data on the keyboard. Example 15.3 demonstrates one such approach, involving two main parts. The first, in Example 15.3a, consists of a sub-routine read_cliffnum() which reads one line from the keyboard and checks the syntax of a single Clifford component. The syntax checking for the numerical coefficient is coded into read_cliffnum() itself. If errors are found, the recovery mechanism is initiated by invoking the sub-routine brexit() from within read_cliffnum(). The syntax checking for the unit of the component is left to the conversion sub-routine spack_unit(). If syntax errors are found in the unit, the recovery mechanism is activated by invoking brexit() from within spack_unit().

The code for the sub-routine read_cliffnum() listed in Example 15.3a prompts the user on line 9 to type a Clifford component on the keyboard in the format described in Section 2.1. Characters typed are read into a buffer using the sub-routine keyboard_input()[1] on line 10. If the string of characters is too long, the recovery mechanism is activated by invoking the sub-routine brexit() on line 15. An attempt to convert the leading characters in the buffer into a pair of floating point numbers is made

---

1 The code is in Example 2.3.

on line 18. If the attempt is unsuccessful, the recovery mechanism is activated[2] on line 23. Use of the sub-routine `brexit ()` is described in detail on page 324 in Chapter 27.

An attempt is made to convert the remainder of the string as a unit by invoking the sub-routine `spack_unit ()` on line 26. If the string cannot be converted due to syntax errors, the recovery mechanism is activated by invoking the sub-routine `brexit ()` from within `spack_unit ()`. If the string is converted successfully a number containing the single component is constructed on line 27 and returned to the code from which `read_cliffnum ()` was invoked.

Example 15.3(a)

```
1:   number read_cliffnum()  /* keyboard entry of Clifford numbers */
2:   {                       /* with implicit syntax check */
3:       char buffer[64];
4:       cliff_comp u_comp;
5:       number u;
6:       double x,y;
7:       int i,n,k;
8:
9:       printf("enter a single Clifford component: ");
10:      k=keyboard_input(64,buffer); /* max 63+NULL, discard extra */
11:      if(k>64)
12:          {
13:          printf("input buffer (%d>64) overflow '%.63s'",k,buffer);
14:          printf(" in routine: read_cliffnum()\n");
15:          brexit(0); /* user restart */
16:          }
17:      printf("%.63s\n",buffer);
18:      i=sscanf(buffer," (%lf,%lf)%n",&x,&y,&n);
19:      if(i!=2)
20:          {
21:          printf("complex number '(x,y)' not found at start of string
                       '%s'",buffer);
22:          printf(" in routine: read_cliffnum()\n");
23:          brexit(0); /* user restart */
24:          }
25:      /* may invoke brexit(0) inside spack_unit() if syntax of unit is
                faulty */
26:      pack_comp(&u_comp,x,y,spack_unit(buffer+n));
27:      u=pack_number(1,&u_comp);
28:      return(u);
29:  }
```

The code listed in Example 15.3b provides a framework using the sub-routine `read _cliffnum()` from Example 15.3a to construct a Clifford number by adding together individual components typed at the keyboard. A numerical context is established in lines 8–10 and 13–15 for 10 attempts at recovery on error before terminating the program.

---

2 As an alternative compare with the Example 2.4 where the syntax is checked explicitly and the recovery mechanism is not used.

The flag value 'k=2' in the second argument to the sub-routine `recovery_level()` selects a recovery strategy whereby no prompt is given and no memory is recycled. The third argument declares to the Clifford numerical suite the location of memory reserved by the user for storage of program state information used to restart the program when a syntax or other error is detected.

Line 22 establishes the location at which the execution will be re-established when an error is encountered, with the invocation to the sub-routine `sigsetjmp()`, storing the information in the memory indicated by the first argument and returning a value of zero. A number is created on line 26 in which to accumulate the total of all components entered on the keyboard. The main code from lines 35 to 42 repeatedly reads components typed on the keyboard until a value of zero is detected on line 35. Non-zero components are added to the total on line 37 inside the loop 'while(){}'. After the loop terminates, the total is printed on line 46.

Example 15.3(b)

```
1:   void demo29() /* keyboard entry of Clifford numbers with error
                      recovery */
2:   {
3:     static sigjmp_buf prog_stack; /* restart location saved here */
4:     int lambda[3];
5:     number u,total;
6:     int status,k,n;
7:
8:     n=10;    /* number of errors tolerated before aborting */
9:     k=2;     /* do not prompt on restart, do not free memory */
10:    recovery_level(n,k,&prog_stack);
11:    printf("you are permitted %d errors\n",n);
12:
13:    create_cliffmem(50);
14:    declare_primal(3,pack_signature(3,lambda,-1));
15:    format_printing("%.2f");
16:
17:    /* give some instructions */
18:    printf("you can use 3-dimensional Clifford numbers\n");
19:    printf("to finish enter: (0,0)e{}\n");
20:
21:    /* program restarts here after error */
22:    status=sigsetjmp(prog_stack,1);
23:    if(status==0) /* first time arrive here */
24:      {
25:        printf("starting %d\n",status);
26:        total=fresh_number();
27:      }
28:    else /* after error arrive here */
29:      {
30:        printf("restarting %d\n",status);
31:      }
32:
33:    /* run main code */
```

(Continued)

Example 15.3(b) (Continued)

```
34:     u=read_cliffnum();
35:     while(is_nonzero(u))
36:       {
37:         total=addto_number(total,u);
38:         printf("number added was: ");
39:         print_number(u,8);
40:         recycle_number(u);
41:         u=read_cliffnum();
42:       }
43:     recycle_number(u);
44:     printf("finished\n");
45:     printf("your total is:\n");
46:     print_number(total,8);
47:     recycle_number(total);
48:
49:     free_cliffmem(0);
50:   }
```

Any syntax error encountered when interpreting characters typed on the keyboard re-establishes the program execution to line 22 with a non-zero return value. The total accumulated at that stage is not affected. After printing a message on line 30, the main code re-starts on line 34, permitting the user to correct the erroneous syntax and continue adding new components to the running total.

The results of running the code in Example 15.3b are shown in Examples 15.3c–15.3g for various scenarios with different typed input. Scenarios (c)–(e) do not have any syntax errors and do not activate the recovery mechanism. Scenario (c) gives a total of zero on line 9 and terminates when a value of zero is the first value typed on line 6.

Example 15.3(c)

```
1:   demo29()
2:   you are permitted 10 errors
3:   you can use 3-dimensional Clifford numbers
4:   to finish enter: (0,0)e{}
5:   starting 0
6:   enter a single Clifford component: (0,0)e{}
7:   finished
8:   your total is:
9:   (0.00+i0.00)e{}
```

Scenario (d) demonstrates initialising a vector in three dimensions on lines 6–11. In this case, all units are of grade 1. The vector is printed on line 15. No errors are encountered.

Example 15.3(d)

```
 1:   demo30()
 2:   you are permitted 10 errors
 3:   you can use 3-dimensional Clifford numbers
 4:   to finish enter: (0,0)e{}
 5:   starting 0
 6:   enter a single Clifford component: (1,2)e{0}
 7:   number added was: (1.00+i2.00)e{0}
 8:   enter a single Clifford component: (3,4)e{1}
 9:   number added was: (3.00+i4.00)e{1}
10:   enter a single Clifford component: (5,6)e{2}
11:   number added was: (5.00+i6.00)e{2}
12:   enter a single Clifford component: (0,0)e{}
13:   finished
14:   your total is:
15:   (1.00+i2.00)e{0}+(3.00+i4.00)e{1}+(5.00+i6.00)e{2}
```

Scenario (e) demonstrates initialising a number containing vector and bivector components. In this case although three components are entered, two have the same unit. These are added together so that the result on line 15 has only two components.

Example 15.3(e)

```
 1:   demo31()
 2:   you are permitted 10 errors
 3:   you can use 3-dimensional Clifford numbers
 4:   to finish enter: (0,0)e{}
 5:   starting 0
 6:   enter a single Clifford component: (1,2)e{0}
 7:   number added was: (1.00+i2.00)e{0}
 8:   enter a single Clifford component: (3,4)e{1,2}
 9:   number added was: (3.00+i4.00)e{1,2}
10:   enter a single Clifford component: (5,6)e{1,2}
11:   number added was: (5.00+i6.00)e{1,2}
12:   enter a single Clifford component: (0,0)e{}
13:   finished
14:   your total is:
15:   (1.00+i2.00)e{0}+(8.00+i10.00)e{1,2}
```

Scenarios (f) and (g) demonstrate activation of the recovery mechanism when errors are encountered. For scenario (f) components without syntax errors are entered on lines 6 and 8, followed on line 10 by a component with invalid syntax. An error message is printed on line 11 and the code is restarted on line 12. The previous component is typed with the correct syntax on line 13, and the total is printed on line 18. The recovery strategy used in this case did not recycle memory so that the running total kept its value through the recovery, and the value printed on line 18 contains the values of all components entered before and after the error.

Example 15.3(f)

```
 1:   demo32()
 2:   you are permitted 10 errors
 3:   you can use 3-dimensional Clifford numbers
 4:   to finish enter: (0,0)e{}
 5:   starting 0
 6:   enter a single Clifford component: (1,2)e{0}
 7:   number added was: (1.00+i2.00)e{0}
 8:   enter a single Clifford component: (3,4)e{1,2}
 9:   number added was: (3.00+i4.00)e{1,2}
10:   enter a single Clifford component: (5,6)e{2,0}
11:   primal unit (0) not > than previous one in string 'e{2,0}' in
                       routine: spack_unit()
12:   restarting 1
13:   enter a single Clifford component: (5,6)e{0,2}
14:   number added was: (5.00+i6.00)e{0,2}
15:   enter a single Clifford component: (0,0)e{}
16:   finished
17:   your total is:
18:   (1.00+i2.00)e{0}+(5.00+i6.00)e{0,2}+(3.00+i4.00)e{1,2}
```

Scenario (g) is full of syntax errors. Error messages are printed for every error, and the code is re-started up to the maximum of 10 times specified by the recovery strategy. On the eleventh error, the recovery mechanism prints a message to indicate that the maximum number of re-starts has been reached and terminates the code.

Example 15.3(g)

```
 1:   demo33()
 2:   you are permitted 10 errors
 3:   you can use 3-dimensional Clifford numbers
 4:   to finish enter: (0,0)e{}
 5:   starting 0
 6:   enter a single Clifford component: (1,2)e
 7:   characters 'e{' not found at start of string 'e' in routine:
                       spack_unit()
 8:   restarting 1
 9:   enter a single Clifford component: {0}
10:   complex number '(x,y)' not found at start of string '{0}' in
                       routine: read_cliffnum()
11:   restarting 2
12:   enter a single Clifford component: (3,4)e{3}
```

```
13:   primal unit (3) not < Clifford dimension (3) in string 'e{3}' in
                    routine: spack_unit()
14:   restarting 3
15:   enter a single Clifford component: (5,6)e{,2}
16:   number not found at offset (2) in string 'e{,2}' in routine:
                    spack_unit()
17:   restarting 4
18:   enter a single Clifford component: (7,8)e{1,1}
19:   primal unit (1) not > than previous one in string 'e{1,1}' in
                    routine: spack_unit()
20:   restarting 5
21:   enter a single Clifford component: (2,3)e{100}
22:   primal unit (100) not < Clifford dimension (3) in string 'e{100}'
                    in routine: spack_unit()
23:   restarting 6
24:   enter a single Clifford component: (3,x)e{2}
25:   complex number '(x,y)' not found at start of string '(3,x)e{2}'
                    in routine: read_cliffnum()
26:   restarting 7
27:   enter a single Clifford component: (5,)e{2}
28:   complex number '(x,y)' not found at start of string '(5,)e{2}'
                    in routine: read_cliffnum()
29:   restarting 8
30:   enter a single Clifford component: (,8)e{}
31:   complex number '(x,y)' not found at start of string '(,8)e{}'
                    in routine: read_cliffnum()
32:   restarting 9
33:   enter a single Clifford component: (0,0)e(1,2)
34:   characters 'e{' not found at start of string 'e(1,2)'
                    in routine: spack_unit()
35:   restarting 10
36:   enter a single Clifford component: {0,0}e{1}
37:   complex number '(x,y)' not found at start of string '{0,0}e{1}'
                    in routine: read_cliffnum()
38:   exiting after (10) restarts
```

Reference

King, K. N. (2008), *C Programming: A Modern Approach*, 2nd edn, W. W. Norton, New York, US-NY.

16

Extension

Operations on Clifford numbers reduce at the lowest level to unary operations on single components or binary operations on pairs of components. The implementation of the code in the Clifford numerical suite provides a separation between the component-level operations and the mechanisms which manage the whole Clifford numbers, (i.e. collections of those components) in the form of trees.

Users may define their own sub-routines which supply component-level operations, and then use them to manipulate whole numbers without the need to perform explicit pruning and grafting surgery within the tree structures. Interfaces are provided for four kinds of manipulation: accumulation, multiplication, transformation, and filtering. The interfaces allow the user to construct additional functionality beyond that already provided, extending the Clifford numerical suite to make it meet more closely their own particular needs.

16.1   Accumulation

The process of combining components into an existing tree is performed by a particular worker routine `merge_comp()`, standing for 'merge component', which takes four arguments as shown in Table 16.1. The particular sub-routines providing the operations used to effect the merge are specified by the user within the argument list of `merge_comp()`. Here those sub-routines are represented by the names `combine()` and `insert()`, but in practice they are defined and named by the user.

The sub-routine `merge_comp()` surgically combines the component 'y_comp' with the number 'x' producing a single entity. As a first step, the location where the component 'y_comp' needs to be attached to the tree holding the number 'x', is identified. If a component already resides at that location, by virtue of having the same unit, the two components are passed to the binary operation `combine()` with the x-component as the first argument and the y-component as the second argument. The resulting component (unless zero) is placed in the tree. If no component with the same unit already exists in the tree, the component 'y_comp' is passed as the first argument to the unary operation `insert()`, and the result (unless zero) is grafted onto the tree. If a result is zero, it is either not inserted, or the original component is pruned from the tree. Pruning is a recursive process which propagates automatically as far towards the root of the tree as necessary.

Numerical Calculations in Clifford Algebra: A Practical Guide for Engineers and Scientists,
First Edition. Andrew Seagar.
© 2023 John Wiley & Sons Ltd. Published 2023 by John Wiley & Sons Ltd.

Table 16.1 Function prototype of worker routine to merge components into trees.

Merge operation	Page

```number merge_comp(number x,```
          ```cliff_comp *y_comp,```
          ```cliff_comp *combine(cliff_comp```
                ```*a, cliff_comp *b),```
          ```cliff_comp *insert(cliff_comp```
                ```*a, cliff_comp *c));```
	377

The sub-routine `merge_comp()` does not make a clone of the original number 'x'. It operates surgically directly on the tree provided so that the original number 'x' is no longer valid after the process is complete. A new token which identifies the tree containing the result of the process is issued to the user through the return value of the routine. The value of the new token may be the same as the old one, if it was not necessary to modify the root of the tree. However, that cannot be assumed.

No attempt should be made to recycle the old number 'x'. That memory has been modified and re-used *in situ*. Any parts of the memory which are no longer needed are recycled from within `merge_comp()`. Attempting to recycle the token for the original number 'x' will probably lead to a system error from which continuation or recovery is not possible.

The implementations for the addition and subtraction of pairs of Clifford numbers within the Clifford numerical suite use `merge_comp()`, repeatedly invoking the sub-routines providing pre-defined unary and binary sub-routines as listed in Table 16.2 in place of the routines `insert()` and `combine()`. The binary operations `addition_to()` and `subtraction_from()` modify their first argument so that they can be applied directly to a component already in the tree.

The user may invoke sub-routine `merge_comp()` directly using `identity()` and `addition_to()` to serve as the routines `insert()` and `combine()`, in order to construct Clifford numbers one component at a time, as in method 1 of Example 2.2a.

Table 16.2 Predefined unary and binary operations for merging components into trees.

`insert()`		Page
`cliff_comp *identity(cliff_comp *a, cliff_comp *c);`	$c \leftarrow a$	353
`cliff_comp *negation(cliff_comp *a, cliff_comp *c);`	$c \leftarrow -a$	379

`combine()`		Page
`cliff_comp *addition_to(cliff_comp *a, cliff_comp *b);`	$a \leftarrow a + b$	320
`cliff_comp *subtraction_from(cliff_comp` `*a, cliff_comp *b);`	$a \leftarrow a - b$	464

For other effects, the user may construct their own operations and pass them as arguments to the routine `merge_comp()` in Table 16.1, to be used in place of the arguments `insert()` and `combine()`. As a guide, the code for the implementation of the binary operation `subtraction_from()` is listed in Example 16.1.

Example 16.1

```
 1:  /* subtraction from component with same unit */
 2:  /* a=a-b */
 3:  /* a and b can point to same component */
 4:  cliff_comp *subtraction_from(a,b)
 5:      cliff_comp *a,*b; /* result into component a */
 6:  {
 7:    bitmap au,bu;
 8:
 9:    au=a->unit;
10:    bu=b->unit;
11:    if(au!=bu)
12:      {
13:        printf("illegal attempt to subtract different units");
14:        printf(" (%d,%d) in routine: subtraction_from()\n",au,bu);
15:        brexit(6); /* no restart */
16:      }
17:    a->coef[0]=a->coef[0]-b->coef[0];
18:    a->coef[1]=a->coef[1]-b->coef[1];
19:    return(a);
20:  }
```

16.2   Multiplication

Multiplication of all types is effected using the three worker routines listed in Table 16.3. The first two manage the distribution of multiplication over addition for a number and a component, from either left or right, and the third does the same for two numbers.

Unlike `merge_comp()`, the sub-routines `mul_comp()` and `comp_mul()` do not perform surgery directly on the tree provided so that the original number 'x' or 'y' remains valid after the process is complete. The sub-routine `mul_number()` provides explicit control, through its fourth (last) argument 'd', over which of the two numbers 'x' and 'y' should be recycled prior to returning from the routine. A zero value of 'd', with no bits set, indicates that neither of the arguments should be recycled. Values 1, 2, and 3 indicate that 'x', 'y,' and both should be recycled, respectively.

The product is formed within the worker routines by invoking whatever sub-routine `multiply()` is passed inside via the argument list of the worker routine. This argument is a pointer to a binary operation which calculates the product of two components. Pre-defined multiplication operations are listed in Table 16.4.

For other types of product, the user may construct their own multiplication routines and pass them as arguments to the worker routines in Table 16.3, to be used in place of the

Table 16.3 Function prototypes of worker routines to multiply numbers and components by applying the distribution of multiplication over addition.

Distribution operation	Page
| ```
number comp_mul(cliff_comp *x_comp,
               cliff_comp *multiply(cliff_comp *a,
               cliff_comp *b, cliff_comp *c),
               number,y);
``` | 334 |
| ```
number mul_comp(number x,
               cliff_comp *multiply(cliff_comp *a,
               cliff_comp *b, cliff_comp *c),
               cliff_comp *y_comp);
``` | 378 |
| ```
number mul_number(number x,
               cliff_comp *multiply(cliff_comp *a,
               cliff_comp *b, cliff_comp *c),
               number y,
               int d);
``` | 378 |

Table 16.4 Pre-defined operations for multiplying two components.

`multiply()`		Page
| ```
cliff_comp *left(cliff_comp *a, cliff_comp *b,
              cliff_comp *c);
``` | $c \leftarrow a \,\lrcorner\, b$ | 370 |
| ```
cliff_comp *right(cliff_comp *a, cliff_comp *b,
              cliff_comp *c);
``` | $c \leftarrow a \,\llcorner\, b$ | 444 |
| ```
cliff_comp *outer(cliff_comp *a, cliff_comp *b,
              cliff_comp *c);
``` | $c \leftarrow a \wedge b$ | 385 |
| ```
cliff_comp *inner(cliff_comp *a, cliff_comp *b,
              cliff_comp *c);
``` | $c \leftarrow (a,b)e_{\emptyset}$ | 355 |
| ```
cliff_comp *regress(cliff_comp *a, cliff_comp *b,
              cliff_comp *c);
``` | $c \leftarrow a \vee b$ | 442 |
| ```
cliff_comp *central(cliff_comp *a, cliff_comp *b,
              cliff_comp *c);
``` | $c \leftarrow ab$ | 327 |

argument `multiply()`. As a guide, the code for the implementation of the multiplication `central()` is listed in Example 16.2.

Example 16.2

```
1:  /* central product of components with different units */
2:  /* c=ab */
3:  /* a,b and c can point to same component */
```
```

```
4:  cliff_comp *central(a,b,c)
5:      cliff_comp *a,*b,*c; /* result into component c */
6:  {
7:     double x,y;
8:     bitmap au,bu;
9:
10:    au=a->unit;
11:    bu=b->unit;
12:    c->unit=(au^bu);         /* unit of product */
13:    switch(u_sign(au,bu))    /* product of units */
14:       {
15:       case -1: /* units multiply to -1, negative complex multiply */
16:          x=  (a->coef[1]*b->coef[1])-(a->coef[0]*b->coef[0]);
17:          y=-((a->coef[0]*b->coef[1])+(a->coef[1]*b->coef[0]));
18:          break;
19:       case  0: /* units multiply to zero */
20:          x=0.0;
21:          y=0.0;
22:          break;
23:       case  1: /* units multiply to +1, positive complex multiply */
24:          x= (a->coef[0]*b->coef[0])-(a->coef[1]*b->coef[1]);
25:          y= (a->coef[0]*b->coef[1])+(a->coef[1]*b->coef[0]);
26:          break;
27:       default:
28:          printf("unrecognised switch value (%d) in routine:
                        central()\n",
29:                 u_sign(au,bu));
30:          brexit(6); /* no restart */
31:          break;
32:       }
33:    c->coef[0]=x;
34:    c->coef[1]=y;
35:    return(c);
36: }
```

16.3   Transformation

The worker routines listed in Table 16.5 provide mechanisms to construct simple transformations which apply functions (unary operations) to the individual components of numbers. The simplest transformation is func_number() which applies *in situ* the function provided by the sub-routine in the second argument to all components of a given number 'x'. The original number is overwritten with the new values. An error occurs if the function changes the unit of the component.

The transformation clonefuncof_number() preserves the input number so that it can be used again and produces the result as a new entity. In contrast to func_number(), when using clonefuncof_number(), it is permissible to use functions which change the unit of the component.

The transformations addfuncto_number() and subfuncfrom_number() behave the same as clonefuncof_number(), adding or subtracting the function of the second number, function(y), to or from the first number 'x'.

Table 16.5 Function prototypes of worker routines to apply transformations to numbers.

Transformation	Page

`number func_number(number x,`
`            cliff_comp *function(cliff_comp *a,`
`            cliff_comp *c));`
	347

`number clonefuncof_number(number x,`
`                cliff_comp *function(cliff_comp *a,`
`                cliff_comp *c));`
	332

`number addfuncto_number(number x,`
`            number y,`
`            cliff_comp *function(cliff_comp *a,`
`            cliff_comp *c));`
	319

`number subfuncfrom_number(number x,`
`                number y,`
`                cliff_comp *function(cliff_comp *a,`
`                cliff_comp *c));`
	461

The sub-routines listed in Table 16.6 provide a variety of pre-defined functions which may be used in place of the argument `function()`.

For other purposes, the user may construct their own functions and pass them to the transformation routines in Table 16.5, to be used in place of the argument `function()`. As a guide, the code for the implementation of the function `half()` is listed in Example 16.3.

Example 16.3

```
 1:   /* half the value of component */
 2:   /* c=0.5*a */
 3:   /* a and c can point to the same component */
 4:   cliff_comp *half(a,c)
 5:       cliff_comp *a,*c; /* result into component c */
 6:   {
 7:       c->unit=a->unit;
 8:       c->coef[0]=a->coef[0]*0.5;
 9:       c->coef[1]=a->coef[1]*0.5;
10:       return(c);
11:   }
```

16.4   Filtration

The worker routines listed in Table 16.7 apply a test (a unary operation) at the level of the individual components within a number and based on the result of the test either set the value of the component to zero or leave it unchanged. Setting the value of a component to zero causes it to be eliminated from the number.

Table 16.6 Pre-defined functions for transforming numbers (a) ignores unit (b) scalars only.

function ()		Page
`cliff_comp *half(cliff_comp *a,` `             cliff_comp *c);`	$c \leftarrow 0.5\,a$	351
`cliff_comp *twice(cliff_comp *a,` `              cliff_comp *c);`	$c \leftarrow 2.0\,a$	467
`cliff_comp *polar(cliff_comp *a,` `              cliff_comp *c);`	$c \leftarrow (r,\theta) \leftarrow (x,y) \leftarrow a$	398
`cliff_comp *degrees(cliff_comp *a,` `                cliff_comp *c);`	$c \leftarrow (r, \frac{180}{\pi}\theta) \leftarrow (r,\theta) \leftarrow a$	339
`cliff_comp *radians(cliff_comp *a,` `                cliff_comp *c);`	$c \leftarrow (r, \frac{\pi}{180}\theta) \leftarrow (r,\theta) \leftarrow a$	428
`cliff_comp *inverse(cliff_comp *a,` `                cliff_comp *c);`	$c \leftarrow 1.0/a$	360
`cliff_comp *log_coef(cliff_comp *a,` `                 cliff_comp *c);`	$c \leftarrow \ln a$ (a)	371
`cliff_comp *negation(cliff_comp *a,` `                 cliff_comp *c);`	$c \leftarrow -a$	379
`cliff_comp *reversion(cliff_comp *a,` `                  cliff_comp *c);`	$c \leftarrow \tilde{a}$	443
`cliff_comp *cartesian(cliff_comp *a,` `                  cliff_comp *c);`	$c \leftarrow (x,y) \leftarrow (r,\theta) \leftarrow a$	326
`cliff_comp *real_part(cliff_comp *a,` `                  cliff_comp *c);`	$c \leftarrow \Re(a)$	433
`cliff_comp *supplement(cliff_comp *a,` `                   cliff_comp *c);`	$c \leftarrow \lvert a$	465
`cliff_comp *involution(cliff_comp *a,` `                   cliff_comp *c);`	$c \leftarrow \overline{a}$	365
`cliff_comp *square_root(cliff_comp *a,` `                    cliff_comp *c);`	$c \leftarrow \sqrt{a}$ (b)	455
`cliff_comp *imaginary_part(cliff_comp *a,` `                       cliff_comp *c);`	$c \leftarrow \Im(a)$	354
`cliff_comp *square_root_coef(cliff_comp *a,` `                         cliff_comp *c);`	$c \leftarrow \sqrt{a}$ (a)	455
`cliff_comp *squareof_latitude(cliff_comp *a,` `                          cliff_comp *c);`	$c \leftarrow \lambda_a\{\Re(a)^2 + \Im(a)^2\}$	456
`cliff_comp *squareof_magnitude(cliff_comp *a,` `                           cliff_comp *c);`	$c \leftarrow \Re(a)^2 + \Im(a)^2$ (a)	456
`cliff_comp *complex_conjugation(cliff_comp *a,` `                            cliff_comp *c);`	$c \leftarrow a^*$	334
`cliff_comp *clifford_conjugation(cliff_comp *a,` `                             cliff_comp *c);`	$c \leftarrow \bar{a}$	329

Table 16.7 Function prototypes of worker routines to apply filters to numbers.

Filter	$[x] \leftarrow 0$	$[x] \leftarrow [x]$	Page
number trim_zero(number x, number z, int d);	$\|[x]\| \leq \|z\|$	$\|[x]\| > \|z\|$	467
number keepone_grade(number x, int n);	grade $\neq$ n	grade == n	370
number cloneone_grade(number x, int n);	grade $\neq$ n	grade == n	333
number onegradeof_number(number x, int n, int d);	grade $\neq$ n	grade == n	385
number keep_filter(number x, int test(cliff_comp *c));	test == 0	test == 1	369
number clone_filter(number x, int test(cliff_comp *c));	test == 0	test == 1	330
number filter_number(number x, int test(cliff_comp *c), int d);	test == 0	test == 1	343

The first filter, trim_zero(), eliminates components [x] in the number 'x' which have a complex magnitude less than or equal to the magnitude of the complex scalar in grade 0 of the number 'z'. The value in 'z' represents the threshold for zero. The trimming is performed *in situ* and replaces the original value of 'x' with its filtered value, effectively recycling the old value.

The third argument 'd' indicates the strategy for the disposal of previous arguments. With the first being recycled always, the only useful values for 'd' are a zero, with no bits set, and a value of two indicating that threshold 'z' should be recycled.

The three filters keepone_grade(), cloneone_grade(), and onegradeof _number() remove all of the components not in the grade specified by the second argument 'n'. These filters differ only in the way they treat the number 'x' to which the filter is applied. The filter keepone_grade() behaves like trim_zero() applying the filter *in situ* and thereby recycling the input number. The filter cloneone_grade() preserves the input by making a clone as the first step and then applying the filter to that. The filter onegradeof_number() invokes either of the other two according to the value of the disposal flag 'd'.

The three filters keep_filter(), clone_filter(), and filter_number() are general purpose, with the test supplied by the user. Like keepone_grade(), clo- neone_grade(), and onegradeof_number(), these filters differ in the way they treat the number 'x' to which they are applied, by either recycling, preserving, or disposing, respectively, of it according to the value of the disposal flag 'd'.

The sub-routines listed in Table 16.8 provide a variety of pre-defined tests which may be used in place of the argument test(). The first two tests determine if a component lacks either the real or imaginary part. Any component extracted from a number that has more than one component has at least one part non-zero. The number zero, which has only one component, is the only number containing a component with both parts zero.

The third and fourth tests are intended for the case of electromagnetic fields F or F_{k_m} rep- resented in the form of Table 10.1, where the magnetic field is associated with only units for spatial dimensions, identified with units e_1, e_2, e_3, and where the electric field is associated with units for both spatial dimensions and the temporal dimension, which is identified with

Table 16.8 Pre-defined tests for filtering numbers.

test()	Page
int no_real(cliff_comp *c);	381
int no_imaginary(cliff_comp *c);	381
int spatial_part(cliff_comp *c);	454
int temporal_part(cliff_comp *c);	466
int grade_zero(cliff_comp *c);	351
int grade_one(cliff_comp *c);	350
int grade_two(cliff_comp *c);	351
int non_scalar(cliff_comp *c);	382
int not_grade_one(cliff_comp *c);	383
int not_grade_two(cliff_comp *c);	384
int null_signature(cliff_comp *c);	384
int positive_signature(cliff_comp *c);	400
int negative_signature(cliff_comp *c);	380

unit e_0. The test spatial_part() detects the components of the magnetic field, and the test temporal_part() detects the components of the electric field.

The fifth to seventh tests determine if components are in grades 0 (scalar), 1, or 2, respectively, and the eight to tenth tests do the opposite. The final three tests determine if the signature of the component is zero, plus, or minus one, respectively.

For other purposes, the user may construct their own tests and pass them as arguments to the filter routines in Table 16.7, to be used in place of the argument test(). The test function should return an integer value of zero for the filter to set components to zero (to block them from passing through the filter), and return a value of one for the filter to leave them untouched (to allow them to pass through the filter). The filters do not behave consistently if the value returned by the test is neither 0 or 1.

As a guide, the code for the implementation of the test spatial_part() is listed in Example 16.4.

Example 16.4

```
 1:   /* test for absence of primal unit e0 */
 2:   /* true (1) if absent */
 3:   int spatial_part(c)
 4:        cliff_comp *c;
 5:   {
 6:     int k;
 7:
 8:     k=(~c->unit)&1; /* 1 = does not have e0 */
 9:     return(k);
10:   }
```

Part III

Application

17

Verification

17.1 Identities

The identities[1] listed in Table 17.1 serve as a useful check on the consistency of the various types of multiplication and other operations supported within Clifford algebra. These identities hold for any number of dimensions, and for any signature: uniformly positive or negative, or mixed positive, negative and zero.

If any of the identities fail for a numerical implementation, it indicates some inconsistency or error in that particular implementation. However, even if all of the identities are honoured that does not guarantee all operations are performing correctly, only consistently.

17.2 Tests

The easiest way to test the identities in Table 17.1 for a numerical implementation is by computing them for a large number of cases and evaluating the error in relation to the expected floating point precision of the machine.

17.2.1 Example Code

The program code listed in Example 17.1a invokes routines which individually calculate each of the identities, subtracting one side from the other and accumulating any error.

The code runs the tests for a chosen dimension 'dim' and a chosen signature 'sig' passed in the argument list on line 1. The signature is established as either all ± 1 on lines 13 and 15 or, if the variable 'sig' has value zero, randomly mixed on lines 14 and 15, and lines 26–51.

All 26 tests are run multiple times '$0 \le i <$ rep' within the loop between lines 19 and 60. The values of the Clifford numbers 'u', 'v', and 'w', and the vector 'a' as used in the tests are generated on lines 21–24, with real and imaginary parts of all coefficients chosen randomly between negative and positive one in a uniform distribution.

The first time the loop is run, for '$i=0$' on line 19, the routines called within the loop print verbose messages. The messages usually show which test is failing, if that is the case. For subsequent passes through the loop, the only message printed is the total error, on line 52.

---

1 Taken mostly from Axelsson (2002).

Numerical Calculations in Clifford Algebra: A Practical Guide for Engineers and Scientists,
First Edition. Andrew Seagar.

Table 17.1 Identities for Clifford numbers u, v, w and vector a.

1	$(u \lrcorner v, w)e_\emptyset = (v, u \wedge w)e_\emptyset$	11b	$(uv, w)e_\emptyset = (u, w\tilde{v})e_\emptyset$
2	$(u \llcorner v, w)e_\emptyset = (u, w \wedge v)e_\emptyset$	12a	$(u, v)e_\emptyset = (\tilde{u}, \tilde{v})e_\emptyset$
3	$u \lrcorner (v \lrcorner w) = (v \wedge u) \lrcorner w$	12b	$(u, v)e_\emptyset = (\bar{u}, \bar{v})e_\emptyset$
4	$u \lrcorner (v \llcorner w) = (u \lrcorner v) \llcorner w$	12c	$(u, v)e_\emptyset = (\hat{u}, \hat{v})e_\emptyset$
5	$au = a \lrcorner u + a \wedge u$	13a	$\widetilde{u \lrcorner v} = \tilde{v} \llcorner \tilde{u}$
6a	$a \lrcorner u = -\hat{u} \llcorner a$	13b	$\widetilde{u \wedge v} = \tilde{v} \wedge \tilde{u}$
6b	$a \lrcorner u = \frac{1}{2}(au - \hat{u}a)$	13c	$\widetilde{uv} = \tilde{v}\tilde{u}$
7a	$a \wedge u = \hat{u} \wedge a$	14a	$\overline{u \lrcorner v} = \bar{u} \lrcorner \bar{v}$
7b	$a \wedge u = \frac{1}{2}(au + \hat{u}a)$	14b	$\overline{u \wedge v} = \bar{u} \wedge \bar{v}$
8	$a \lrcorner (u \wedge v) = (a \lrcorner u) \wedge v + \hat{u} \wedge (a \lrcorner v)$	14c	$\overline{uv} = \bar{u}\bar{v}$
9	$a \lrcorner (uv) = (a \lrcorner u)v + \hat{u}(a \lrcorner v)$	15a	$** [u]_k = (-1)^{k(n-k)} \lambda_N [u]_k$
10	$a \wedge (uv) = (a \lrcorner u)v + \hat{u}(a \wedge v)$	15b	$u \wedge *v = (u, v)e_N$
11a	$(uv, w)e_\emptyset = (v, \tilde{u}w)e_\emptyset$	16	$*(u \vee v) = (*u) \wedge (*v)$

The peak memory used is printed on line 59 to indicate whether there is any problem in some routine failing to recycle properly all memory used on a temporary basis.

Example 17.1(a)

```
 1:   double test_identities(lambda,dim,sig,rep)
 2:        int *lambda;
 3:        int dim,sig,rep;
 4:   {
 5:       number a,u,v,w;
 6:       double error,total;
 7:       int mem[2][3];
 8:       int n,i;
 9:
10:       n=power_of_two(dim);
11:       printf("tests for dimension %d, coefficients %d,
                     signature ",dim,n);
12:       random_seed(0xABCDEF);
13:       if(sig!=0) { pack_signature(dim,lambda,sig);
                      printf("%d.\n",sig);   }
14:       else       { random_signature(dim,lambda,1);
                      printf("mixed.\n"); }
15:       declare_primal(dim,lambda);
16:       get_mem_usage(mem);
17:       printf("memory peak (%d structural)
                     (%d data)\n",mem[0][1],mem[1][1]);
18:       total=0.0;
19:       for(i=0;i<rep;i++) /* repeat tests this many times */
20:          {
21:              a=random_vector();
22:              u=random_number(); /* also u=random_vector();*/
```

```
23:          v=random_number(); /*        v=random_vector();*/
24:          w=random_number(); /*        w=random_vector();*/
25:          error=0.0;
26:          if(sig==0) {random_signature(dim,lambda,i);}
                        error=error+check1(i,u,v,w);
27:          if(sig==0) {random_signature(dim,lambda,i);}
                        error=error+check2(i,u,v,w);
28:          if(sig==0) {random_signature(dim,lambda,i);}
                        error=error+check3(i,u,v,w);
29:          if(sig==0) {random_signature(dim,lambda,i);}
                        error=error+check4(i,u,v,w);
30:          if(sig==0) {random_signature(dim,lambda,i);}
                        error=error+check5(i,a,u);
31:          if(sig==0) {random_signature(dim,lambda,i);}
                        error=error+check6a(i,a,u);
32:          if(sig==0) {random_signature(dim,lambda,i);}
                        error=error+check6b(i,a,u);
33:          if(sig==0) {random_signature(dim,lambda,i);}
                        error=error+check7a(i,a,u);
34:          if(sig==0) {random_signature(dim,lambda,i);}
                        error=error+check7b(i,a,u);
35:          if(sig==0) {random_signature(dim,lambda,i);}
                        error=error+check8(i,a,u,v);
36:          if(sig==0) {random_signature(dim,lambda,i);}
                        error=error+check9(i,a,u,v);
37:          if(sig==0) {random_signature(dim,lambda,i);}
                        error=error+check10(i,a,u,v);
38:          if(sig==0) {random_signature(dim,lambda,i);}
                        error=error+check11a(i,u,v,w);
39:          if(sig==0) {random_signature(dim,lambda,i);}
                        error=error+check11b(i,u,v,w);
40:          if(sig==0) {random_signature(dim,lambda,i);}
                        error=error+check12a(i,u,v);
41:          if(sig==0) {random_signature(dim,lambda,i);}
                        error=error+check12b(i,u,v);
42:          if(sig==0) {random_signature(dim,lambda,i);}
                        error=error+check12c(i,u,v);
43:          if(sig==0) {random_signature(dim,lambda,i);}
                        error=error+check13a(i,u,v);
44:          if(sig==0) {random_signature(dim,lambda,i);}
                        error=error+check13b(i,u,v);
45:          if(sig==0) {random_signature(dim,lambda,i);}
                        error=error+check13c(i,u,v);
46:          if(sig==0) {random_signature(dim,lambda,i);}
                        error=error+check14a(i,u,v);
47:          if(sig==0) {random_signature(dim,lambda,i);}
                        error=error+check14b(i,u,v);
48:          if(sig==0) {random_signature(dim,lambda,i);}
                        error=error+check14c(i,u,v);
49:          if(sig==0) {random_signature(dim,lambda,i);}
                        error=error+check15a(i,u);
50:          if(sig==0) {random_signature(dim,lambda,i);}
                        error=error+check15b(i,u,v);
```

(Continued)

Example 17.1(a) (Continued)

```
51:           if(sig==0) {random_signature(dim,lambda,i);}
                         error=error+check16(i,u,v);
52:           printf("run %3d, sum of errors (26 tests)% 5.2e ",i,error);
53:           total=total+error;
54:           recycle_number(a);
55:           recycle_number(u);
56:           recycle_number(v);
57:           recycle_number(w);
58:           get_mem_usage(mem);
59:           printf("memory peak (%d structural)
                         (%d data)\n",mem[0][1],mem[1][1]);
60:       }
61:     printf("\n");
62:     total=total/(26*rep); /* average error per identity */
63:     return(total);
64:   }
```

17.2.2 Example Output

Example 17.1b lists the results of running the code in Example 17.1a for seven dimensions with positive, negative, and mixed signature, with ten repetitions. Individual errors are printed for positive signatures in lines 6–31, most of which are slightly above the double precision accuracy of the underlying data type. The total error for all tests in the first run through the loop is printed on line 32. The total error for all other runs is printed on lines 33–41. If the value of the error is identically zero, the logarithm is reported (for convenience) as −99.

Example 17.1(b)

```
1:    application4()
2:    Enter dimension (1-32) for test: 7
3:    Enter number of repetitions (>0): 10
4:    tests for dimension 7, coefficients 128, signature 1.
5:    memory peak (0 structural) (0 data)
6:    1     log error -13.7,   error 1.811536e-14
7:    2     log error -13.6,   error 2.486900e-14
8:    3     log error -13.6,   error 2.527115e-14
9:    4     log error -13.5,   error 3.275894e-14
10:   5     log error -14.7,   error 2.210667e-15
11:   6a    log error -99.0,   error 0.000000e+00
12:   6b    log error -14.8,   error 1.527770e-15
13:   7a    log error -15.0,   error 1.001174e-15
14:   7b    log error -14.9,   error 1.335390e-15
15:   8     log error -13.6,   error 2.280436e-14
16:   9     log error -13.1,   error 7.901540e-14
17:   10    log error -13.1,   error 8.329679e-14
18:   11a   log error -13.4,   error 4.176387e-14
19:   11b   log error -14.2,   error 6.883383e-15
```

```
20:   12a  log error -99.0,   error 0.000000e+00
21:   12b  log error -99.0,   error 0.000000e+00
22:   12c  log error -99.0,   error 0.000000e+00
23:   13a  log error -99.0,   error 0.000000e+00
24:   13b  log error -13.9,   error 1.163384e-14
25:   13c  log error -13.4,   error 4.415526e-14
26:   14a  log error -99.0,   error 0.000000e+00
27:   14b  log error -99.0,   error 0.000000e+00
28:   14c  log error -99.0,   error 0.000000e+00
29:   15a  log error -99.0,   error 0.000000e+00
30:   15b  log error -99.0,   error 0.000000e+00
31:   16   log error -14.0,   error 9.533020e-15
32:   run   0, sum of errors (26 tests) 4.06e-13 memory peak
            (895 structural) (903 data)
33:   run   1, sum of errors (26 tests) 3.77e-13 memory peak
            (895 structural) (903 data)
34:   run   2, sum of errors (26 tests) 4.26e-13 memory peak
            (895 structural) (903 data)
35:   run   3, sum of errors (26 tests) 5.60e-13 memory peak
            (895 structural) (903 data)
36:   run   4, sum of errors (26 tests) 4.03e-13 memory peak
            (895 structural) (903 data)
37:   run   5, sum of errors (26 tests) 4.71e-13 memory peak
            (895 structural) (903 data)
38:   run   6, sum of errors (26 tests) 4.11e-13 memory peak
            (895 structural) (903 data)
39:   run   7, sum of errors (26 tests) 4.13e-13 memory peak
            (895 structural) (903 data)
40:   run   8, sum of errors (26 tests) 5.65e-13 memory peak
            (895 structural) (903 data)
41:   run   9, sum of errors (26 tests) 4.87e-13 memory peak
            (895 structural) (903 data)
42:
43:   tests for dimension 7, coefficients 128, signature -1.
44:   memory peak (895 structural) (903 data)
45:   1    log error -14.0,   error 8.950904e-15
46:   2    log error -13.3,   error 5.024296e-14
47:   3    log error -13.4,   error 4.293741e-14
48:   4    log error -13.6,   error 2.248622e-14
49:   5    log error -14.7,   error 1.831290e-15
50:   6a   log error -99.0,   error 0.000000e+00
51:   6b   log error -14.8,   error 1.456722e-15
52:   7a   log error -15.0,   error 1.001174e-15
53:   7b   log error -14.9,   error 1.153632e-15
54:   8    log error -13.6,   error 2.280436e-14
55:   9    log error -13.0,   error 9.197668e-14
56:   10   log error -13.1,   error 8.141948e-14
57:   11a  log error -13.4,   error 4.158345e-14
58:   11b  log error -13.6,   error 2.383233e-14
59:   12a  log error -99.0,   error 0.000000e+00
60:   12b  log error -99.0,   error 0.000000e+00
61:   12c  log error -99.0,   error 0.000000e+00
62:   13a  log error -99.0,   error 0.000000e+00
63:   13b  log error -13.9,   error 1.163384e-14
```

(Continued)

Example 17.1(b) (Continued)

```
64:   13c  log error -13.3,   error 4.946399e-14
65:   14a  log error -99.0,   error 0.000000e+00
66:   14b  log error -99.0,   error 0.000000e+00
67:   14c  log error -99.0,   error 0.000000e+00
68:   15a  log error -99.0,   error 0.000000e+00
69:   15b  log error -99.0,   error 0.000000e+00
70:   16   log error -14.0,   error 9.533020e-15
71:   run   0, sum of errors (26 tests) 4.62e-13 memory peak
                 (895 structural) (903 data)
72:   run   1, sum of errors (26 tests) 4.82e-13 memory peak
                 (895 structural) (903 data)
73:   run   2, sum of errors (26 tests) 4.17e-13 memory peak
                 (895 structural) (903 data)
74:   run   3, sum of errors (26 tests) 5.97e-13 memory peak
                 (895 structural) (903 data)
75:   run   4, sum of errors (26 tests) 4.58e-13 memory peak
                 (895 structural) (903 data)
76:   run   5, sum of errors (26 tests) 4.51e-13 memory peak
                 (895 structural) (903 data)
77:   run   6, sum of errors (26 tests) 4.12e-13 memory peak
                 (895 structural) (903 data)
78:   run   7, sum of errors (26 tests) 4.63e-13 memory peak
                 (895 structural) (903 data)
79:   run   8, sum of errors (26 tests) 4.85e-13 memory peak
                 (895 structural) (903 data)
80:   run   9, sum of errors (26 tests) 4.88e-13 memory peak
                 (895 structural) (903 data)
81:
82:   tests for dimension 7, coefficients 128, signature mixed.
83:   memory peak (895 structural) (903 data)
84:   sig[00-0-0-] 1    log error -15.3,   error 4.742875e-16
85:   sig[0+0-0+-] 2    log error -15.5,   error 3.140185e-16
86:   sig[-00-0--] 3    log error -14.4,   error 4.117949e-15
87:   sig[++-+00-] 4    log error -14.1,   error 8.314463e-15
88:   sig[-0++---] 5    log error -14.6,   error 2.352357e-15
89:   sig[-+00+++] 6a   log error -99.0,   error 0.000000e+00
90:   sig[00+0-0+] 6b   log error -15.0,   error 1.044557e-15
91:   sig[0++00-0] 7a   log error -14.8,   error 1.528448e-15
92:   sig[-00--0-] 7b   log error -14.8,   error 1.554126e-15
93:   sig[-+--00+] 8    log error -13.8,   error 1.574691e-14
94:   sig[000++0-] 9    log error -13.5,   error 2.826323e-14
95:   sig[--++-+0] 10   log error -13.0,   error 9.237736e-14
96:   sig[0++0--+] 11a  log error -14.3,   error 5.329071e-15
97:   sig[0++00-0] 11b  log error -15.0,   error 9.930137e-16
98:   sig[++0--++] 12a  log error -99.0,   error 0.000000e+00
99:   sig[-0--0--] 12b  log error -99.0,   error 0.000000e+00
100:  sig[++0-0-+] 12c  log error -99.0,   error 0.000000e+00
101:  sig[+0+--+0] 13a  log error -99.0,   error 0.000000e+00
102:  sig[0+-0+00] 13b  log error -14.1,   error 8.719029e-15
103:  sig[+00+-+-] 13c  log error -13.5,   error 2.852170e-14
104:  sig[++-0++0] 14a  log error -99.0,   error 0.000000e+00
105:  sig[0+0--00] 14b  log error -99.0,   error 0.000000e+00
106:  sig[0--+-0-] 14c  log error -99.0,   error 0.000000e+00
107:  sig[+-++00+] 15a  log error -99.0,   error 0.000000e+00
```

```
108:    sig[0--0--+] 15b  log error -16.7,   error 2.081668e-17
109:    sig[+-0-+-0] 16   log error -99.0,   error 0.000000e+00
110:    run   0, sum of errors (26 tests) 2.00e-13 memory peak
                (895 structural) (903 data)
111:    run   1, sum of errors (26 tests) 1.26e-13 memory peak
                (895 structural) (903 data)
112:    run   2, sum of errors (26 tests) 1.65e-13 memory peak
                (895 structural) (903 data)
113:    run   3, sum of errors (26 tests) 2.61e-13 memory peak
                (895 structural) (903 data)
114:    run   4, sum of errors (26 tests) 2.29e-13 memory peak
                (895 structural) (903 data)
115:    run   5, sum of errors (26 tests) 2.10e-13 memory peak
                (895 structural) (903 data)
116:    run   6, sum of errors (26 tests) 1.30e-13 memory peak
                (895 structural) (903 data)
117:    run   7, sum of errors (26 tests) 3.01e-13 memory peak
                (895 structural) (903 data)
118:    run   8, sum of errors (26 tests) 1.98e-13 memory peak
                (895 structural) (903 data)
119:    run   9, sum of errors (26 tests) 2.23e-13 memory peak
                (895 structural) (903 data)
120:
121:    total time 2.367039e+01(s)
122:    time per identity: 3.034665e-02(s)
        error per identity: 1.445815e-14
123:    statistics for structural heap:
124:    initial size (256)
125:    peak size     (895), as % of initial size (350%)
126:    full size     (1031), expansion count (8)
127:    free count    (1031), free start (5)
128:    statistics for data heap:
129:    initial size (256)
130:    peak size     (903), as % of initial size (353%)
131:    full size     (1031), expansion count (8)
132:    free count    (1031), free start (0)
```

The results for negative and mixed signatures are listed in the same fashion between lines 45 and 80 and between lines 84 and 119. In the case of the mixed signatures, the signature for each test in the first run through the loop is printed from lines 84 to 109.

Figure 17.1 shows the average error per identity as a function of the number of dimensions. The upper curve is the case when the Clifford numbers 'u', 'v', 'w' used in the tests are non-zero in all components. The lower curve is the case when 'u', 'v', 'w' are non-zero only in grade 1, i.e. they are Clifford vectors, as when the code in the comments at the end of lines 22–24 of Example 17.1a is brought into use. The errors increase as the number of dimensions increases due to the larger number of arithmetic operations required for each Clifford calculation.

For full Clifford numbers a central multiplication operation in n dimensions requires 4×2^{2n} real-valued multiplications and about half that number of additions. In principle, the number of rounding errors accumulated increases with dimension in the ratio of $2^2 = 4$.

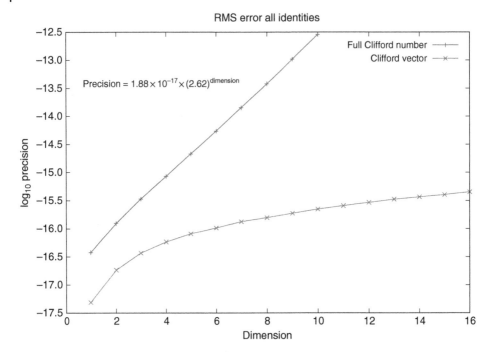

Figure 17.1 Numerical error as function of dimension for full Clifford numbers.

If the errors are independent, then the total error for the whole operation increases as some-where in the vicinity of the square root of that. The factor 2.62 found in practice here as the slope of the line in Figure 17.1, is about 30% higher.

For Clifford vectors, there are fewer operations, only $4n^2$ multiplications and half as many additions. As a consequence the lower curve in Figure 17.1 increases much more slowly than the upper curve, the curve for full Clifford numbers.

The accumulation of errors for full numbers behaves exponentially with a constant as the base raised to the dimension as exponent, given by the formula adjacent to the upper line in Figure 17.1. The constants in the formula are obtained by numerically fitting a straight line as a chord between dimensions 3 and 10.

In contrast, the accumulation of errors for vectors behaves as a power law with the dimension as the base raised to a constant power. This relationship is shown in Figure 17.2 where the vector data are replotted with logarithm on both abscissa and ordinate. Fitting a straight line to the data as a chord between dimensions 9 and 16 gives the constants as shown in the formula in the figure.

In principle the computational effort, as shown in Figure 17.3, increases with dimension in direct proportion to the number of operations. Ignoring any memory accessing, the number of arithmetic operations increases by a factor of $2^2 = 4$. The factor found in practice, taken as the slope of the line in Figure 17.3, is 4.49. This is 12% higher than the expected value.

The computational effort for vectors, the lower line in Figure 17.3, increases more slowly. The increase to be expected is in the order of the dimension to the power of 2. In Figure 17.4, the chord spanning the upper eight data points fits closer to a power of 3.20.

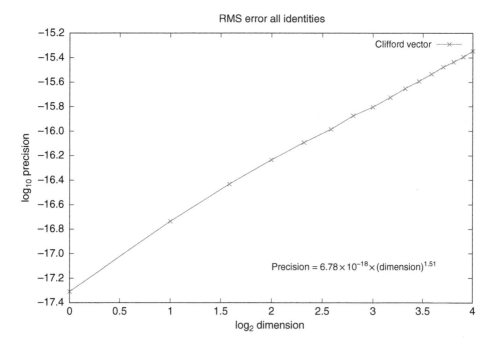

Figure 17.2 Numerical error as function of dimension for vectors.

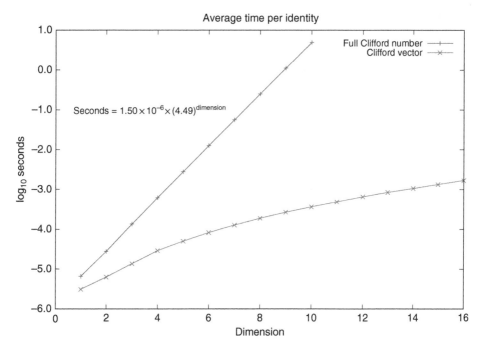

Figure 17.3 Computational time as function of dimension for full Clifford numbers.

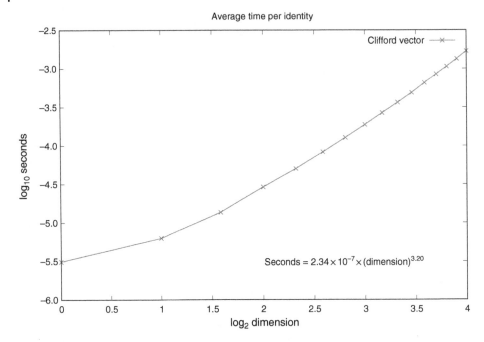

Figure 17.4 Computational time as function of dimension for vectors.

Reference

Axelsson, A. (2002), Transmission Problems for Dirac's and Maxwell's Equations with Lipschitz Interfaces, PhD thesis, Australian National University, Canberra, Australia.

18

Lines Not Parallel

Two lines which are not parallel intersect if they lie in a common plane, but fail to intersect if they have no plane in common. In the first case, as in Figure 18.1a, the shortest line segment which connects the two lines is of length zero and identifies a single point in space. In the second case, as in Figure 18.1b, the shortest line segment connecting the two lines is non-zero in length and identifies two points in space, one on each line.

Identifying the shortest line segment between two non-parallel lines in space is taken here as a simple geometrical problem on which to demonstrate the use of Grassmann's inner and outer products. Since both of these products operate consistently in any number of dimensions[1], the solutions given here can be used in all dimensions $n \geq 2$. Although the code used to demonstrate these solutions numerically is applied only in three dimensions, it also applies directly to higher dimensions simply by declaring a greater number of dimensions at the outset and providing higher dimensional data.

18.1 Theory

18.1.1 Common Plane

The intersection of two lines can be calculated by solving a system of linear equations involving Cartesian (or other) components of vectors. A linear system of two interdependent equations arises when Grassmann's inner product is used. When using Grassmann's outer product, the linear system reduces to two independent equations.

Consider the point $\mathbf{x}$ of intersection for the two line segments $\mathbf{b} - \mathbf{a}$ and $\mathbf{s} - \mathbf{r}$ as shown in Figure 18.2. The two segments are in a common plane and are not parallel because otherwise there is no proper solution. The plane may be embedded in two, three, or any higher number of dimensions n. As a consequence $n \geq 2$ Cartesian or other components are employed to identify points in the plane, as in

$$\mathbf{x} = x = x_0 e_0 + x_1 e_1 + \cdots + x_{n-1} e_{n-1} \tag{18.1}$$

---

1 Unlike the vector cross product.

Numerical Calculations in Clifford Algebra: A Practical Guide for Engineers and Scientists,
First Edition. Andrew Seagar.
© 2023 John Wiley & Sons Ltd. Published 2023 by John Wiley & Sons Ltd.

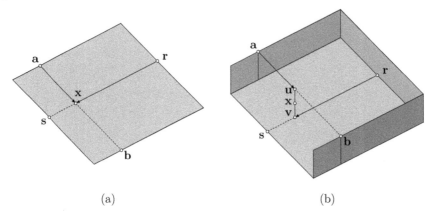

(a)   (b)

Figure 18.1   Point **x** closest to two lines **b** − **a** and **s** − **r** (a) with common plane, (b) with no common plane.

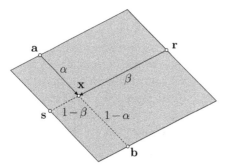

Figure 18.2   Intersection of two lines in *n* dimensions at point **x**.

A solution for the intersection **x** is sought in terms of two scalar parameters $\alpha = (\mathbf{x} - \mathbf{a})/(\mathbf{b} - \mathbf{a})$ and $\beta = (\mathbf{x} - \mathbf{r})/(\mathbf{s} - \mathbf{r})$, which specify the proportion of each segment from its beginning, **a** or **r**, to the intersection **x**:

$$\mathbf{x} = \begin{cases} (1 - \alpha)\,\mathbf{a} + \alpha\,\mathbf{b} \\ (1 - \beta)\,\mathbf{r} + \beta\,\mathbf{s} \end{cases} \tag{18.2}$$

Eliminating the point **x** produces a single vector equation:

$$\alpha\,(\mathbf{b} - \mathbf{a}) - \beta\,(\mathbf{s} - \mathbf{r}) = \mathbf{r} - \mathbf{a} \tag{18.3}$$

From this stage, there are two alternative methods, one using the inner product and one using the outer product.

18.1.1.1   Inner Product

Multiplying segments **s** − **r** and **b** − **a** by Eq. (18.3) using the dot[2] version of the inner product splits the single vector equation into two equations, each of which describes a relationship in the direction of one of the segments:

$$\begin{cases} \alpha\,(\mathbf{b} - \mathbf{a}) \cdot (\mathbf{s} - \mathbf{r}) - \beta\,(\mathbf{s} - \mathbf{r}) \cdot (\mathbf{s} - \mathbf{r}) = (\mathbf{r} - \mathbf{a}) \cdot (\mathbf{s} - \mathbf{r}) \\ \alpha\,(\mathbf{b} - \mathbf{a}) \cdot (\mathbf{b} - \mathbf{a}) - \beta\,(\mathbf{s} - \mathbf{r}) \cdot (\mathbf{b} - \mathbf{a}) = (\mathbf{r} - \mathbf{a}) \cdot (\mathbf{b} - \mathbf{a}) \end{cases} \tag{18.4}$$

---

2 It is convenient here, for clarity, to write the inner product $(\mathbf{p}, \mathbf{q})e_\emptyset$ as $\mathbf{p} \cdot \mathbf{q}$, the vectorial 'dot' version of the inner product. This is possible because all of the quantities involved are vectors and is desirable because otherwise the parentheses used extensively to group terms can be confused with the parentheses of the inner product.

The two equations form a linear system with scalar unknowns and scalar coefficients, which can be solved by matrix inversion. The matrix is full except in the special case when the lines are perpendicular.

18.1.1.2 Outer Product

Using the outer product, each of the segments $\mathbf{s} - \mathbf{r}$ and $\mathbf{b} - \mathbf{a}$ is multiplied by Eq. (18.3) yielding in the first instance the linear system:

$$\begin{cases} \alpha\,(\mathbf{b} - \mathbf{a}) \wedge (\mathbf{s} - \mathbf{r}) \; - \; \beta\,(\mathbf{s} - \mathbf{r}) \wedge (\mathbf{s} - \mathbf{r}) \; = \; (\mathbf{r} - \mathbf{a}) \wedge (\mathbf{s} - \mathbf{r}) \\ \alpha\,(\mathbf{b} - \mathbf{a}) \wedge (\mathbf{b} - \mathbf{a}) \; - \; \beta\,(\mathbf{s} - \mathbf{r}) \wedge (\mathbf{b} - \mathbf{a}) \; = \; (\mathbf{r} - \mathbf{a}) \wedge (\mathbf{b} - \mathbf{a}) \end{cases} \tag{18.5}$$

At first sight, this appears much the same as Eq. (18.4). However, unlike the inner product, the outer product of a vector with itself is zero (see Section 5.1.1), so the matrix is not full. Two of the terms vanish, leaving

$$\begin{cases} \alpha\,(\mathbf{b} - \mathbf{a}) \wedge (\mathbf{s} - \mathbf{r}) = (\mathbf{r} - \mathbf{a}) \wedge (\mathbf{s} - \mathbf{r}) \\ -\beta\,(\mathbf{s} - \mathbf{r}) \wedge (\mathbf{b} - \mathbf{a}) = (\mathbf{r} - \mathbf{a}) \wedge (\mathbf{b} - \mathbf{a}) \end{cases} \tag{18.6}$$

or alternatively:

$$\begin{cases} \alpha = (\mathbf{r} - \mathbf{a}) \wedge (\mathbf{s} - \mathbf{r}) \; / \; (\mathbf{b} - \mathbf{a}) \wedge (\mathbf{s} - \mathbf{r}) \\ \beta = (\mathbf{r} - \mathbf{a}) \wedge (\mathbf{b} - \mathbf{a}) \; / \; (\mathbf{b} - \mathbf{a}) \wedge (\mathbf{s} - \mathbf{r}) \end{cases} \tag{18.7}$$

Although the numerator and denominator are bivectors, the division makes sense because all of the vectors involved lie in the same plane, meaning that the numerator and denominator are simply scalar multiples of the same bivector. Either of the two equations gives the solution without need for any matrix inversion.

For the numerical calculation of the two ratios in Eq. (18.7), the numerator and denominator can both be reduced to scalars by inner multiplication with the denominator[3]. The vectorial dot product cannot be used here because the entities involved are bivectors not vectors. For example

$$\alpha = \frac{\big((\mathbf{r} - \mathbf{a}) \wedge (\mathbf{s} - \mathbf{r}),\, (\mathbf{b} - \mathbf{a}) \wedge (\mathbf{s} - \mathbf{r})\big)\, e_\emptyset}{\big((\mathbf{b} - \mathbf{a}) \wedge (\mathbf{s} - \mathbf{r}),\, (\mathbf{b} - \mathbf{a}) \wedge (\mathbf{s} - \mathbf{r})\big)\, e_\emptyset} \tag{18.8}$$

18.1.1.3 Geometrical Interpretation

The geometrical interpretation of the solution in Eq. (18.7) is as the ratio of two areas. In Figure 18.3a, the numerator is the small white parallelogram with sides parallel to $\mathbf{r} - \mathbf{a}$

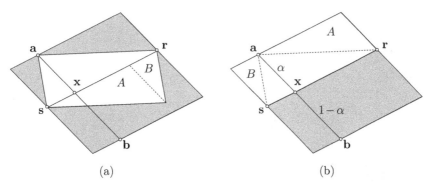

(a)   (b)

Figure 18.3 Solution for scalar parameter α as ratio of two areas (a) as per equation, (b) geometrically reconfigured.

---

3 This will not work in general for two arbitrary bivectors.

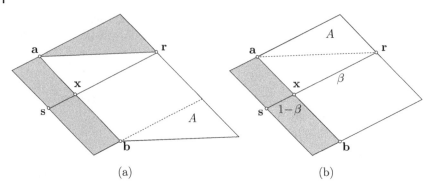

(a)　　　　　　　　　　　(b)

Figure 18.4 Solution for scalar parameter β as a ratio of two areas (a) as per equation, (b) geometrically reconfigured.

and $\mathbf{s} - \mathbf{a}$, sitting on top of the large grey (denominator) parallelogram with sides parallel to $\mathbf{b} - \mathbf{a}$ and $\mathbf{s} - \mathbf{r}$.

In Figure 18.3b, the smaller parallelogram has been reconfigured by moving triangles A and B without changing the total area. By inspection the ratio of the area of the smaller parallelogram in relation to the larger is exactly the same as the ratio α, the distance of $\mathbf{x}$ from $\mathbf{a}$ in relation to the distance of $\mathbf{b}$ from $\mathbf{a}$. Figure 18.4 shows a similar construction for the solution β as the ratio of the areas of two parallelograms.

18.1.2　No Plane in Common

The minimum separation of two lines with no plane in common can be calculated by finding the shortest line segment which connects the two lines, shown as $\mathbf{u} - \mathbf{v}$ in Figure 18.5.

A solution is sought in terms of two scalar parameters $\alpha = (\mathbf{u} - \mathbf{a})/(\mathbf{b} - \mathbf{a})$ and $\beta = (\mathbf{v} - \mathbf{r})/(\mathbf{s} - \mathbf{r})$, which specify the proportion of each segment from its beginning, $\mathbf{a}$ or $\mathbf{r}$, to the points $\mathbf{u}$ and $\mathbf{v}$:

$$\begin{cases} \mathbf{u} = (1 - \alpha)\,\mathbf{a} + \alpha\,\mathbf{b} \\ \mathbf{v} = (1 - \beta)\,\mathbf{r} + \beta\,\mathbf{s} \end{cases} \tag{18.9}$$

with midpoint:

$$\mathbf{x} = \tfrac{1}{2}(\mathbf{u} + \mathbf{v}) \tag{18.10}$$

Figure 18.5 Minimum separation of two lines in n dimensions at points $\mathbf{u}$ and $\mathbf{v}$.

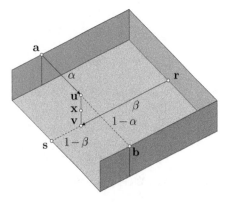

18.1.2.1 Inner Product

The squared length of the segment:

$$\mathbf{u} - \mathbf{v} = (1 - \alpha)\mathbf{a} + \alpha\mathbf{b} - (1 - \beta)\mathbf{r} - \beta\mathbf{s} = \alpha(\mathbf{b} - \mathbf{a}) - \beta(\mathbf{s} - \mathbf{r}) + (\mathbf{a} - \mathbf{r}) \tag{18.11}$$

is proportional to the Grassmann's inner product[4] $(\mathbf{u} - \mathbf{v}) \cdot (\mathbf{u} - \mathbf{v})$. Differentiating that with respect to the parameters β and α and seeking the points at which the derivatives are zero then gives a pair of equations:

$$\begin{cases} \frac{\partial}{\partial \beta}(\mathbf{u} - \mathbf{v}) \cdot (\mathbf{u} - \mathbf{v}) = -(\mathbf{s} - \mathbf{r}) \cdot (\mathbf{u} - \mathbf{v}) + (\mathbf{u} - \mathbf{v}) \cdot (-)(\mathbf{s} - \mathbf{r}) = -2(\mathbf{u} - \mathbf{v}) \cdot (\mathbf{s} - \mathbf{r}) = 0 \\ \frac{\partial}{\partial \alpha}(\mathbf{u} - \mathbf{v}) \cdot (\mathbf{u} - \mathbf{v}) = (\mathbf{b} - \mathbf{a}) \cdot (\mathbf{u} - \mathbf{v}) + (\mathbf{u} - \mathbf{v}) \cdot (\mathbf{b} - \mathbf{a}) = 2(\mathbf{u} - \mathbf{v}) \cdot (\mathbf{b} - \mathbf{a}) = 0 \end{cases} \tag{18.12}$$

which expand on elimination of $\mathbf{u} - \mathbf{v}$, using Eq. (18.11), as follows:

$$\begin{cases} \alpha(\mathbf{b} - \mathbf{a}) \cdot (\mathbf{s} - \mathbf{r}) - \beta(\mathbf{s} - \mathbf{r}) \cdot (\mathbf{s} - \mathbf{r}) + (\mathbf{a} - \mathbf{r}) \cdot (\mathbf{s} - \mathbf{r}) = 0 \\ \alpha(\mathbf{b} - \mathbf{a}) \cdot (\mathbf{b} - \mathbf{a}) - \beta(\mathbf{s} - \mathbf{r}) \cdot (\mathbf{b} - \mathbf{a}) + (\mathbf{a} - \mathbf{r}) \cdot (\mathbf{b} - \mathbf{a}) = 0 \end{cases} \tag{18.13}$$

This pair of equations exhibits the form of a linear system of equations much as in Eq. (18.4), the first case for the intersection of two lines:

$$\begin{cases} \alpha(\mathbf{b} - \mathbf{a}) \cdot (\mathbf{s} - \mathbf{r}) - \beta(\mathbf{s} - \mathbf{r}) \cdot (\mathbf{s} - \mathbf{r}) = (\mathbf{r} - \mathbf{a}) \cdot (\mathbf{s} - \mathbf{r}) \\ \alpha(\mathbf{b} - \mathbf{a}) \cdot (\mathbf{b} - \mathbf{a}) - \beta(\mathbf{s} - \mathbf{r}) \cdot (\mathbf{b} - \mathbf{a}) = (\mathbf{r} - \mathbf{a}) \cdot (\mathbf{b} - \mathbf{a}) \end{cases} \tag{18.14}$$

and requires solution by matrix inversion in the same way as Eq. (18.4). It should not be surprising that one solution for the case where the two planes are in common is the same as the case where they are not, since the former is just a special case of the latter. However, the numerical values when calculating the solution using Eq. (18.14) are different than when using Eq. (18.4) because the vector $\mathbf{r} - \mathbf{a}$ traverses from one plane to the other and takes different values in the two sets of equations.

18.1.2.2 Solution

Solution of the linear system in Eq. (18.14) can be achieved by constructing an auxiliary two-dimensional[5] geometric problem which involves finding the coordinates α, β of a point $\mathbf{p}_0$ in terms of basis vectors $\mathbf{p}_1$ and $\mathbf{p}_2$.

The construction is effected by multiplying each of the lines in Eq. (18.14) by different primal units:

$$\begin{cases} \alpha(\mathbf{b} - \mathbf{a}) \cdot (\mathbf{s} - \mathbf{r})e_0 - \beta(\mathbf{s} - \mathbf{r}) \cdot (\mathbf{s} - \mathbf{r})e_0 = (\mathbf{r} - \mathbf{a}) \cdot (\mathbf{s} - \mathbf{r})e_0 \\ \alpha(\mathbf{b} - \mathbf{a}) \cdot (\mathbf{b} - \mathbf{a})e_1 - \beta(\mathbf{s} - \mathbf{r}) \cdot (\mathbf{b} - \mathbf{a})e_1 = (\mathbf{r} - \mathbf{a}) \cdot (\mathbf{b} - \mathbf{a})e_1 \end{cases} \tag{18.15}$$

then adding the two equations, producing the vectorial equation:

$$\alpha\, \mathbf{p}_1 - \beta\, \mathbf{p}_2 = \mathbf{p}_0 \tag{18.16}$$

where the vectors $\mathbf{p}_0$, $\mathbf{p}_1$, and $\mathbf{p}_2$ are defined as follows:

$$\begin{cases} \mathbf{p}_1 = (\mathbf{b} - \mathbf{a}) \cdot (\mathbf{s} - \mathbf{r})e_0 + (\mathbf{b} - \mathbf{a}) \cdot (\mathbf{b} - \mathbf{a})e_1 \\ \mathbf{p}_2 = (\mathbf{s} - \mathbf{r}) \cdot (\mathbf{s} - \mathbf{r})e_0 + (\mathbf{s} - \mathbf{r}) \cdot (\mathbf{b} - \mathbf{a})e_1 \\ \mathbf{p}_0 = (\mathbf{r} - \mathbf{a}) \cdot (\mathbf{s} - \mathbf{r})e_0 + (\mathbf{r} - \mathbf{a}) \cdot (\mathbf{b} - \mathbf{a})e_1 \end{cases} \tag{18.17}$$

---

4 Here written as the dot product of vectors.
5 The general n-dimensional problem is treated in Chapter 20.

Multiplying Eq. (18.16) together with first $\mathbf{p}_2$ and then $\mathbf{p}_1$ using the outer product:

$$\begin{cases} \alpha\,\mathbf{p}_1 \wedge \mathbf{p}_2 - \beta\,\mathbf{p}_2 \wedge \mathbf{p}_2 = \mathbf{p}_0 \wedge \mathbf{p}_2 \\ \alpha\,\mathbf{p}_1 \wedge \mathbf{p}_1 - \beta\,\mathbf{p}_2 \wedge \mathbf{p}_1 = \mathbf{p}_0 \wedge \mathbf{p}_1 \end{cases} \tag{18.18}$$

directly eliminates two of the terms:

$$\begin{cases} \alpha\,\mathbf{p}_1 \wedge \mathbf{p}_2 = \mathbf{p}_0 \wedge \mathbf{p}_2 \\ -\beta\,\mathbf{p}_2 \wedge \mathbf{p}_1 = \mathbf{p}_0 \wedge \mathbf{p}_1 \end{cases} \tag{18.19}$$

and the solution emerges in explicit form as the ratio of bivectors:

$$\begin{cases} \alpha = \mathbf{p}_0 \wedge \mathbf{p}_2 \,/\, \mathbf{p}_1 \wedge \mathbf{p}_2 \\ \beta = \mathbf{p}_0 \wedge \mathbf{p}_1 \,/\, \mathbf{p}_1 \wedge \mathbf{p}_2 \end{cases} \tag{18.20}$$

These particular bivectors are both directly proportional to the pseudo-scalar for the two-dimensional auxiliary problem, so the division makes sense and is straightforward, with no need to resort to using the inner product as in Eq. (18.8).

18.2 Practice

18.2.1 Example Code

The code listed in Example 18.1a implements Eq. (18.8) to perform the intersection of the two line segments in arrays 'a[2]' and 'r[2]'. Lines 6–8 calculate the differences between the specified points, and lines 10 and 11 implement the left-hand members of the inner products in Eq. (18.8). Lines 12 and 13 implement the inner products themselves and line 14 calculates the ratio, which is then passed back to the calling routine.

Example 18.1(a)

```
1:    number intersect_lines(a,r)
2:         number a[2], r[2];
3:    {
4:      number ra,ba,sr,alpha,denom;
5:
6:      ra=sub_number(r[0],a[0],0);
7:      sr=sub_number(r[1],r[0],0);
8:      ba=sub_number(a[1],a[0],0);
9:
10:     alpha=outer_mul(ra,sr,1);
11:     denom=outer_mul(ba,sr,3);
12:     alpha=inner_mul(alpha,denom,1);
13:     denom=inner_mul(denom,denom,3);
14:     alpha=central_mul(alpha,inverseof_component(denom,1),3);
15:     return(alpha); /* distance of x along line a */
16:   }
```

Calculation of the point of intersection as a vector is achieved by the code listed in Example 18.1b, which implements the top line of Eq. (18.2).

Example 18.1(b)

```
1:   number pointon_line(a,alpha)
2:        number a[2];
3:        number alpha;
4:   {
5:     number x;
6:
7:     x=sub_number(scalar_number(1.0,0.0),alpha,1);    /* 1-alpha */
8:     x=add_number(central_mul(x,a[0],1),central_mul(alpha,a[1],0),3);
9:     return(x);
10:  }
```

The code listed in Example 18.1c calculates the shortest distance between two lines using Eqs. (18.17) and (18.20). Lines 7–9 are the same as in Example 18.1a. Lines 11–18 construct the auxiliary problem, and lines 23–25 form the outer products and take their ratio.

Example 18.1(c)

```
1:   number connect_lines(a,r)
2:        number a[2], r[2];
3:   {
4:     number ra,ba,sr,alpha,denom;
5:     number e0,e1,p0,p1,p2;
6:
7:     ra=sub_number(r[0],a[0],0);
8:     sr=sub_number(r[1],r[0],0);
9:     ba=sub_number(a[1],a[0],0);
10:
11:    e0=primal_vector(1.0,0.0,0);
12:    e1=primal_vector(1.0,0.0,1);
13:    p0=add_number(central_mul(inner_mul(ra,sr,0),e0,1),
14:                  central_mul(inner_mul(ra,ba,0),e1,1),3);
15:    p1=add_number(central_mul(inner_mul(ba,sr,0),e0,1),
16:                  central_mul(inner_mul(ba,ba,0),e1,1),3);
17:    p2=add_number(central_mul(inner_mul(sr,sr,0),e0,3),
18:                  central_mul(inner_mul(sr,ba,0),e1,3),3);
19:    recycle_number(ra);
20:    recycle_number(sr);
21:    recycle_number(ba);
22:
23:    alpha=outer_mul(p0,p2,1);
24:    denom=outer_mul(p1,p2,3);
25:    alpha=central_mul(alpha,inverseof_component(denom,1),3);
26:
27:    return(alpha); /* distance of x along line a */
28:  }
```

Note that implicit recycling is not used within the invocations to `inner_mul()` in lines 13–18. Recycling is employed explicitly on lines 19–21 instead. Although it would appear possible to recycle vector 'ra' on line 14 where it is last used as an argument to `inner_mul()`, that is dangerous. The problem is that the 'c' programming language (King, 2008) does not stipulate the order in which the parameters to a sub-routine are evaluated. The first argument to the invocation of `add_number()` on line 13 is not necessarily evaluated before the second argument on line 14. It is important that the vector 'ra' is not recycled where it is used in one of these arguments before being accessed in the other. Since the order is unknown, it is not possible to recycle in either.

As a practical example, the code listed in Example 18.1d demonstrates the intersection of the lines as shown in Figure 18.2 as well as calculation of the shortest line segment connecting two lines, as in Figure 18.5. The coordinates for the endpoints of the lines are established to their initial values in lines 4–7. All four of these points occupy the same plane $z = x - y$.

Example 18.1(d)

```
 1:    void application5(argc,argv) /* intersection of lines */
 2:          int argc; char **argv;
 3:    {
 4:        double point_a[3]={10.0,35.0,-25.0};
 5:        double point_b[3]={40.0,5.0,35.0};
 6:        double point_r[3]={53.333,36.667,16.666};
 7:        double point_s[3]={13.333,16.666,-3.333};
 8:        number a[2],r[2];
 9:        number alpha,beta,u,v,x,offset;
10:        int lambda[3];
11:
12:        recovery_level(0,0,0);
13:        create_cliffmem(50);
14:        declare_primal(3,pack_signature(3,lambda,-1));
15:        format_printing("%.3f#r");
16:
17:        a[0]=pack_R3vector(point_a);  a[1]=pack_R3vector(point_b);
18:        r[0]=pack_R3vector(point_r);  r[1]=pack_R3vector(point_s);
19:
20:        alpha=intersect_lines(a,r);
21:        x=pointon_line(a,alpha);
22:
23:        printf("fractional distance of intersection x from point a
                   to point b is: ");
24:        print_number(alpha,8);
25:        printf("coordinates of intersection are: ");
26:        print_vector(x);
27:        printf("\n");
28:        recycle_number(alpha);
29:        recycle_number(x);
30:
31:        offset=primal_vector(5.0,0.0,2);  /* 5z */
32:        printf("add +/- offset to lines: ");
```

```
33:     print_vector(offset);
34:
35:     a[0]=add_number(a[0],offset,1);   a[1]=add_number(a[1],offset,1);
36:     r[0]=sub_number(r[0],offset,1);   r[1]=sub_number(r[1],offset,3);
37:
38:     alpha=connect_lines(a,r);
39:     u=pointon_line(a,alpha);
40:     beta=connect_lines(r,a);
41:     v=pointon_line(r,beta);
42:
43:     printf("fractional distance of point u from point a
                    to point b is: ");
44:     print_number(alpha,8);
45:     printf("coordinates of point are: ");
46:     print_vector(u);
47:     printf("fractional distance of point v from point r
                    to point s is: ");
48:     print_number(beta,8);
49:     printf("coordinates of point are: ");
50:     print_vector(v);
51:     printf("\n");
52:
53:     recycle_number(alpha);   recycle_number(beta);
54:     recycle_number(u);       recycle_number(v);
55:     recycle_array(2,a);      recycle_array(2,r);
56:
57:     free_cliffmem(1);
58:  }
```

A three-dimensional Clifford context with negative signature is established in lines 12–15. The four points are packed into two arrays on lines 17 and 18. The intersection is invoked at line 20, and the coordinates of the intersection are calculated on line 21. The results are printed in lines 23–27. Memory for some of the variables in use is recycled on lines 28 and 29.

The lines are moved away from their common plane by offsetting one by $+5z$ and the other by $-5z$ on lines 35 and 36. The two closest points are then calculated in lines 38–41, and the results are printed in lines 43–51. Memory for variables remaining in use is recycled in lines 53–55. Finally on line 57, the bulk memory allocated to the Clifford context is abandoned.

18.2.2 Example Output

The results of running the code in Example 18.1 are listed in Example 18.2. The value of the parameter α is printed on line 2, and the coordinates of the point of intersection is printed on line 3.

After offsetting the two lines, the parameters α and β and their corresponding points **u** and **v** on each line are printed in lines 6–9.

Example 18.2

```
 1:  application5()
 2:  fractional distance of intersection x from point a
        to point b is: (0.444)e{}
 3:  coordinates of intersection are: [23.333, 21.667, 1.667]
 4:
 5:  add +/- offset to lines: [0.000, 0.000, 5.000]
 6:  fractional distance of point u from point a
        to point b is: (0.333)e{}
 7:  coordinates of point are: [20.000, 25.000, 0.000]
 8:  fractional distance of point v from point r
        to point s is: (0.750)e{}
 9:  coordinates of point are: [23.333, 21.667, -3.333]
10:
11:  statistics for structural heap:
12:  initial size (50)
13:  peak size    (19), as % of initial size (38%)
14:  full size    (50), expansion count (0)
15:  free count   (50), free start (6)
16:  statistics for data heap:
17:  initial size (50)
18:  peak size    (33), as % of initial size (66%)
19:  full size    (50), expansion count (0)
20:  free count   (50), free start (11)
```

Reference

King, K. N. (2008), *C Programming: A Modern Approach*, 2nd edn, W. W. Norton, New York, US-NY.

19

Perspective Projection

Often, illustrations of simple three-dimensional objects such as a cube or tetrahedron can be drawn in projection onto two dimensions by eye, either without attempting to incorporate the effects of perspective, or including those effects in an artistic manner, not resorting to the calculation of individual coordinates. As the objects become more complicated, it becomes harder to make them appear visually correct. Perspective projection is used here as a simple example of the application of the Clifford numerical suite for performing a geometric operation.

19.1 Theory

Perspective projection can be interpreted as the shadows of objects cast by the sun onto a plane. In Figure 19.1, the sun stands vertically above the point P in the horizontal plane $x_3 y_3$ (small rectangle). Rays from the sun pass through points x_3, y_3, z_3 in three-dimensional space casting shadows on the tilted shadow-plane $x_2 y_2$ (large rectangle). The shadow-plane is tilted so that its normal n is pointing towards the sun. The orientation of the axes x_2, y_2 within the shadow-plane is determined by specifying the upward direction 'zenith'. In the figure, the zenith is in the direction of the vertical axis z_3. This usually makes sense, although it is also possible to choose a different direction. The y_2 axis in the shadow-plane follows the direction of the shadow of the zenith. The x_2 axis in the shadow-plane is directed perpendicular to both the y_2 axis and the normal n, so as to form a right-handed system x_2, y_2, n.

19.2 Practice

19.2.1 Example Code

Perspective projection is achieved within the Clifford numerical suite in two steps. The first step, which only need be done once, is to determine the basis vectors x_2, y_2 for the shadow-plane from the vectors towards the sun and the zenith. The code listed in Example 19.1a does those calculations on lines 8 and 9, saving the basis vectors and the normal as an array of numbers in lines 10–12. In order to calculate x_2 from y_2 and n, the

Numerical Calculations in Clifford Algebra: A Practical Guide for Engineers and Scientists,
First Edition. Andrew Seagar.
© 2023 John Wiley & Sons Ltd. Published 2023 by John Wiley & Sons Ltd.

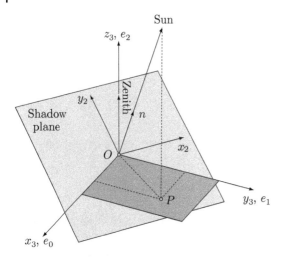

Figure 19.1 Shadow-plane $x_2 y_2$, onto which points x_3, y_3, z_3 in space are projected by rays from the sun.

approach demonstrated by Grassmann (1877) with outer multiplication and supplement is used instead of using the cross product from vector calculus.

Example 19.1(a)

```
 1:   number *orient(sun,zenith,plane)  /* set orientation of
                                            projection plane */
 2:       number sun,zenith;             /* as normal to the sun */
 3:       number plane[3];               /* y-axis is in direction of
                                            shadow of the zenith */

 4:   {
 5:       number n,x2,y2;
 6:
 7:       n=normalise_vector(sun,0);
 8:       y2=normalise_vector(outerspace_vector(zenith,n,0),1);
 9:       x2=supplementof_number(outer_mul(y2,n,0),1);
10:       plane[0]=x2;
11:       plane[1]=y2;
12:       plane[2]=n;
13:       return(plane);
14:   }
```

The second step is to cast shadows of any chosen points in x_3, y_3, z_3 space onto the shadow-plane. The code listed in Example 19.1b achieves this using primarily the `innerspace_vector()` and `outerspace_vector()` vector splitting operators, which on lines 11 and 12 separate the vector $r = xe_0 + ye_1 + ze_2$ into components z_0 (as variable 'z0') towards the sun, and r_0 (as variable 'r0') in the shadow-plane.

Perspective is introduced in lines 13 and 14 by scaling the length of 'r0' according to the distance to the sun. The perspective scale factor is in principle a scalar (in grade 0), but in the program variable of that name, there are some non-zero components in other grades, left there by imprecise numerical cancellation to machine precision. It is possible to filter them out, although they have very little effect on the final result. The use of outer multiplication rather than central multiplication on line 14 filters some of them out. The sub-routine `onegradeof_number()` can be used to clean out all such cancellation errors when that is important (not in this application).

After applying the perspective, the resulting vector is split (again using the `innerspace _vector()` operator) into components along the x_2 and y_2 axes (as variables 'x2' and 'y2'), from which the signed magnitudes are extracted by inner multiplication, all in lines 15 and 16.

Example 19.1(b)

```
 1:   number *shadow(r,sun,plane,xy)  /* cast shadow of point on plane */
 2:          number r,sun;
 3:          number plane[3],xy[2];
 4:   {
 5:     number n,x2,y2;
 6:     number z0,r0,perspective;
 7:
 8:     x2=plane[0];
 9:     y2=plane[1];
10:     n=plane[2];
11:     z0=innerspace_vector(r,n,0);  /* component of r along normal */
12:     r0=outerspace_vector(r,n,0);  /* component of r at right angle
                                         to normal */
13:     perspective=central_mul(sun,inverseof_vector
                                 (sub_number(z0,sun,1),1),2);
14:     r0=outer_mul(r0,perspective,3);
15:     xy[0]=inner_mul(innerspace_vector(r0,x2,0),x2,1);
16:     xy[1]=inner_mul(innerspace_vector(r0,y2,1),y2,1);
17:     return(xy);
18:   }
```

A concrete example applying perspective projection to the 12 faces of a dodecahedron is listed in the code of Examples 19.2a–19.2c. The main routine in Example 19.2a establishes a three-dimensional context within the Clifford numerical suite, with negative signature, on lines 6–9.

Example 19.2(a)

```
 1:   void application6(argc,argv)  /* perspective drawing */
 2:          int argc; char **argv;
 3:   {
 4:     int lambda[3];
 5:
 6:     recovery_level(0,0,0);
 7:     create_cliffmem(50);
 8:     declare_primal(3,pack_signature(3,lambda,-1));
 9:     format_printing("%.3f#r");
10:
11:     plot_dodecahedron();
12:
13:     free_cliffmem(1);
14:   }
```

The sub-routine `plot_dodecahedron()` in Example 19.2b calculates the orientation of the projection plane on line 32 from the chosen values of zenith and sun on lines 31 and 30 and passes the coordinates of the corners of the faces to the sub-routine `project_tile()` on line 36. Six corners are used to describe each face rather than five, with the first and last being the same. Adjacent pairs of points can then be taken to represent each of the five edges when plotting.

Example 19.2(b)

```
 1:   /* constants for dodecahedron coordinates */
 2:   #define A -0.5*(sqrt(5.0)-1.0)
 3:   #define B 0.5*(sqrt(5.0)-1.0)
 4:   #define C 0.5*(3.0-sqrt(5.0))
 5:   #define D -0.5*(3.0-sqrt(5.0))
 6:   #define E 1.0
 7:   #define F -1.0
 8:   /*-----------------------------------------------------*/
 9:   void plot_dodecahedron(void)
10:   {
11:      double faces[12][6][3]=
12:         {{{C,0,E},{D,0,E},{A,B,B},{0,E,C},{B,B,B},{C,0,E}},
13:          {{D,0,E},{C,0,E},{B,A,B},{0,F,C},{A,A,B},{D,0,E}},
14:          {{C,0,F},{D,0,F},{A,A,A},{0,F,D},{B,A,A},{C,0,F}},
15:          {{D,0,F},{C,0,F},{B,B,A},{0,E,D},{A,B,A},{D,0,F}},
16:          {{0,E,D},{0,E,C},{B,B,B},{E,C,0},{B,B,A},{0,E,D}},
17:          {{0,E,C},{0,E,D},{A,B,A},{F,C,0},{A,B,B},{0,E,C}},
18:          {{0,F,D},{0,F,C},{A,A,B},{F,D,0},{A,A,A},{0,F,D}},
19:          {{0,F,C},{0,F,D},{B,A,A},{E,D,0},{B,A,B},{0,F,C}},
20:          {{E,C,0},{E,D,0},{B,A,B},{C,0,E},{B,B,B},{E,C,0}},
21:          {{E,D,0},{E,C,0},{B,B,A},{C,0,F},{B,A,A},{E,D,0}},
22:          {{F,C,0},{F,D,0},{A,A,A},{D,0,F},{A,B,A},{F,C,0}},
23:          {{F,D,0},{F,C,0},{A,B,B},{D,0,E},{A,A,B},{F,D,0}}};
24:
25:      double a[3];
26:      number plane[3];
27:      number sun,zenith;
28:      int i;
29:
30:      sun=pack_R3vector(pack_R3array(30.0,60.0,80.0,a));
            /* sun is here    */
31:      zenith=pack_R3vector(pack_R3array(0.0,0.0,1.0,a));
            /* z=0,0,1 is up */
32:      orient(sun,zenith,plane);
            /* establish projection plane */
33:
34:      for(i=0;i<12;i++)
35:         {
36:            project_tile(6,faces[i],sun,plane);
37:         }
38:      recycle_number(sun);
39:      recycle_number(zenith);
40:      recycle_array(3,plane);
41:   }
```

The coordinates themselves are calculated by projection using the sub-routine `shadow()` at line 14 and 24 within sub-routine `project_tile()` in Example 19.2c. The first pair of coordinates is calculated separately, giving the option (by deleting lines 19 and 20) to express the remaining coordinates as distances from the first. That may be of benefit for some plotting utilities.

The coordinates in the shadow-plane are extracted and scaled in lines 16 and 17 and 26 and 27 and then printed on lines 18 and 28.

Example 19.2(c)

```
 1:    void project_tile(n,t,sun,plane)  /* project 3D coordinates
                                             onto 2D shadow plane */
 2:         int n;
 3:         double (*t)[3];
 4:         number sun;
 5:         number *plane;
 6:    {
 7:       number xy[2];
 8:       number r;
 9:       double x0,y0,x,y,s;
10:       int i;
11:
12:       s=25.0; /* scale factor */
13:       r=pack_R3vector(t[0]);
14:       shadow(r,sun,plane,xy);
15:       recycle_number(r);
16:       x0=s*get_real(xy[0],0,1);
17:       y0=s*get_real(xy[1],0,1);
18:       printf("%.4f %.4f\n",x0,y0);
19:       x0=0.0;   /* use lines 19,20 for absolute coordinates */
20:       y0=0.0;   /* delete lines 19,20 for coordinates relative
                        to first */
21:       for(i=1;i<n;i++)
22:          {
23:             r=pack_R3vector(t[i]);
24:             shadow(r,sun,plane,xy);
25:             recycle_number(r);
26:             x=s*get_real(xy[0],0,1);
27:             y=s*get_real(xy[1],0,1);
28:             printf("%.4f %.4f\n",x-x0,y-y0);
29:          }
30:       printf("\n");
31:    }
```

19.2.2 Example Output

Figure 19.2 shows the result of plotting the coordinates produced by running the code listed in Examples 19.1 and 19.2. On the left side of the figure, all 12 faces are plotted as transparent. On the right, the front faces are plotted as opaque. There is no automatic hidden line removal here, so it is done by manually selecting only the front faces.

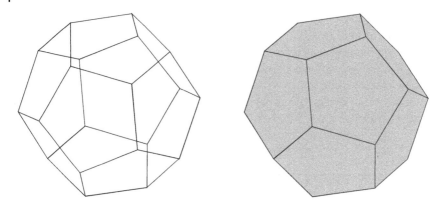

Figure 19.2 Dodecahedron with perspective projection.

Reference

Grassmann, H. (1877), 'Der ort der Hamilton'schen quaternionen in der Ausdehnungslehre', *Mathematische Annalen* **12**, 375–386.

20

Linear Systems

Linear systems are often solved by matrix methods. Grassmann showed how to do it using the outer product. The method is not especially efficient, being much the same as Cramer's method[1].

20.1 Theory

In his book, Grassmann (1995) takes a system of linear equations as in Eq. (20.1), with unknown values x_k and known scalar[2] coefficients a_{jk}, and multiplies each row by a different primal unit:

$$\begin{cases} a_{11}x_1 + a_{12}x_2 + \cdots + a_{1n}x_n = a_{10} \\ a_{21}x_1 + a_{22}x_2 + \cdots + a_{2n}x_n = a_{20} \\ \cdots + \cdots + \cdots + \cdots = \cdots \\ a_{n1}x_1 + a_{n2}x_2 + \cdots + a_{nn}x_n = a_{n0} \end{cases} \quad (20.1)$$

That produces the system (20.2) in which each column has a common factor x_k but different coefficients and primal units:

$$\begin{cases} a_{11}x_1 e_0 + a_{12}x_2 e_0 + \cdots + a_{1n}x_n e_0 = a_{10}e_0 \\ a_{21}x_1 e_1 + a_{22}x_2 e_1 + \cdots + a_{2n}x_n e_1 = a_{20}e_1 \\ \cdots + \cdots + \cdots + \cdots = \cdots \\ a_{n1}x_1 e_{n-1} + a_{n2}x_2 e_{n-1} + \cdots + a_{nn}x_n e_{n-1} = a_{n0}e_{n-1} \end{cases} \quad (20.2)$$

Leaving aside the common factors, Grassmann defined vectors p_k from each column as in Eq. (20.3):

$$\begin{cases} p_1 = a_{11}e_0 + a_{21}e_1 + \cdots + a_{n1}e_{n-1} \\ p_2 = a_{12}e_0 + a_{22}e_1 + \cdots + a_{n2}e_{n-1} \\ \cdots = \cdots + \cdots + \cdots + \cdots \\ p_n = a_{1n}e_0 + a_{2n}e_1 + \cdots + a_{nn}e_{n-1} \\ p_0 = a_{10}e_0 + a_{20}e_1 + \cdots + a_{n0}e_{n-1} = q_0 \end{cases} \quad (20.3)$$

---

1 Published earlier by MacLaurin (Hedman 1999).
2 For linear systems in which the coefficients are non-scalar Clifford numbers, see Chapters 13 and 25.

Numerical Calculations in Clifford Algebra: A Practical Guide for Engineers and Scientists,
First Edition. Andrew Seagar.
© 2023 John Wiley & Sons Ltd. Published 2023 by John Wiley & Sons Ltd.

Then he rewrote the original linear system (20.1) as a linear combination of the vectors[3]:

$$p_1 x_1 + p_2 x_2 + \cdots + p_n x_n = q_0 \tag{20.4}$$

Grassmann developed the solution for the first of the unknowns by outer multiplying both sides with the vectors associated with all unknowns except the first, as on the first line of Eq. (20.5). Within the parentheses on the left-hand side, every vector meets itself from outside the parentheses, except for the first one, which does not appear outside the parentheses. Outer multiplying a vector by itself produces zero, as in Figure 5.2. The only term inside the parentheses that survives the multiplication is the first, and the only unknown that survives is also the first, as in the second line of Eq. (20.5). With a single multiplication, Grassmann has eliminated all but one of the unknowns:

$$
\begin{aligned}
(p_1 x_1 + p_2 x_2 + \cdots + p_n x_n) \wedge p_2 \wedge p_3 \ldots \wedge p_n &= q_0 \wedge p_2 \wedge p_3 \ldots \wedge p_n \\
x_1 \, (p_1 \wedge p_2 \wedge p_3 \ldots \wedge p_n) &= q_0 \wedge p_2 \wedge p_3 \ldots \wedge p_n \\
x_1 = (q_0 \wedge p_2 \wedge p_3 \ldots \wedge p_n) \, / \, (p_1 \wedge p_2 \wedge p_3 \ldots \wedge p_n) &
\end{aligned}
\tag{20.5}
$$

In the second line of Eq. (20.5), the term in parentheses has a coefficient which takes the value of the determinant of the linear system and has a unit which is the unit of the pseudo-scalar (element of volume) for n dimensions. The right-hand side has the same unit but a different coefficient.

The solution is completed in the third line multiplying both sides by the inverse of the unit pseudo-scalar for n dimensions and the inverse of the determinant (i.e. dividing by the contents of the parentheses on the left-hand side). Pseudo-scalars have inverses as long as the signature contains no zeros. Determinants also have inverses only when non-zero.

It is perhaps somewhat surprising when the linear system is written in the form of vectors as in Eq. (20.4) that there are effectively only two steps in order to obtain the solution for one of the variables. The solution for other variables can be found in the same manner, leaving the vector associated with that variable, now x_k in Eq. (20.6), absent from the multiplication in the first line:

$$
\begin{aligned}
(p_1 x_1 + p_2 x_2 + \cdots + p_n x_n) \wedge p_1 \wedge p_2 \cdots \wedge p_{k-1} \wedge p_{k+1} \cdots \wedge p_n &= q_0 \wedge p_1 \wedge p_2 \cdots \wedge p_{k-1} \wedge p_{k+1} \cdots \wedge p_n \\
x_k \, (p_k \wedge p_1 \wedge p_2 \cdots \wedge p_{k-1} \wedge p_{k+1} \cdots \wedge p_n) &= q_0 \wedge p_1 \wedge p_2 \cdots \wedge p_{k-1} \wedge p_{k+1} \cdots \wedge p_n \\
x_k = (q_0 \wedge p_1 \wedge p_2 \cdots \wedge p_{k-1} \wedge p_{k+1} \cdots \wedge p_n) \, / \, (p_k \wedge p_1 \wedge p_2 \cdots \wedge p_{k-1} \wedge p_{k+1} \cdots \wedge p_n) & \\
= (p_1 \wedge p_2 \cdots \wedge p_{k-1} \wedge q_0 \wedge p_{k+1} \cdots \wedge p_n) \, / \, (p_1 \wedge p_2 \cdots \wedge p_{k-1} \wedge p_k \wedge p_{k+1} \cdots \wedge p_n) &
\end{aligned}
$$

$$\tag{20.6}$$

Reordering the factors in the numerator and denominator as in the final line either has no effect or changes the sign of both, leaving in either case the overall result unchanged. In the final form, the denominator is the determinant, and the numerator differs from that by substituting the known data vector q_0 in place of the factor p_k.

The solution of a linear system of equations is often interpreted geometrically as the point at the intersection of a number of (hyper) planes. The solution as expressed in Eq. (20.6) is interpreted geometrically as the ratio of pairs of (hyper) volumes.

---

3 Grassmann's p_0 is relabelled here as q_0 to distinguish the data vector clearly from the others.

20.2 Practice

Equation (20.7) serves as a test case for solving a linear system using the method described in Section 20.1:

$$
\begin{pmatrix}
1 & 2 & 4 & -1 \\
3 & -1 & -1 & 2 \\
-4 & 1 & 3 & 0 \\
1 & 2 & 3 & 4
\end{pmatrix}
\begin{pmatrix}
x_1 \\
x_2 \\
x_3 \\
x_4
\end{pmatrix}
=
\begin{pmatrix}
-4 \\
5 \\
-8 \\
3
\end{pmatrix}
\tag{20.7}
$$

20.2.1 Example Code

The code listed in Examples 20.1a and 20.1b demonstrates the use of Grassmann's method of outer multiplication for the solution of the 4×4 linear system in Eq. (20.7). The sub-routine `lin_solve()` in Example 20.1a implements Grassmann's method as in Eq. (20.6) to solve the $n \times n$ linear system 'pn' with data vector 'qn[0]'.

Example 20.1(a)

```
1:    number lin_solve(n,pn,qn)
2:        int n;
3:        number *pn,*qn;
4:    {
5:      number d,c;
6:      double m;
7:      int i;
8:
9:      d=cloneof_number(pn[0]);
10:     for(i=1;i<n;i++)                    /* on left of data vector */
11:       {
12:         qn[i]=outer_mul(d,qn[0],0);
13:         d=outer_mul(d,pn[i],1);
14:       }
15:     m=get_real(magnitudeof_number(d,0),0,1);
16:     if(m==0.0)
17:       {
18:         printf("attempt to invert matrix with determinant (%.5e) ",m);
19:         printf("in routine: solve()\n");
20:         brexit(0);
21:       }
22:     c=inverseof_component(d,1); /* inverse of determinant */
23:     qn[n-1]=central_mul(qn[n-1],c,1);
24:     d=cloneof_number(pn[n-1]);
25:     for(i=n-2;i>=0;i--)               /* on right of data vector */
26:       {
27:         qn[i]=central_mul(outer_mul(qn[i],d,1),c,1);
28:         d=outer_mul(pn[i],d,2);
29:       }
30:     recycle_number(c);
31:     return(d);
32:   }
```

The loop from lines 10 to 14 handles the multiplication on the left- hand side of the data vector q_0 as variable 'qn[0]', and the loop from lines 25 to 29 handles all of the multiplication on the right-hand side. The determinant is calculated twice, both on line 13 and 28 as a by-product of the two loops. Its inverse under central multiplication is calculated on line 22 and used in the second loop.

The code in Example 20.1b runs the linear system test case. The coefficients of the linear system are first defined in lines 5–8, and the vector of data values on the right-hand side is defined on line 9. Lines 14–17 establish a four-dimensional context for Clifford numbers with all negative signatures. The four vectors p_1 to p_4 are generated in lines 20–23 as in Eq. (20.3) (as variables 'p[0]' to 'p[3]' with the indices decremented by one to conform with the conventions of the 'c' programming language), and the vector of data values is constructed on line 24. Solution is effected by line 26. The values of the determinant and the solution are printed on lines 28–32.

Example 20.1(b)

```
 1:   void application7(argc,argv)  /* solution of 4x4 linear system */
 2:        int argc; char**argv;
 3:   {
 4:       int lambda[4];
 5:       double a[4][4]={{ 1.0,  2.0,  4.0,-1.0},
 6:                       { 3.0,-1.0,-1.0,  2.0},
 7:                       {-4.0,  1.0,  3.0,  0.0},
 8:                       { 1.0,  2.0,  3.0,  4.0}};
 9:       double b[4]={-4.0,5.0,-8.0,3.0};  /* data vector */
10:       number p[4];                      /* vectors */
11:       number q[4];                      /* data->unknowns */
12:       number d;                         /* determinant */
13:
14:       recovery_level(0,0,0);
15:       create_cliffmem(50);
16:       declare_primal(4,pack_signature(4,lambda,-1));
17:       format_printing("%.5f");
18:
19:       transpose(4,a[0]);    /* convert from row to column major */
20:       p[0]=pack_RNvector(4,a[0]);             /* set up data */
21:       p[1]=pack_RNvector(4,a[1]);
22:       p[2]=pack_RNvector(4,a[2]);
23:       p[3]=pack_RNvector(4,a[3]);
24:       q[0]=pack_RNvector(4,b);
25:
26:       d=lin_solve(4,p,q);                    /* invert matrix */
27:
28:       printf("determinant = "); print_number(d,8);
29:       printf("q0 = ");  print_number(q[0],8);
30:       printf("q1 = ");  print_number(q[1],8);
31:       printf("q2 = ");  print_number(q[2],8);
32:       printf("q3 = ");  print_number(q[3],8);
33:
34:       recycle_array(4,q);
35:       recycle_array(4,p);
36:       recycle_number(d);
37:
38:       free_cliffmem(0);
39:   }
```

20.2.2 Example Output

The results from running the code in Examples 20.1a and 20.1b are listed in Example 20.1c. The determinant is printed on line 2. The values of the unknowns are printed on lines 3–6. Substitution of the values into Eq. (20.7) verifies the solution as correct.

Example 20.1(c)

```
1:   application7()
2:   determinant = (-97.00000+i0.00000)e{0,1,2,3}
3:   q0 = (1.00000+i0.00000)e{}
4:   q1 = (2.00000+i0.00000)e{}
5:   q2 = (-2.00000+i0.00000)e{}
6:   q3 = (1.00000+i0.00000)e{}
```

References

Grassmann, H. (1995), The extension theory of 1844, *in* 'A New Branch of Mathematics: The Ausdehnungslehre of 1844, and Other Works', Open Court, Chicago, US-IL, pp. 2–295. (Translated by Lloyd C. Kannenberg).

Hedman, B. A. (1999), 'An earlier date for "Cramer's rule', *Historica Mathematica* **26**(4), 365–368.

21

Fast Fourier Transform

Standard fast Fourier transform (FFT) algorithms support at best complex signals (functions of time) or waveforms (functions of space). In Chapter 23, it is of interest to use the FFT to calculate derivatives of an electromagnetic field in order to simulate wave propagation using the finite difference time domain (FDTD) method. For that application, the signal is a Clifford bivector in four dimensions (both space and time), making it necessary to construct a Fourier transform which operates at least on bivectors, and preferably on all other Clifford entities as well.

21.1 Theory

In general, the spectrum $S(\omega)$ is calculated from a signal $s(t)$ by the Fourier transform as follows:

$$S(\omega) = \int s(t) e^{-i\omega t} \, dt \tag{21.1}$$

When the signal $s(t)$ is bandlimited and periodic, the same spectral values are obtained[1] if the original signal is replaced by a new function:

$$s'(t) = \frac{1}{N} \sum_{n=0}^{N-1} s_n \, \delta(t - n\Delta t) \tag{21.2}$$

consisting of N samples $s_n = \int s(t) \, \delta(t - n\Delta t) \, dt$ covering one cycle of the signal, equally spaced by Δt in time, provided the value of N is large enough. Here the function δ is the Dirac delta function, and N is restricted to a power of 2.

The situation for $N = 8$ samples is shown in Figure 21.1. In this case, the maximum frequency which can be sampled without error is shown by the short sinewave, and the minimum non-zero frequency is shown by the long sinewave.

Table 21.1 gives expressions for the limits on the frequencies and periods which can be properly reconstructed for a chosen number of samples N spaced at interval Δt. Angular frequencies ω are supported over the range $-\frac{\pi}{\Delta t} \le \omega \le \frac{\pi}{\Delta t}$, separated by $\Delta \omega = \frac{2\pi}{N\Delta t}$.

---

1 This is not obvious.

Numerical Calculations in Clifford Algebra: A Practical Guide for Engineers and Scientists, First Edition. Andrew Seagar.
© 2023 John Wiley & Sons Ltd. Published 2023 by John Wiley & Sons Ltd.

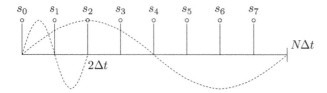

Figure 21.1 Sampled periodic signal.

Table 21.1 Limits on frequencies and periods for signal in terms of sampling parameters.

minimum sinusoidal period	$2\Delta t$	maximum sinusoidal period	$N\Delta t$
maximum frequency	$f = \frac{1}{2\Delta t}$	minimum frequency	$\Delta f = \frac{1}{N\Delta t}$
maximum angular frequency	$\omega = 2\pi f = \frac{\pi}{\Delta t}$	minimum angular frequency	$\Delta\omega = 2\pi\Delta f = \frac{2\pi}{N\Delta t}$

Figure 21.2 Periodic spectrum of periodic signal.

Values of the discrete spectral lines, $S(\omega)$, where $\omega = m\Delta\omega$, are calculated from the samples as follows:

$$S(\omega) = \int s'(t)e^{-i\omega t}\,dt = \int \frac{1}{N}\sum_{n=0}^{N-1} s_n\,\delta(t - n\Delta t)e^{-i\omega t}\,dt = \frac{1}{N}\sum_{n=0}^{N-1} s_n\,e^{-i\omega n\Delta t} \tag{21.3}$$

Spectral values are cyclic with a period of twice the maximum frequency:

$$S(\omega) = S(\omega + \tfrac{2\pi}{\Delta t})$$

$$e^{-i\omega t} = e^{-i(\omega + \frac{2\pi}{\Delta t})t} = e^{-i\omega t}e^{-i\frac{2\pi}{\Delta t}t} = e^{-i\omega t}e^{-i\frac{2\pi}{\Delta t}n\Delta t} = e^{-i\omega t}e^{-i2\pi n} = e^{-i\omega t}\times 1 \tag{21.4}$$

so that $S(-\omega) = S(\frac{2\pi}{\Delta t} - \omega)$, as shown in Figure 21.2.

Rather than calculating the negative frequency spectral components $-\frac{\pi}{\Delta t} \le \omega < 0$ directly it is often convenient to calculate the duplicates found at $\frac{\pi}{\Delta t} \le \omega < \frac{2\pi}{\Delta t}$.

21.2 Practice

21.2.1 Example Code

The code shown in Example 21.1a implements an FFT according to Eq. (21.3), using a recursive algorithm with decimation in time, for an array of samples of any kinds of Clifford numbers. The spectrum is calculated for the signal in array 'a [n]'. The result is returned in array 'b [n]'. Line 31 implements the negative complex exponential for the forward transform[2].

---

2 Following Cooley and Tukey (1965).

The results are stored in order *ABCD*: positive frequencies *A* to *B* from low to high as in Figure 21.2, followed by negative frequencies *C* to *D* from largest negative to smallest negative as the 'copy of negative frequency' on the right-hand side in Figure 21.2.

Example 21.1(a)

```
 1:   number *transform(n,a,b)  /* fast Fourier transform of array
                                       of numbers */
 2:        int n;                   /* input array a[n] is recycled */
 3:        number *a,*b;            /* length n must be power of 2  */
 4:   {
 5:     cliff_comp exp_omega,half;
 6:     double theta;
 7:     number x;
 8:     int k,n2;
 9:
10:     if(not_powerof_two(n)) /* not power of two */
11:        {
12:          printf("length of array (%d) not a power of two in
                      routine: transform()\n",n);
13:          brexit(0); /* user restart */
14:        }
15:     pack_comp(&half,0.5,0.0,0);
16:     n2=n/2;
17:     if(n>2) /* divide and conquer */
18:        {
19:          for(k=0;k<n2;k++)
20:             {
21:               b[k]=mul_comp(a[2*k],central,&half);
                       /* cosine weighted data */
22:               b[k+n2]=mul_comp(a[2*k+1],central,&half);
                       /* sine weighted data */
23:               recycle_number(a[2*k]);
24:               recycle_number(a[2*k+1]);
25:             }
26:          transform(n2,b,a);     /* destroys b, overwrites a */
27:          transform(n2,&b[n2],&a[n2]);
28:          for(k=0;k<n2;k++)
29:             {
30:               theta=2.0*M_PI*k/n; /* one sample = 2pi/n */
31:               pack_comp(&exp_omega,cos(theta),-sin(theta),0);
32:               b[k]=add_number(a[k],mul_comp(a[n2+k],
                       central,&exp_omega),2);
33:             }
34:          for(k=n2;k<n;k++)
35:             {
36:               theta=2.0*M_PI*k/n; /* one sample = 2pi/n */
37:               pack_comp(&exp_omega,cos(theta),-sin(theta),0);
38:               b[k]=add_number(a[k-n2],mul_comp(x=a[k],
                       central,&exp_omega),3);
39:               recycle_number(x);
40:             }
41:        }
```

(Continued)

Example 21.1(a) (Continued)

```
42:     else /* n=2 */
43:        {
44:           b[0]=mul_comp(x=add_number(a[0],a[1],0),central,&half);
45:           recycle_number(x);
46:           b[1]=mul_comp(x=sub_number(a[0],a[1],3),central,&half);
47:           recycle_number(x);
48:        }
49:     return(b);
50:  }
```

The code listed in Example 21.1b shows the inverse transform. The signal is reconstructed from the spectrum in array 'a [n]'. The result is returned in array 'b [n]'. The algorithm differs from the forward transform only in the sign of the complex exponential, on line 28, and a scale factor proportional to the number of samples. The scale factor appears as the value 'half' at line 15 of Example 21.1a, where it is used on lines 21 and 22. However, it is absent from the corresponding lines 20 and 21 of Example 21.1b. Cumulative application of the factor of one half produces the value $\frac{1}{N}$ in Eq. (21.3).

Example 21.1(b)

```
 1:   number *inverse_transform(n,a,b)  /* fast inverse Fourier
                                           transform of array of numbers */
 2:        int n;                        /* input array a[n] is recycled */
 3:        number *a,*b;                 /* length n must be power of 2   */
 4:   {
 5:        cliff_comp exp_omega;
 6:        double theta;
 7:        number x;
 8:        int k,n2;
 9:
10:        if(not_powerof_two(n))  /* not power of two */
11:           {
12:              printf("length of array (%d) not a power of two in
                 routine: inverse_transform()\n",n);
13:              brexit(0); /* user restart */
14:           }
15:        n2=n/2;
16:        if(n>2)  /* divide and conquer */
17:           {
18:              for(k=0;k<n2;k++)
19:                 {
20:                    b[k]=a[2*k];         /* cosine weighted data */
21:                    b[k+n2]=a[2*k+1];    /* sine weighted data */
22:                 }
23:              inverse_transform(n2,b,a);    /* destroys b, overwrites a */
24:              inverse_transform(n2,&b[n2],&a[n2]);
25:              for(k=0;k<n2;k++)
```

```
26:                {
27:                    theta=2.0*M_PI*k/n; /* one sample = 2pi/n */
28:                    pack_comp(&exp_omega,cos(theta),sin(theta),0);
29:                    b[k]=add_number(a[k],mul_comp(a[n2+k],
                            central,&exp_omega),2);
30:                }
31:            for (k=n2;k<n;k++)
32:                {
33:                    theta=2.0*M_PI*k/n; /* one sample = 2pi/n */
34:                    pack_comp(&exp_omega,cos(theta),sin(theta),0);
35:                    b[k]=add_number(a[k-n2],mul_comp(x=a[k],
                            central,&exp_omega),3);
36:                    recycle_number(x);
37:                }
38:            }
39:        else /* n=2 */
40:            {
41:                b[0]=add_number(a[0],a[1],0);
42:                b[1]=sub_number(a[0],a[1],3);
43:            }
44:        return(b);
45:    }
```

Application of the forward transform from Example 21.1a to calculate the spectrum of a pair of signals is demonstrated in Example 21.2a. The number of samples taken for the signals is defined in line 9, and a Clifford two-dimensional context is established from line 10 to 13. All multiplications in the Fourier transform are with complex scalars, so it doesn't matter whether a positive or negative signature is used. The two dimensions[3] are used here to store the two signals, effectively in parallel, as two different vector units.

Input and output arrays of Clifford numbers 'a[n]' and 'b[n]' are allocated in lines 16 and 17, and single cycle bipolar signals[4] of square and triangular shape are placed in the input array from line 25 to 41. The square signal is assigned vector unit e_0, and the triangular signal is assigned the vector unit e_1.

The transform, on both signals simultaneously, is carried on line 43, and the first 10 spectral values are printed for checking from line 46 to 49.

Application of the inverse transform is also demonstrated in Example 21.2a. The spectra for two single cosine signals are created in lines 53–57. The signal with the vector unit e_0 is the 256th harmonic (one cycle has 16 samples), and the signal with the vector unit e_1 is the 512th harmonic (one cycle has eight samples). The harmonics are set up with half value in the positive frequency, $\frac{1}{2}e^{i\omega t}$, and half value in the negative frequency $\frac{1}{2}e^{-i\omega t}$ so that on reconstruction the sum of the two delivers $\cos \omega t$.

---

3 Strictly speaking a Clifford number of one dimension would be sufficient. In two dimensions, there are four components; therefore, it would be possible to accommodate four signals. However, because the implementation is sparse, not storing non-zero components, it doesn't make much difference whether a context of one, two, or more dimensions is used.

4 Signals are normally understood as functions of time, and waves are understood as functions of space. The Fourier transform applies equally well to both.

The inverse transform, on both signals simultaneously, is carried on line 59, and the first 16 samples of the signals are printed for checking from line 62 to 65.

Example 21.2(a)

```
1:   void application8(argc,argv) /* fast Fourier transform */
2:        int argc; char **argv;
3:   {
4:      int lambda[2];
5:      number *a, *b;
6:      int i,n,n1,n2,n3;
7:      bitmap e0,e1;
8:
9:      n=4096;   /* samples in temporal dimension */
10:     recovery_level(0,0,0);
11:     create_cliffmem(3*n);
12:     declare_primal(2,pack_signature(2,lambda,-1));
13:     format_printing("%.5f");
14:
15:     /* allocate array memory here */
16:     a=(number *)malloc(n*sizeof(number));
17:     b=(number *)malloc(n*sizeof(number));
18:
19:     /* calculate bitmaps for units one time only */
20:     e0=spack_unit("e{0}");
21:     e1=spack_unit("e{1}");
22:
23:     /* calculate spectrum of square wave and triangle wave
                simultaneously */
24:     /* square wave has unit e{0}, triangle wave has unit e{1} */
25:     n1=n/4; n2=2*n1; n3=3*n1;
26:     a[0]=add_number(simple_number(0.0,0.0,e0),
27:                     simple_number(0.0,0.0,e1),3);
28:     for(i=1;i<n1;i++)
29:       { a[i]=add_number(simple_number(1.0,0.0,e0),
30:                         simple_number(4.0*i/n,0.0,e1),3); }
31:     for(i=n1;i<n2;i++)
32:       { a[i]=add_number(simple_number(1.0,0.0,e0),
33:                         simple_number(2.0-4.0*i/n,0.0,e1),3); }
34:     a[n2]=add_number(simple_number(0.0,0.0,e0),
35:                      simple_number(0.0,0.0,e1),3);
36:     for(i=n2+1;i<n3;i++)
37:       { a[i]=add_number(simple_number(-1.0,0.0,e0),
38:                         simple_number(2.0-4.0*i/n,0.0,e1),3); }
39:     for(i=n3;i<n;i++)
40:       { a[i]=add_number(simple_number(-1.0,0.0,e0),
41:                         simple_number(4.0*i/n-4.0,0.0,e1),3); }
42:     transform(n,a,b); /* transform vectors in array a[n] */
43:
44:     printf("spectral components\n");
45:     for(i=0;i<10;i++) /* print first ten spectral values */
46:       {
47:         printf("%2d: ",i); print_number(b[i],8);
48:       }
```

```
49:    recycle_array(n,b);
50:
51:    /* calculate signals for two single harmonics */
52:    fresh_array(n,a);
53:    recycle_number(a[256]);    a[256]=simple_number(0.5,0.0,e0);
54:    recycle_number(a[n-256]);  a[n-256]=simple_number(0.5,0.0,e0);
55:    recycle_number(a[512]);    a[512]=simple_number(0.5,0.0,e1);
56:    recycle_number(a[n-512]);  a[n-512]=simple_number(0.5,0.0,e1);
57:
58:    inverse_transform(n,a,b); /* inverse transform vectors
                                     in array a[n] */
59:
60:    printf("\nreconstructed cosine waves\n");
61:    for(i=0;i<16;i++) /* print first sixteen samples */
62:       {
63:       printf("%2d: ",i); print_number(b[i],8);
64:       }
65:    recycle_array(n,b);
66:
67:    free((void *)a);
68:    free((void *)b);
69:    free_cliffmem(0);
70:    }
```

21.2.2 Example Output

The results of running the code listed in Example 21.2a is given in Example 21.2b, with the spectral components from lines 3 to 12, and the reconstructed signals from lines 15 to 30.

As constructed in Example 21.2a, both the square- and triangular-shaped signals are odd functions. As a consequence only sinusoidal components of the results are non-zero; these being found in the imaginary part of the complex coefficients. In addition to having an odd symmetry about the origin, the signals as constructed have an odd symmetry about the middle of the cycle. That means of all the sinusoidal harmonics, only the odd ones $1, 3, 5 \ldots$ are non-zero. These symmetries are properly reflected in the results.

Example 21.2(b)

```
1:  application8()
2:  spectral components
3:    0: (0.00000+i0.00000)e{}
4:    1: (0.00000-i0.63662)e{0}+(0.00000-i0.40528)e{1}
5:    2: (0.00000+i0.00000)e{}
6:    3: (0.00000-i0.21221)e{0}+(-0.00000+i0.04503)e{1}
7:    4: (0.00000+i0.00000)e{}
8:    5: (0.00000-i0.12732)e{0}+(0.00000-i0.01621)e{1}
9:    6: (0.00000+i0.00000)e{}
10:   7: (0.00000-i0.09094)e{0}+(-0.00000+i0.00827)e{1}
11:   8: (0.00000+i0.00000)e{}
```

(Continued)

Example 21.2(b) (Continued)

```
12:    9:  (0.00000-i0.07073)e{0}+(0.00000-i0.00500)e{1}
13:
14:   reconstructed cosine waves
15:    0:  (1.00000+i0.00000)e{0}+(1.00000+i0.00000)e{1}
16:    1:  (0.92388+i0.00000)e{0}+(0.70711-i0.00000)e{1}
17:    2:  (0.70711-i0.00000)e{0}+(-0.00000+i0.00000)e{1}
18:    3:  (0.38268-i0.00000)e{0}+(-0.70711+i0.00000)e{1}
19:    4:  (-0.00000+i0.00000)e{0}+(-1.00000+i0.00000)e{1}
20:    5:  (-0.38268+i0.00000)e{0}+(-0.70711+i0.00000)e{1}
21:    6:  (-0.70711+i0.00000)e{0}+(0.00000-i0.00000)e{1}
22:    7:  (-0.92388+i0.00000)e{0}+(0.70711-i0.00000)e{1}
23:    8:  (-1.00000+i0.00000)e{0}+(1.00000+i0.00000)e{1}
24:    9:  (-0.92388+i0.00000)e{0}+(0.70711-i0.00000)e{1}
25:   10:  (-0.70711+i0.00000)e{0}+(-0.00000+i0.00000)e{1}
26:   11:  (-0.38268+i0.00000)e{0}+(-0.70711+i0.00000)e{1}
27:   12:  (0.00000-i0.00000)e{0}+(-1.00000+i0.00000)e{1}
28:   13:  (0.38268+i0.00000)e{0}+(-0.70711+i0.00000)e{1}
29:   14:  (0.70711-i0.00000)e{0}+(0.00000-i0.00000)e{1}
30:   15:  (0.92388-i0.00000)e{0}+(0.70711-i0.00000)e{1}
```

Inspection of analytical results for the Fourier series (Spiegel 1968) of signals with square and triangular shapes indicates that the non-zero harmonics are proportional to the reciprocal of the frequency for the square shape, and proportional to the square of the reciprocal of the frequency for the triangular shape. The FFT can never exactly match these results because it assumes some bandlimit, whereas square and triangular shapes have no bandlimit.

From the results, ratios of the harmonics $1:3$, $1:5$, $1:7$, and $1:9$ for the square shape are 2.9999529, 5.0001571, 7.0004399, and 9.0007069, and for the triangular shape are 9.0002221, 25.001851, 49.006046, and 81.056, respectively, much as expected. The deviation from the analytical results can be reduced by increasing the number of samples, thereby sampling the signals more finely and raising the bandlimit.

Expected values for the reconstructed signals are the cosines of $0°$, $22\frac{1}{2}°$, $45°$, $67\frac{1}{2}°$, and $90°$ which are $1, \frac{1}{2}\sqrt{2+\sqrt{2}}, \frac{1}{2}\sqrt{2}, \frac{1}{2}\sqrt{2-\sqrt{2}}$, and 0, respectively. The samples of the lower-frequency signal contains all of these values, whereas the samples of the higher-frequency signal has only half of them. Checking the values numerically verifies that the results are as expected.

References

Cooley, J. W. and Tukey, J. W. (1965), 'An algorithm for the machine calculation of complex Fourier series', *Mathematics of Computation* **19**, 297–301.

Spiegel, M. R. (1968), *Mathematical Handbook of Formulas and Tables*, Schaum's outline series, McGraw–Hill, Cambridge, UK.

22

Hertzian Dipole

The simplest source of electromagnetic waves in three dimensions is the Hertzian dipole. Although its radiation characteristics are somewhat less than desirable[1] it can be realised in practice (Hertz 1894), it can play a useful role as a viable antenna in situations where space is limited, and (conveniently) its fields can be expressed in closed form.

22.1 Theory

Figure 22.1 shows a Hertzian dipole of (infinitesimal) length h and cross-sectional area A standing vertically along the z-axis, carrying current $I_k = J_k A$ uniformly distributed throughout the cylindrical region Ω_0 and varying sinusoidally in time. Conservation of charge is honoured by temporary storage of charge in the reservoirs of thickness Δh at the ends of the dipole, throughout the alternating cycle of current, according to need.

The electromagnetic field is given as magnetic and electric phasors in spherical coordinates r, θ, ϕ by

$$
\begin{cases}
H_k(\mathbf{r}) = -\dfrac{I_k h}{r^2} B_{-k}(\mathbf{r}) \left[1 + ikr\right] (r \sin \theta) \, \hat{\phi} \\[2mm]
E_k(\mathbf{r}) = -\eta \dfrac{I_k h}{r^2} B_{-k}(\mathbf{r}) \left[\left(\dfrac{2}{ikr} + 2\right)(r \cos \theta) \, \hat{\mathbf{r}} + \left(\dfrac{1}{ikr} + 1 + ikr\right)(r \sin \theta) \, \hat{\theta}\right]
\end{cases}
\tag{22.1}
$$

where

$$
B_{-k}(\mathbf{r}) = -\dfrac{1}{4\pi r} e^{-ikr}
\tag{22.2}
$$

is the Bessel potential in three dimensions for negative exponents, and where $\eta = \sqrt{\mu}/\sqrt{\epsilon}$ is[2] the intrinsic impedance of the region through which the wave is travelling.

---

1 The radiation resistance is too low so that a high current must be used to achieve a useful amount of radiation. However, there are then high losses due to low electrical efficiency, and a significant amount of the power is wasted as heat.

2 For complex numbers $\sqrt{a/b}$ does not always give the same result as $\sqrt{a}/\sqrt{b}$.

Numerical Calculations in Clifford Algebra: A Practical Guide for Engineers and Scientists,
First Edition. Andrew Seagar.
© 2023 John Wiley & Sons Ltd. Published 2023 by John Wiley & Sons Ltd.

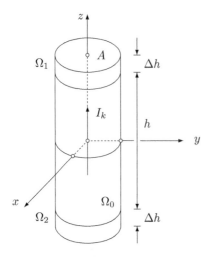

Figure 22.1 Vertical current source I_k within region Ω_0 of length h and cross-sectional area A, standing vertically at the origin, capped at the ends with charge reservoirs Ω_1, Ω_2.

The two fields can be expressed in a common metric when rescaled by their respective material properties:

$$
\begin{cases}
\sqrt{\bar{\mu}}H_k(\mathbf{r}) = -\dfrac{\sqrt{\mu I_k}h}{r^2} B_{-k}(\mathbf{r})\left[1 + ikr\right](r\sin\theta)\,\hat{\phi} \\[2ex]
\sqrt{\epsilon}E_k(\mathbf{r}) = -\dfrac{\sqrt{\mu I_k}h}{r^2} B_{-k}(\mathbf{r})\left[\left(\dfrac{2}{ikr} + 2\right)(r\cos\theta)\,\hat{\mathbf{r}} + \left(\dfrac{1}{ikr} + 1 + ikr\right)(r\sin\theta)\,\hat{\theta}\right]
\end{cases}
$$

$$(22.3)$$

and combined as the single bivector electromagnetic field:

$$
\mathcal{F}_{k(\mathrm{Hz})}(\mathbf{r}, I_k h) = -\frac{\sqrt{\mu I_k}h}{r^2} B_{-k}(\mathbf{r})\left\{\left[1 + ikr\right](r\sin\theta)\hat{\phi}\,\sigma - i\left[\left(\frac{2}{ikr} + 2\right)(r\cos\theta)\,\hat{\mathbf{r}}\right]\right.
$$
$$
\left. + \left(\frac{1}{ikr} + 1 + ikr\right)(r\sin\theta)\,\hat{\theta}\,e_0\right\}
$$

$$(22.4)$$

When evaluating the fields for the Hertzian dipole numerically using Eqs. (22.1), (22.3), or (22.4), there is no need to involve spherical coordinates. The term $(r\sin\theta)\hat{\phi}$ is constructed by projecting vector $\mathbf{r}$ onto the xy plane and then rotating by 90° in that plane. Similarly, the term $(r\sin\theta)\hat{\theta}$ is constructed by projecting vector $\mathbf{r}$ onto the xy plane and then rotating by θ towards the $-z$ axis. The term $(r\cos\theta)\hat{\mathbf{r}}$ is constructed by projecting vector $\mathbf{r}$ onto the z axis and then rotating by angle θ towards the xy plane. All of these calculations can be achieved using the projection and rotation operations in Table 6.1 without the need for angles or for trigonometric functions.

22.2 Practice

22.2.1 Example Code

The code listed in Example 22.1a provides a routine to calculate the Bessel potential, and the code listed in Example 22.1b produces the field from a Hertzian dipole. The code for the

Bessel potential is a straightforward implementation of Eq. (22.2) for the case of wavenumbers k without the minus sign. For negative wavenumbers, the negative sign should be applied to the argument before invoking the routine.

Example 22.1(a)

```
1:   number bessel_potential3D(k,r)  /* Bessel potential in 3D */
2:           cliff_comp *k;              /* use negative k for negative
                                           exponent */
3:           double r;                   /* r is radial distance */
4:   {
5:      cliff_comp ikr;
6:      double s;
7:      number bp;
8:
9:      if(r<=0.0)
10:        {
11:         printf("radius (%e) not > (0) in routine:
                      bessel_potential3D()\n",r);
12:         brexit(0); /* user restart */
13:        }
14:      s=-0.25/(M_PI*r);
15:      pack_comp(&ikr,-r*k->coef[1],r*k->coef[0],0);
16:      if(ikr.coef[0]!=0.0)
17:        {
18:         s=s*exp(ikr.coef[0]); /* attenuation for lossy media */
19:        }
20:      bp=scalar_number(s*cos(ikr.coef[1]),s*sin(ikr.coef[1]));
21:      return(bp);
22:   }
```

The code for the dipole field constructs the result from several parts. The terms preceding the brackets in Eq. (22.3) are combined in lines 12–16. Components for the terms involving complex numbers are set up in lines 20 and 21 and then used when needed in lines 50, 58, and 61. Terms involving trigonometric functions are calculated from line 28 to 46. There are two cases, depending on whether the point r is above or below the plane $z = 0$, in order to avoid problems (division by zero) at the poles. Lines 28 and 29 project the radial vector onto the z axis and the xy plane. Lines 31, and 35 or 44 construct unit vectors in order to rotate the projections into the directions of the spherical unit vectors $\hat{\phi}$ and $\hat{\theta}$ which follow the trigonometric terms. Lines 32, and 36 to 37 or 45 to 46 apply the rotations.

Example 22.1(b)

```
1:   number hertzian_dipole(Ih,mu,k,r)  /* Hertzian Dipole field */
2:           cliff_comp *Ih,*mu,*k;
3:           number r;
4:   {
5:      cliff_comp s,minus_k,ikr,inv_ikr,one,x,y;
6:      double rlen;
```

(Continued)

Example 22.1(b) (Continued)

```
 7:    number bk,xunit,yunit,zunit,runit,ntheta,nphi;;
 8:    number rsintheta,rcostheta,rsintheta_uphi,rsintheta_utheta,
             rcostheta_ur;
 9:    number hfield,efield,f;
10:
11:    /* -sqrt(mu)Ih/(r^2) B(-k,r) */
12:    rlen=get_real(magnitudeof_vector(r,0),0,1);
13:    rscale(-1.0/(rlen*rlen),Ih,&s);
14:    bk=mul_comp(f=bessel_potential3D(negation(k,&minus_k),rlen),
15:                central,
16:                central(&s,square_root(mu,&x),&x));
17:    recycle_number(f);
18:
19:    pack_comp(&one,1.0,0.0,0);                          /* 1.0 */
20:    pack_comp(&ikr,-rlen*k->coef[1],rlen*k->coef[0],0); /* ikr */
21:    inverse(&ikr,&inv_ikr);                             /* 1/ikr */
22:
23:    runit=normalise_vector(r,0);
           /* unit radial vector */
24:    xunit=simple_number(1.0,0.0,spack_unit("e{1}"));
           /* unit x vector */
25:    yunit=simple_number(1.0,0.0,spack_unit("e{2}"));
           /* unit y vector */
26:    zunit=simple_number(1.0,0.0,spack_unit("e{3}"));
           /* unit z vector */
27:
28:    rsintheta=outerspace:vector(r,zunit,0);
           /* projection r on xy plane */
29:    rcostheta=innerspace:vector(r,zunit,0);
           /* projection r on z axis */
30:
31:    nphi=normalise_vector(add_number(xunit,yunit,2),1);
           /* for 90 phi rotation */
32:    rsintheta_uphi=rotate_number(rsintheta,xunit,nphi,6);
           /* direction of u_phi vector */
33:    if(get_real(inner_mul(zunit,runit,0),0,1)<0.0)
           /* positive z component */
34:       {
                  /* rotate from positive pole */
35:        ntheta=normalise_vector(add_number(zunit,runit,2),1);
                  /* for theta rotation */
36:        rsintheta_utheta=rotate_number(rsintheta,zunit,ntheta,1);
                  /* of u_theta vector  */
37:        rcostheta_ur=rotate_number(rcostheta,zunit,ntheta,7);
                  /* of u_radial vector */
38:       }
39:    else
                  /* negative z component, rotate from negative pole */
40:       {
41:        zunit=negativeof_number(zunit,1);
42:        rsintheta=negativeof_number(rsintheta,1);
43:        rcostheta=negativeof_number(rcostheta,1);
44:        ntheta=normalise_vector(add_number(zunit,runit,2),1);
                  /* for theta rotation */
```

```
45:            rsintheta_utheta=rotate_number(rsintheta,zunit,ntheta,1);
                             /* of u_theta vector  */
46:            rcostheta_ur=rotate_number(rcostheta,zunit,ntheta,7);
                             /* of u_radial vector */
47:       }
48:    /* sqrt(mu) H */
49:    hfield=central_mul(bk,
50:                        mul_comp(rsintheta_uphi,central,
                           addition(&one,&ikr,&x)),
51:                        2);
52:    recycle_number(rsintheta_uphi);
53:
54:    /* sqrt(epsilon) E */
55:    efield=central_mul(bk,
56:                        addto_number(mul_comp(rcostheta_ur,
57:                                     central,
58:                                     rscale(2.0,addition
                                            (&inv_ikr,&one,&x),&x)),
59:                            f=mul_comp(rsintheta_utheta,
60:                                       central,
61:                                       addition(addition
                                          (&inv_ikr,&one,&y),
                                          &ikr,&y))),
62:                        3);
63:    recycle_number(f);
64:    recycle_number(rcostheta_ur);
65:    recycle_number(rsintheta_utheta);
66:
67:    /* sqrt(mu) H sigma - i sqrt(epsilon) E e0 */
68:    f=subfrom_number(mul_comp(hfield,central,
                       pack_comp(&x,-1.0,0.0,14)),    /* -e1e2e3 */
69:                     bk=mul_comp(efield,central,
                       pack_comp(&y,0.0,1.0,1)));  /* ie0      */
70:    f=keepone_grade(f,2); /* eliminate non-exact zeros
                                outside grade 2 */
71:    recycle_number(bk);
72:    recycle_number(hfield);
73:    recycle_number(efield);
74:
75:    return(f);
76:  }
```

The earlier calculations are brought together for the magnetic field in lines 49–51, and for the electric field in lines 55–62. Finally, the Clifford electromagnetic field is constructed from the magnetic and electric fields in lines 68 and 69.

Although in principle, the result is wholly contained within grade 2, inexact numerical calculation can leave non-zero values at the level of machine precision in other grades. The filter on line 70 is included to eliminate any of those which may occur.

Application of the code in Example 22.1b to calculate the dipole field along the x axis is listed in Example 22.1c. A four-dimensional Clifford context with negative signature is established in lines 12–15. The parameters defining the wave propagation are set up in lines 17–23 and printed in lines 24–30. For this example, the electric permeability and permittivity are set to free space values, so the speed of propagation is the speed of light in vacuum. The value of the angular frequency is chosen to give a wavelength of half a metre.

Example 22.1(c)

```
 1:   void application9(argc,argv)  /* Hertzian dipole field on x-axis */
 2:        int argc; char **argv;
 3:   {
 4:      double H[3][2], E[3][2];
 5:      double xyz[3],one[2];
 6:      int lambda[4];
 7:      cliff_comp Ih,mu,k,epsilon,v,inv_v,temp;
 8:      double omega,wlen;
 9:      number f,r,m;
10:      int i;
11:
12:      recovery_level(0,0,0);
13:      create_cliffmem(50);
14:      declare_primal(4,pack_signature(4,lambda,-1));
                       /* four dimensions for EM fields */
15:      format_printing("%.3e");
16:
17:      omega=4.0*M_PI*C0;                       /* angular frequency */
18:      pack_comp(&Ih,600.0,0.0,0);              /* dipole moment=600.0 */
19:      pack_comp(&mu,1.0*MU0,0.0,0);            /* permeability */
20:      pack_comp(&epsilon,1.0*EPS0,0.0,0);      /* permittivity */
21:      inverse(central(square_root(&mu,&temp),
                  square_root(&epsilon,&v),&inv_v),&v);
22:      rscale(omega,&inv_v,&k);                 /* wavenumber */
23:      wlen=2.0*M_PI*v.coef[0]/omega;           /* wavelength */
24:      printf("# permittivity    = ");  print_comp(&epsilon);
           printf("(F/m)\n");
25:      printf("# permeability    = ");  print_comp(&mu);
              printf("(H/m)\n");
26:      printf("# velocity        = ");  print_comp(&v);
               printf("(m/s)\n");
27:      printf("# angular frequency = ");  printf(" %.3e",omega);
           printf("(radians/s)\n");
28:      printf("# frequency       = ");  printf(" %.3e",0.5*omega/M_PI);
           printf("(Hz)\n");
29:      printf("# wavenumber      = ");  print_comp(&k);
                  printf("(1/m)\n");
30:      printf("# wavelength      = ");  printf(" %.3e",wlen);
           printf("(m)\n");
31:
32:      one[0]=1.0; one[1]=0.0;
33:      for(i=4;i<257;i++)
34:         {
35:            xyz[0]=i*0.0625*wlen;  /* field location */
36:            xyz[1]=0.0;
37:            xyz[2]=0.0;                /* =j*0.0625*wlen */
38:            r=pack_R03vector(xyz);
39:            f=hertzian_dipole(&Ih,&mu,&k,r);
40:            /* f=inverseof_emfield(f,1);  add this for inverse field */
41:            m=magnitudeof_emfield(f,0);
```

```
42:              unpack_emfield(f,one,one,H,E);
43:              printf("%.5f    ",xyz[0]);
44:              printf("%.3e    ",E[2][0]);      /* Ez real */
45:              printf("%.3e    ",E[2][1]);      /* Ez imag */
46:              printf("%.3e    ",H[1][0]);      /* Hy real */
47:              printf("%.3e    ",H[1][1]);      /* Hy imag */
48:              printf("%.3e\n",get_real(m,0,1)); /* magnitude */
49:              recycle_number(f);
50:              recycle_number(r);
51:           }
52:       free_cliffmem(1);
53:    }
```

The field is calculated on line 39 within the loop from line 33 to 51, at distances along the *x* axis between $\frac{1}{4}$ and 16 wavelengths ($\frac{1}{8}$ and 8 m) away from the dipole. The particular locations at which the field is calculated are generated in lines 35–37. The electric and magnetic fields are extracted on line 42 (each scaled by the square root of its material property), and the real and imaginary parts of the *y* component of the magnetic field and of the *z* component of the electric field are printed on lines 44–47. The magnitude of the whole field is calculated on line 41 and printed on line 48.

22.2.2 Example Output

Figure 22.2 shows the real and imaginary parts of the *y* component of the magnetic field and the *z* component of the electric field as well as the magnitude of the whole field.

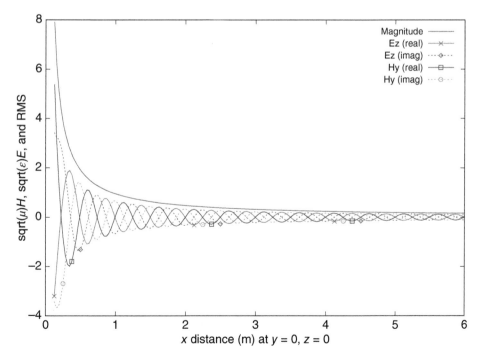

Figure 22.2 Components of Hertzian dipole field.

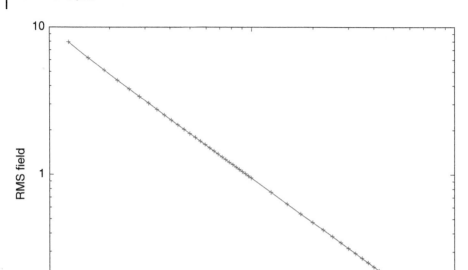

Figure 22.3 Magnitude of Hertzian dipole field.

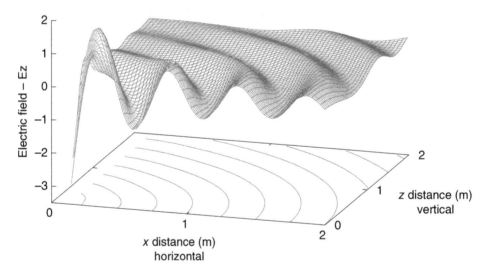

Figure 22.4 Hertzian dipole field in vertical plane.

The real parts of the electric and magnetic fields are almost opposite in value, especially further from the dipole. The imaginary parts are also almost opposite in value.

Figure 22.3 shows the magnitude of the whole field plotted on a log–log scale. The line is almost straight, more so away from the dipole, with a slope asymptotic to −1. This is

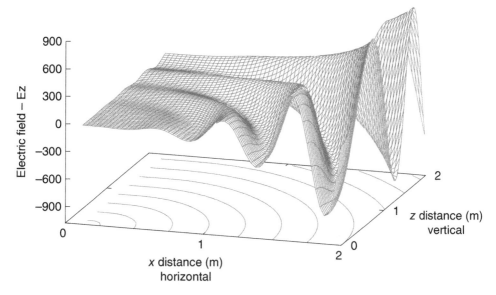

Figure 22.5 Inverse of Hertzian dipole field in vertical plane.

consistent with the expectation that the field decrease as $1/r$ in the far field (many wavelengths distant from the dipole).

Figure 22.4 shows the real part of the z component of the electric field in the xz (vertical) plane through the origin. The contour lines beneath the surface show where the waves pass through zero. In order to generate the data for this result, it is straightforward to nest the loop 'for(i=...) {}' spanning from line 33 to 51 inside another loop 'for(j=0; j<64;j++) {}' which increments the z location 'xyz[2]=j*0.0625*wlen;' on line 37.

Figure 22.5 shows the real part of the z component of the temporal part E' of the inverse of the electromagnetic field in the xz (vertical) plane through the origin, calculated as in Section 10.1.7 by including the code written as a comment on line 40.

Reference

Hertz, H. (1894), Untersuchungen über die ausbreitung der elecktrischen kraft, *in 'Gesammelte Werke on Heinrich Hertz'*, Vol. 2, J.A. Barth, Leipzig, Germany.

23

Finite Difference Time Domain

The finite difference time domain (FDTD) method was introduced by Yee (1966) as a numerical method for simulating the propagation of electromagnetic waves. Of all methods for simulating electromagnetic wave propagation, this is the simplest to implement.

23.1 Theory

The propagation of electromagnetic waves was first described by Maxwell (1873a, 1873b). Written for a plane wave travelling in the x direction within a uniform region of space void of sources, using a four-dimensional Clifford number context with all negative signature, Maxwell's equations take the form of a single first-order ordinary differential equation:

$$\mathcal{D}\mathcal{F} = 0 \tag{23.1}$$

where:

$$\begin{cases} \mathcal{D} = \frac{\partial}{\partial x}e_1 - \frac{i}{c}\frac{\partial}{\partial t}e_0 \\ \mathcal{F} = \sqrt{\mu}H_z e_3 \sigma - i\sqrt{\epsilon}E_y e_2 e_0 \end{cases} \tag{23.2}$$

The spatiotemporal differential operator $\mathcal{D}$ is the Dirac (1928) operator, and the function $\mathcal{F}$ is the electromagnetic field, as in Section 10.1.2. In this case, H_z and E_y are the magnetic and electric fields in the z and y directions, respectively, μ and ϵ are the magnetic permeability and the electric permittivity, $\sigma = -e_1 e_2 e_3$ is the negatively signed unit of volume, and c is the velocity of propagation.

23.1.1 Analytical Solution

The magnetic and electric fields in Eq. (23.2) can be functions $H_z = f$ and $E_y = \eta f$ of any shape f as long as that function's profiles in space and time are inter-related in the form (Seagar 2018):

$$f = f(x - ct) \tag{23.3}$$

Here $\eta = \sqrt{\mu}/\sqrt{\epsilon}$ is the intrinsic impedance, and the velocity c can take any positive or negative value. The solution represents any shape which moves with a constant velocity $c = \frac{dx}{dt}$ along the x spatial dimension either in the positive or negative direction.

Numerical Calculations in Clifford Algebra: A Practical Guide for Engineers and Scientists,
First Edition. Andrew Seagar.
© 2023 John Wiley & Sons Ltd. Published 2023 by John Wiley & Sons Ltd.

23.1.2 Series Solution

If the value of the field F is known everywhere at one time t, it is possible to calculate the value of the field at a later time $t + h$ numerically, using the FDTD method. The method involves exploiting the relationship between the temporal and spatial parts of the differential operator D, to calculate temporal derivatives from spatial derivatives. The temporal derivatives can then be used to extend the field as a Taylor series to a later instant in time. The whole process can be repeated as many times as necessary.

Utilising the general knowledge of the analytical solution as $H_z = \eta E_y = f(x, t)$, reduces Maxwell's equations (23.1) to the simpler equation:

$$Df = 0 \tag{23.4}$$

for the scalar function f. The foreknowledge that the analytic solution $f(x, t) = f(x - ct)$ is left unstated, so as not to prejudice the numerical solution, which may indeed not be so perfect.

At this stage, it becomes the differential operator which is of most interest and not the function to which it is applied. So rather than carrying the function f in the equations that follow it is omitted, on the understanding that whatever result is obtained must be applied to f on both sides of the equal sign. Equation (23.4) then gives

$$e_1 \frac{\partial}{\partial x} = \frac{ie_0}{c} \frac{\partial}{\partial t} \Leftrightarrow \begin{cases} \frac{\partial}{\partial x} = \frac{ie_0 e_1}{c} \frac{\partial}{\partial t} \\ \frac{\partial}{\partial t} = ie_0 e_1 c \frac{\partial}{\partial x} \end{cases} \tag{23.5}$$

Care must be taken in Clifford algebra to honour the rules in Eq. (5.19) when re-ordering expressions; otherwise, it is possible accidentally to introduce spurious minus signs.

Applying the temporal derivative twice gives

$$\frac{\partial^2}{\partial t^2} = (ie_0 e_1 c)^2 \frac{\partial^2}{\partial x^2} = c^2 \frac{\partial^2}{\partial x^2} \tag{23.6}$$

Repeated application leads to expressions for any order of derivative:

$$\begin{cases} \frac{\partial^{2n}}{\partial t^{2n}} = c^{2n} \frac{\partial^{2n}}{\partial x^{2n}} \\ \frac{\partial^{2n+1}}{\partial t^{2n+1}} = ie_0 e_1 c^{2n+1} \frac{\partial^{2n+1}}{\partial x^{2n+1}} \end{cases} \tag{23.7}$$

The Taylor series for extending a differentiable function f, some distance h along the dimension of time t at a fixed point x in space takes the form:

$$f(t + h) = f(t) + h \frac{\partial f(t)}{\partial t} + \frac{h^2}{2!} \frac{\partial^2 f(t)}{\partial t^2} + \frac{h^3}{3!} \frac{\partial^3 f(t)}{\partial t^3} + \dots \tag{23.8}$$

It would be somewhat more precise here to write the function f as a function in two dimensions, $f(x, t)$. However, that just makes all of the equations more cumbersome. Instead, the function is written in the abbreviated form $f(t)$ with x missing when x is a fixed value and is written as $f(x)$ with t missing when t is a fixed value.

Using Eq. (23.7), the temporal derivatives can be converted to equivalent expressions in terms of spatial derivatives:

$$f(t + h) = f(t) + h(ie_0 e_1)c \frac{\partial f(x)}{\partial x} + \frac{h^2}{2!} c^2 \frac{\partial^2 f(x)}{\partial x^2} + \frac{h^3}{3!}(ie_0 e_1)c^3 \frac{\partial^3 f(x)}{\partial x^3} + \dots$$

$$= f(t) + hc(ie_0 e_1) \frac{\partial f(x)}{\partial x} + \frac{(hc)^2}{2!} \frac{\partial^2 f(x)}{\partial x^2} + \frac{(hc)^3}{3!}(ie_0 e_1) \frac{\partial^3 f(x)}{\partial x^3} + \dots \tag{23.9}$$

Figure 23.1 Sinusoidal waveform.

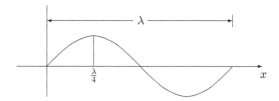

The Taylor series may or may not converge, depending on the value of the interval of time h and the behaviour of the derivatives.

23.1.3 Analytical Example

Suppose at time t, the waveform takes a sinusoidal shape:

$$f(x) = \sin(\frac{2\pi x}{\lambda}) \tag{23.10}$$

as shown in Figure 23.1. The first six spatial derivatives are the following:

$$\frac{\partial f(x)}{\partial x} = \frac{2\pi}{\lambda} \cos(\frac{2\pi x}{\lambda}) \qquad \frac{\partial^2 f(x)}{\partial x^2} = -(\frac{2\pi}{\lambda})^2 \sin(\frac{2\pi x}{\lambda}) \qquad \frac{\partial^3 f(x)}{\partial x^3} = -(\frac{2\pi}{\lambda})^3 \cos(\frac{2\pi x}{\lambda})$$

$$\frac{\partial^4 f(x)}{\partial x^4} = (\frac{2\pi}{\lambda})^4 \sin(\frac{2\pi x}{\lambda}) \qquad \frac{\partial^5 f(x)}{\partial x^5} = (\frac{2\pi}{\lambda})^5 \cos(\frac{2\pi x}{\lambda}) \qquad \frac{\partial^6 f(x)}{\partial x^6} = -(\frac{2\pi}{\lambda})^6 \sin(\frac{2\pi x}{\lambda})$$

$$\tag{23.11}$$

Choosing the point $x = \frac{\lambda}{4}$ gives the specific values:

$$\frac{\partial f}{\partial x} = 0 \qquad \frac{\partial^2 f}{\partial x^2} = -(\frac{2\pi}{\lambda})^2 \qquad \frac{\partial^3 f}{\partial x^3} = 0$$

$$\frac{\partial^4 f}{\partial x^4} = (\frac{2\pi}{\lambda})^4 \qquad \frac{\partial^5 f}{\partial x^5} = 0 \qquad \frac{\partial^6 f}{\partial x^6} = -(\frac{2\pi}{\lambda})^6 \tag{23.12}$$

Using these values for the derivatives, and the value $f(\frac{\lambda}{4}) = 1$ for the function itself, the temporal Taylor series at the point $x = \frac{\lambda}{4}$ is

$$f(t+h) = 1 - \frac{(hc)^2}{2!}(\frac{2\pi}{\lambda})^2 + \frac{(hc)^4}{4!}(\frac{2\pi}{\lambda})^4 - \frac{(hc)^6}{6!}(\frac{2\pi}{\lambda})^6 + \dots$$

$$= 1 - \frac{1}{2!}(\frac{2\pi hc}{\lambda})^2 + \frac{1}{4!}(\frac{2\pi hc}{\lambda})^4 - \frac{1}{6!}(\frac{2\pi hc}{\lambda})^6 + \dots \tag{23.13}$$

$$= 1 - \frac{1}{2!}(h\omega)^2 + \frac{1}{4!}(h\omega)^4 - \frac{1}{6!}(h\omega)^6 + \dots$$

The series in Eq. (23.13) is recognised (Spiegel 1968) as the Taylor series expansion for the cosine function $\cos(\frac{2\pi hc}{\lambda})$ or $\cos(h\omega)$, where ω is the angular frequency. This particular series is convergent for all values of h, so the temporal expansion can be calculated for all forward values of time. However, using the series to do that may require many terms. If a faster convergence of the series is required, then the value of h may be restricted. For example $h < \frac{1}{\omega} = \frac{\lambda}{2\pi c}$ ensures that the power terms decrease as well as the factorial terms. This is equivalent to restricting the forward propagation of the wave hc to the fraction $\frac{1}{2\pi} \approx 0.159$ of a wavelength.

23.1.4 Numerical Derivatives

Calculating derivatives numerically can be achieved in various ways. None of them is wholly reliable. The method employed here involves fitting a set of trigonometric basis functions to the field and then differentiating those.

The differentiation is carried in the frequency domain by applying a filter equivalent to a differential operator. The Fourier transform (as in Chapter 21) is used to calculate the spectral values and to return to the spatial domain once the filter has been applied.

The spectrum[1] $S(\omega)$ of a waveform $s(x)$ is

$$S(\omega) = \int s(x)e^{-i\omega x}\, dx \qquad (23.14)$$

so that the spectrum of the differential of the waveform is

$$S'(\omega) = \int \left[\tfrac{d}{dx}s(x)\right]e^{-i\omega x}\, dx \qquad (23.15)$$

In order to find the relation between the pair of spectra S and S', we differentiate the product:

$$\tfrac{d}{dx}\left[s(x)e^{-i\omega x}\right] = \left[\tfrac{d}{dx}s(x)\right]e^{-i\omega x} + s(x)(-i\omega)e^{-i\omega x}$$

$$\left[\tfrac{d}{dx}s(x)\right]e^{-i\omega x} = \tfrac{d}{dx}\left[s(x)e^{-i\omega x}\right] - s(x)(-i\omega)e^{-i\omega x}$$

$$\int \left[\tfrac{d}{dx}s(x)\right]e^{-i\omega x}\, dx = \int \tfrac{d}{dx}\left[s(x)e^{-i\omega x}\right]\, dx - \int s(x)(-i\omega)e^{-i\omega x}\, dx$$

$$\int_a^b \left[\tfrac{d}{dx}s(x)\right]e^{-i\omega x}\, dx = \left[s(x)e^{-i\omega x}\right]_a^b + i\omega \int_a^b s(x)e^{-i\omega x}\, dx \qquad (23.16)$$

Here the waveform $s(x)$ is periodic, sampled at N points equally spaced by Δx over a single wavelength λ. In this case, it is convenient to take the limits of integration over one wavelength of the function, $\lambda = N\Delta x$:

$$\int_0^\lambda \left[\tfrac{d}{dx}s(x)\right]e^{-i\omega x}\, dx = \left[s(x)e^{-i\omega x}\right]_0^\lambda + i\omega \int_0^\lambda s(x)e^{-i\omega x}\, dx$$

$$\int_0^\lambda \left[\tfrac{d}{dx}s(x)\right]e^{-i\omega x}\, dx = \left[s(\lambda)e^{-i\omega \lambda} - s(0)e^{-i\omega 0}\right] + i\omega \int_0^\lambda s(x)e^{-i\omega x}\, dx$$

$$= i\omega \int_0^\lambda s(x)e^{-i\omega x}\, dx \qquad (23.17)$$

The value of the term in brackets on the right-hand side is zero because *both* the waveform $s(x)$ and the complex exponential function are periodic, taking the same values at each end of an integral number of wavelengths. For the exponential function, for any spectral line frequency $\omega = m\Delta\omega$:

$$e^{-i\omega \lambda} = e^{-i(m\Delta\omega)(N\Delta x)} = e^{-i(m2\pi/N\Delta x)(N\Delta x)} = e^{-im2\pi} = [e^{-i2\pi}]^m = [1]^m = 1 = e^0 = e^{-i\omega 0} \qquad (23.18)$$

Comparison of Eq. (23.17) with Eqs. (23.14) and (23.15) indicates that the relationship between the spectrum of the waveform $S(\omega)$ and the spectrum of the derivative of the waveform $S'(\omega)$ is

$$S'(\omega) = i\omega S(\omega) \qquad (23.19)$$

In order to calculate the derivative of the waveform, it is therefore possible to calculate its spectrum (using the Fourier transform), multiply by $i\omega$, then calculate the waveform

---

1 This is the same Fourier transform as in Eq. (21.1). However, here the function is a wave (a function of space x) rather than a signal (a function of time t). Some of the symbols have been changed to respect that distinction.

corresponding to that spectrum (using the inverse Fourier transform). For derivatives $\frac{d^n}{dx^n}$ higher than the first, a power $(i\omega)^n$ is used instead:

$$S^{(n)}(\omega) = (i\omega)^n S(\omega) \tag{23.20}$$

23.2 Practice

23.2.1 Example Code

The key components for the code used to implement the FDTD algorithm, i.e. spatial differentiation (Eq. (23.20)), spatial to temporal conversion (Eq. (23.7)), and Taylor extension (Eq. (23.9)) are listed in Examples 23.1b, 23.1c and 23.1d, respectively, whereas the code responsible for setting the context for a particular problem and tying all the other parts together is listed in Example 23.1a.

In Example 23.1a, the number of samples of the waveform is chosen on line 28, then in lines 29–32 a four-dimensional Clifford context with all negative signature is established. Arrays of Clifford numbers are created for the values of the field and its first four spatial derivatives in lines 35–37, including an extra array needed occasionally for temporary work space.

The material properties for the space in which the wave propagates are defined in lines 40 and 41, and the velocity is calculated from those. Note that the square root is taken for each factor on line 42 before multiplying, rather than doing it the other way around. This is necessary because in general the material properties can both be any complex value and in that case multiplying before rooting sometimes gives the wrong result.

The shape of the pulse is defined in lines 51–60. Lines 67–70 reverse the direction of propagation if that choice was made in the arguments passed into the routine.

The spatial sampling Δx chosen in line 84 is based on the distance covered by the simulation, which is from $x = 0$ to $x = 2\pi$, and on the number of samples.

The time step chosen in line 91 is $h = \frac{\Delta x}{2c}$ which ensures the wave cannot travel more than one-half of the distance between samples in one time step. This is slightly over the maximum step $h = \frac{\Delta x}{\pi c}$ needed to ensure that the Taylor series has decreasing power terms for the shortest wavelength $\lambda_{min} = 2\Delta x$ supported by the sampling interval Δx.

Lines 95–103 are used to select either a short or long simulation. The short one takes about five seconds and the long one about three hours on a typical contemporary desktop PC.

The main work of the simulation is initiated repeatedly by line 118 in the loop from lines 116 to 123.

Example 23.1(a)

```
 1:   void application11(argc,argv)  /* finite difference time domain */
 2:         int argc; char** argv;
 3:   {
 4:      FILE *fp1;
 5:      int lambda[4];
 6:      cliff_comp mu,epsilon,v,h,w,t,temp;
 7:      number *a,*b,*d4;
```

(Continued)

Example 23.1(a) (Continued)

```
 8:    int i,n,width,pulse,nstep,tstep,distance;
 9:    bitmap unit;
10:
11:    pulse=0;
12:    distance=0;
13:    if(argc==3)
14:      {
15:        printf("%s %s %s\n",argv[0],argv[1],argv[2]);
16:        sscanf(argv[1],"%d",&pulse);
17:        sscanf(argv[2],"%d",&distance);
18:      }
19:    if(pulse<0||pulse>6||distance<-2||distance==0||
          distance>2||argc!=3)
20:      {
21:        printf("useage: application 11 pulse distance\n");
22:        printf(" pulse=0(square), 1(triangle), 2(quadratic),");
23:        printf(" 3(cubic), 4(sine), 5(gaussian), 6(bell)\n");
24:        printf(" distance=1(one cycle), 2(2048 cycles),");
25:        printf(" -1(backwards one cycle),
                    -2(backwards 2048 cycles)\n");
26:        brexit(0);
27:      }
28:    n=256;   /* samples in spatial dimension */
29:    recovery_level(0,0,0);
30:    create_cliffmem(12*n+2);
31:    declare_primal(4,pack_signature(4,lambda,-1));
       /* four dimensions for EM fields */
32:    format_printing("%.5f");
33:
34:    /* allocate array memory here */
35:    a=(number *)malloc(n*sizeof(number));
       /* array of numbers for waveform */
36:    b=(number *)malloc(n*sizeof(number));
       /* extra array of numbers */
37:    d4=(number *)malloc(4*n*sizeof(number));
       /* array of numbers for 4 derivatives */
38:
39:    /* set values of material properties */
40:    pack_comp(&mu,1.0,0.0,0);                /* permeability=1.0 */
41:    pack_comp(&epsilon,1.0,0.0,0);           /* permittivity=1.0 */
42:    inverse(central(square_root(&mu,&temp),
                square_root(&epsilon,&v),&v),&v);
43:    printf("# permittivity = ");  print_comp(&epsilon);
           printf("(F/m)\n");
44:    printf("# permeability = ");  print_comp(&mu);
           printf("(H/m)\n");
45:    printf("# velocity     = ");  print_comp(&v);
           printf("(m/s)\n");
46:
47:    /* set width of pulse in samples */
48:    width=n/4;
49:
50:    /* set up pulse shape */
51:    fresh_array(n,a);
```

```
52:    switch(pulse)
53:      {
54:      case 0: constant_pulse(n,a,width,&mu,&epsilon);      break;
55:      case 1: linear_pulse(n,a,width,&mu,&epsilon);        break;
56:      case 2: quadratic_pulse(n,a,width,&mu,&epsilon);     break;
57:      case 3: cubic_pulse(n,a,width,&mu,&epsilon);         break;
58:      case 4: cosine_pulse(n,a,width,&mu,&epsilon);        break;
59:      case 5: gaussian_pulse(n,a,width,&mu,&epsilon);      break;
60:      case 6: bell_pulse(n,a,width,&mu,&epsilon);          break;
61:      default:
62:        printf("pulse shape (%d) outside range (0-6) ",pulse);
63:        printf("in routine: application11()\n");
64:        brexit(0);
65:        break;
66:      }
67:    if(distance<0) /* reverse direction of propagation */
68:      {
69:      reverse_Hfield(n,a);
70:      distance=-distance;
71:      }
72:
73:    /* open file to save intermediate results */
74:    fp1=fopen("output.lis","w");
75:
76:    /* show and save initial pulse */
77:    printf("# starting pulse\n");
78:    unit=spack_unit("e{0,2}"); /* E field */
79:    printf_onecoefficientof_array("%.5e#i",n,a,unit);
       /* imaginary part */
80:    printf("\n");
81:    printto_file(fp1,0,n,a,unit,1);
82:
83:    /* choose spatial sampling, w; (w = delta x, not omega) */
84:    pack_comp(&w,2.0*M_PI/n,0.0,0);
85:    printf("# %d samples at separation of ",n); print_comp(&w);
86:    printf(" = ");
87:    print_comp(central(pack_comp(&temp,(double)n,0.0,0),&w,&temp));
88:    printf(" total spatial extent\n");
89:
90:    /* use time step h=w/2v */
91:    central(&w,inverse(central(pack_comp(&h,2.0,0.0,0),
         &v,&h),&h),&h);
92:    printf("# one time step = "); print_comp(&h); printf("\n");
93:
94:    /* set length of time in steps */
95:    switch(distance)
96:       {
97:       case 1:             /* five seconds on my PC */
98:         tstep=(2*n);      /* (1) for one cycle of 2*n=512 steps */
99:         nstep=8;          /* (1) save after each 8 steps */
100:        break;
101:       case 2:            /* three hours on my PC */
102:        tstep=2048*(2*n); /* (2) for 2048 cycles of 2*n=512 steps */
103:        nstep=32*(2*n);   /* (2) save after each 32 cycles */
```

(Continued)

Example 23.1(a) (Continued)

```
104:         break;
105:      default:
106:         printf("distance (%d) outside range (0-1) ",distance);
107:         printf("in routine: application11()\n");
108:         brexit(0);
109:         break;
110:      }
111:      central(pack_comp(&t,(double)tstep,0.0,0),&h,&t);
112:      printf("# %d repetitions of %d steps ",tstep/nstep,nstep);
113:      printf("= %d steps over time period of ",tstep); print_comp(&t);
                printf("(s)\n");
114:
115:      /* this is where the main work happens */
116:      for(i=1;i<1+tstep/nstep;i++)
         /* repeat loop tstep/nstep times */
117:         {
118:            propagate_multi(n,a,b,&w,&h,&v,nstep,4,d4);
                   /* propagate nstep time steps */
119:            printf("# pulse after %d time steps\n",i*nstep);
                   /* show */
120:            printf_onecoefficientof_array("%.5e#i",n,a,unit);
                   /* imaginary part */
121:            printf("\n");
122:            printto_file(fp1,i*nstep,n,a,unit,1);          /* save */
123:         }
124:
125:      /* cleanup */
126:      fclose(fp1);           /* close file */
127:      recycle_array(n,a);  /* recycle numbers back into cliffmem */
128:      free((void *)a);     /* free storage allocated by malloc */
129:      free((void *)b);     /* free storage allocated by malloc */
130:      free((void *)d4);    /* free storage allocated by malloc */
131:      free_cliffmem(1);    /* free storage allocated to cliffmem */
132:   }
```

The code for spatial differentiation in Example 23.1b is less than wholly straightforward because it is written to minimise the number of Fourier transforms. Although these are fast Fourier transforms, they still consume a considerable amount of time.

An initial transform is taken on line 15 to convert the waveform into the frequency domain. The outermost loops from line 25 to 36 and from line 44 to 55 process positive and negative frequency components, respectively. Lines 18–24 and 37–43 deal with the zero and maximum frequency components, respectively. The different orders of differentiation are calculated firstly on lines 29 and 48, and then repeatedly for higher orders, one from the previous, in the loops from line 31 to 35 and line 50 to 54.

Once all the frequency domain filters have been applied, the inverse transforms are calculated in lines 59–65 in order to return the results into the spatial domain.

Example 23.1(b)

```
 1:   /* calculate derivatives 1 to k from of waveform or signal a[n] */
 2:   number *multi_integer_differentiate(n,a,dk,k,ds)
 3:        int n;      /* length n must be power of 2 */
 4:        number *a;  /* input array a[n] is recycled before return */
 5:        number *dk; /* output array [k*n] */
 6:        int k;      /* the order of highest derivative */
 7:        double ds;  /* the distance between samples */
 8:   {
 9:     cliff_comp factor;
10:     double d_omega;
11:     number *in,*out;
12:     int i,j,n2;
13:
14:     out=dk+(k-1)*n;
15:     transform(n,a,out);
16:     d_omega=2.0*M_PI/(n*ds);      /* scale for real distance */
17:     n2=n/2;
18:     a[0]=scalar_number(0.0,0.0);
19:     out=dk;
20:     for(j=1;j<k;j++)
21:        {
22:          out[0]=scalar_number(0.0,0.0);
23:          out=out+n;
24:        }
25:     for(i=1;i<n2;i++)              /* i is angular frequency omega */
26:        {
27:          in=dk+(k-1)*n;
28:          pack_comp(&factor,0.0,i*d_omega,0);    /* i*delta_omega */
29:          a[i]=mul_comp(in[i],central,&factor);  /* apply filter */
30:          in=a; out=dk;
31:          for(j=1;j<k;j++)
32:             {
33:               out[i]=mul_comp(in[i],central,&factor);
                  /* apply filter */
34:               in=out; out=out+n;
35:             }
36:        }
37:     a[n2]=scalar_number(0.0,0.0);
38:     out=dk;
39:     for(j=1;j<k;j++)
40:        {
41:          out[n2]=scalar_number(0.0,0.0);
42:          out=out+n;
43:        }
44:     for(i=n2+1;i<n;i++)            /* n-i is angular freqency omega */
45:        {
46:          in=dk+(k-1)*n;
47:          pack_comp(&factor,0.0,-(n-i)*d_omega,0); /* -i*delta_omega */
48:          a[i]=mul_comp(in[i],central,&factor);    /* apply filter */
49:          in=a; out=dk;
50:          for(j=1;j<k;j++)
```

(Continued)

Example 23.1(b) (Continued)

```
51:              {
52:                   out[i]=mul_comp(in[i],central,&factor);
                      /* apply filter */
53:                   in=out;  out=out+n;
54:              }
55:         }
56:    out=dk+(k-1)*n;
57:    in=out-n;
58:    recycle_array(n,out);
59:    for(j=1;j<k;j++)
60:         {
61:            inverse_transform(n,in,out);
62:            out=in;
63:            in=in-n;
64:         }
65:    inverse_transform(n,a,dk);
66:    return(dk);
67:  }
```

Once all of the spatial derivatives have been calculated, they are converted to the temporal derivatives using the code listed in Example 23.1c. This is a straightforward complex scaling of all derivatives, in the loop from line 18 to 27, at all points on the waveform from line 20 to 24.

Example 23.1(c)

```
 1:   /* calculate k time derivatives from spatial derivatives */
 2:   number *dkft_global(n,a,b,d,k,w,v)
 3:        int n,k;
 4:        number *a,*b,*d;
 5:        cliff_comp *w,*v;
 6:  {
 7:    cliff_comp ie0e1v,ie0e1vk;
 8:    number *dk;
 9:    number x;
10:    int i,j;
11:
12:    clone_array(n,a,b);
13:    /* get spatial derivatives */
14:    multi_integer_differentiate(n,b,d,k,w->coef[0]);
15:    dk=d;
16:    pack_comp(&ie0e1v,-(v->coef[1]),v->coef[0],3);  /* (i*e0*e1*v) */
17:    identity(&ie0e1v,&ie0e1vk);            /* (i*e0*e1*v)^k) with k=1 */
18:    for(j=0;j<k;j++)                       /* loop for k derivatives */
19:         {
20:            for(i=0;i<n;i++)
21:              {
22:                dk[i]=comp_mul(&ie0e1vk,central,x=dk[i]);
23:                recycle_number(x);
24:              }
```

```
25:            dk=dk+n;
26:            central(&ie0e1v,&ie0e1vk,&ie0e1vk);      /* (i*e0*e1*v)^k */
27:        }
28:    return(d);
29:  }
```

After calculating the temporal derivatives, the Taylor expansion is evaluated by the code listed in Example 23.1d. Four derivatives give an accurate result with the values chosen here for other parameters, such as spatial sampling and time step. If only three derivatives are used, the results are visibly less accurate.

Example 23.1(d)

```
1:   number taylor4(h,f,df1,df2,df3,df4)
                /* calculate Taylor series at f(t+h) */
2:         cliff_comp *h;
                /* for step size h */
3:         number f,df1,df2,df3,df4;
                /* using f(t) and first four derivatives */
4:   {
5:     cliff_comp temp;
6:     number sum,x;
7:
8:     sum=cloneof_number(df4);
9:     sum=mul_comp(x=sum,central,central(h,pack_comp(&temp,1.0/
                                       4.0,0.0,0),&temp));
10:    recycle_number(x);
11:    sum=addto_number(sum,df3);
12:    sum=mul_comp(x=sum,central,central(h,pack_comp(&temp,1.0/
                                       3.0,0.0,0),&temp));
13:    recycle_number(x);
14:    sum=addto_number(sum,df2);
15:    sum=mul_comp(x=sum,central,central(h,pack_comp(&temp,1.0/
                                       2.0,0.0,0),&temp));
16:    recycle_number(x);
17:    sum=addto_number(sum,df1);
18:    sum=mul_comp(x=sum,central,h);
19:    recycle_number(x);
20:    sum=addto_number(sum,f);
21:    return(sum);
22:  }
```

23.2.2 Example Output

Figure 23.2 depicts the scenario established here for the purposes of experimentation. The spatial coordinate x is of infinite extent, and the shaded pulse in the middle is one of an infinite series of pulses to both left and right. As the series of pulses propagate towards the right, the one visible in the window between 0 and L eventually disappears from view, and the one out of range to the left moves into view.

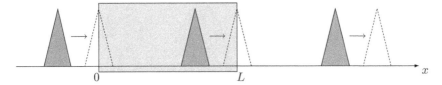

Figure 23.2   Observation window 0 to *L* for infinite series of travelling pulses.

The purpose of this approach is to make it possible to track the propagation of the pulses over a long (in principle unlimited) period of time using only a small amount of computer memory for the finite spatial window, here from $x = 0$ to $x = L = 2\pi = 256\Delta x$.

It takes the pulses two time steps to cover one unit of distance (from x to $x + \Delta x$) so that it takes 512 time steps for any pulse to move away and be replaced by the one on its left. This cycle is repeated every 512 time steps. All pulses are identical; so it is possible in practice to observe the effects on one pulse as it travels over long distances (much greater than L) and long periods of time.

Figure 23.3 shows the result of running the FDTD algorithm for a Gaussian pulse with unit height and a half-width of 20 distance units ($20\Delta x$), starting at time $t = 0$ in the centre of the spatial axis at $x = L/2 = 128\Delta x$. The velocity is positive, so the pulse propagates towards the right. The value plotted on the vertical scale is the electric field. Increasing values of time are plotted from front to back in the figure so that as the pulse propagates towards the right it appears further back in the figure.

The lines on the plane beneath the pulse are contour lines drawn where the height of the pulse takes values of 0.1, 0.3, 0.5, 0.7, and 0.9. These are straight lines because the velocity is constant.

A Gaussian pulse is smooth, with no discontinuities in value or derivative. That means it can be quite well represented in the spatial domain using only a limited number of

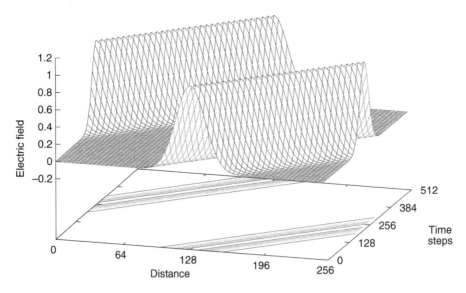

Figure 23.3   Gaussian pulse travelling one cycle.

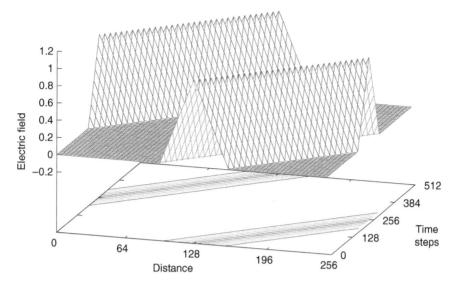

Figure 23.4 Triangular pulse travelling one cycle.

samples, and well represented in the frequency domain using only the first few terms of a Taylor series. In this case, five terms are used: the value of the pulse itself and the first four derivatives.

Figure 23.4 shows the same experiment repeated for a triangular pulse of the same height and half-width. A triangular pulse is not smooth. Although it has no discontinuities in value, it does have three discontinuities in the first (and all higher) derivatives. That means it cannot be as well represented as the Gaussian pulse using the same number of samples in the spatial dimension or terms in the Taylor series, which could lead to deterioration in the shape of the pulse as is travels in space and time. For the single cycle shown, there is no visible deterioration. Longer times are shown in Figures 23.6–23.8.

Figure 23.5 shows the same experiment repeated for a square pulse of the same height and full-width of 40 distance units. A square pulse is neither smooth nor continuous. It has two discontinuities in value and in the first (and all higher) derivatives. This means it cannot be as well represented as the triangular pulse using the same number of samples in the spatial dimension or terms in the Taylor series. Nevertheless, for the single cycle, it is hard to see any visible deterioration in the shape.

Figure 23.6 shows the same experiment as in Figure 23.3 continued for a longer time. In this case, the Gaussian pulse travels across the window 2048 times. The pulse is plotted only once at the same location for every cycle so that it appears to be stationary. As a result, the figure shows the pulse moving along the time axis towards the rear, but apparently frozen in space. This makes it easy to observe the change in the shape of the pulse over long periods of time.

Visible inspection of the figure doesn't show any deterioration in the shape of the pulse. Inspecting the actual numerical values does show a little deterioration, but it is not very significant even after one million iterations of the algorithm.

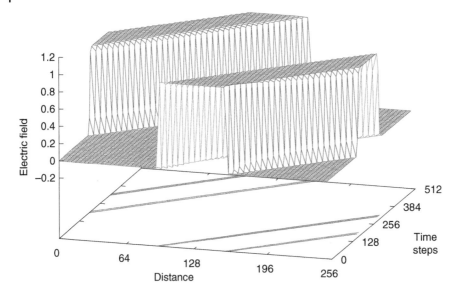

Figure 23.5   Square pulse travelling one cycle.

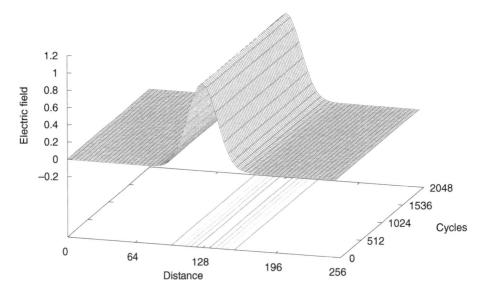

Figure 23.6   Gaussian pulse travelling 2048 cycles.

Figure 23.7 shows the triangular pulse travelling the longer distance. In this case, it is possible to see an initial deterioration in the shape of the pulse from the first to the second instant of time. The forward-most pulse, at $t = 0$ has a sharp peak on top. This is the initial shape, before sampling, before using the spectrum to calculate the derivatives, and before making the Taylor series expansion. The first derivative is not continuous. As a consequence the spatial derivatives are not wholly accurate, the Taylor series is calculated somewhat in error, and none of the later pulses have the same sharp peak.

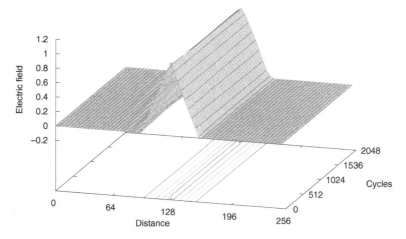

Figure 23.7 Triangular pulse travelling 2048 cycles.

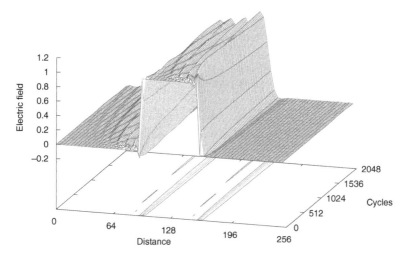

Figure 23.8 Square pulse travelling 2048 cycles.

Figure 23.8 shows the square pulse travelling the longer distance. In this case, deterioration in the shape of the pulse is absolutely clear. As time and distance increases, the slopes of the leading and trailing edges of the pulse become less steep, with a more rounded initial onset. The leading and trailing edges are followed by a damped oscillation, which grows in amplitude as the pulse travels. The oscillation shows on the contour plot beneath the pulse as the extra line trailing the main group. The main group of contours are not parallel. They spread as time increases, as the slopes of the leading and trailing edges lessen.

The oscillations trailing the pulse are seen more clearly in Figure 23.9, which shows the pulse after 1024 cycles at time step 524 288. The pulse has a shape in the same general form as that for an under-damped ($\zeta < 1$) second-order differential system with one variable. The differential system here is first order but has two variables: space and time.

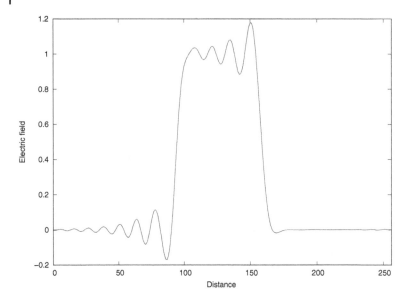

Figure 23.9 Square pulse shape after 1024 cycles.

The oscillations trailing the pulse are consistent with a causal system. Oscillations appearing in front of the pulse are non-causal and would indicate that the simulation was not a good representation of the physical world.

More accurate results can be achieved if more terms are used in the Taylor series, if the pulse is sampled more finely, and if the time steps are reduced. However, all of those mean more computational time.

References

Dirac, P. A. M. (1928), 'The quantum theory of the electron', *Proceedings of the Royal Society. Series A, Containing Papers of a Mathematical and Physical Character* **117**(778), 610–624.

Maxwell, J. C. (1873a), *A Treatise on Electricity and Magnetism*, Vol. 1, Clarendon Press, Oxford, UK.

Maxwell, J. C. (1873b), *A Treatise on Electricity and Magnetism*, Vol. 2, Clarendon Press, Oxford, UK.

Seagar, A. (2018), 'First order wavelike phenomena with electromagnetic and acoustic applications', *IEEE Transactions on Antennas and Propagation* **57**(7), 3526–3531.

Spiegel, M. R. (1968), *Mathematical Handbook of Formulas and Tables*, Schaum's outline series, McGraw–Hill, Cambridge, UK.

Yee, K. (1966), 'Numerical solution of initial boundary value problems involving Maxell's equations in isotropic media', *IEEE Transactions on Antennas and Propagation* **14**(3), 302–307.

24

Cauchy Extension

Cauchy (1841) discovered that an analytic function of a complex variable in the two dimensions of the complex plane can be reproduced everywhere inside a closed boundary from the values of the function on the boundary. Under the right conditions, the extension can be made either to the inner region or the outer region.

The same kind of extension is possible for electromagnetic fields in regions without sources and with uniform and constant material properties when represented in Clifford algebra (McIntosh 1989). The mathematical functions representing the electromagnetic fields are in this case termed monogenic rather than analytic. The monogenic functions behave as a generalisation of complex analytic functions to dimensions higher than two.

24.1   Background

Solutions to problems involving boundary element methods (Brebbia, Telles & Wrobel 1984) yield, in the first instance, field values on some boundary. Often, it is also of interest to evaluate the field away from the boundary, as in the case of the electromagnetic field around an antenna or inside a microwave cavity.

Normally, there is an explicit integral relationship between the field values on the boundary and the field value at any chosen point away from the boundary. The integral usually has one factor, such as a Green's function (Shahpari & Seagar 2020) or a Cauchy kernel, which contains a difference of two terms in the denominator. As the point at which the field is evaluated approaches the boundary, these two terms approach the same value, so division by something close to zero occurs and the value of the Cauchy kernel or Green's function approaches infinity. This makes direct numerical calculation of the field close to or on the boundary problematic. Far from the boundary, there is usually no problem.

Numerical Calculations in Clifford Algebra: A Practical Guide for Engineers and Scientists,
First Edition. Andrew Seagar.
© 2023 John Wiley & Sons Ltd. Published 2023 by John Wiley & Sons Ltd.

24.2 Theory

24.2.1 Two Dimensions

In two dimensions, the complex value of an analytic function is extended from its value $f(z)$ on a closed boundary Σ to the region inside by the Cauchy extension integral (Cauchy 1841, Smithies 1997):

$$f(u) = \frac{1}{2\pi i} \int_\Sigma \frac{f(z)}{z-u}\, dz = \int_\Sigma E_*(z-u)n(z)f(z)\,|dz| \tag{24.1}$$

where $E_*(z-u) = 1/2\pi(z-u)$ is the Cauchy kernel, and $(1/i)dz = n(z)|dz|$ is the product of the complex representation of the unit normal to the boundary $n(z)$ with the element of arc length $|dz|$.

The first integral in Eq. (24.1) is expressed in the standard form. The second integral is an alternative form which conforms better to the more general higher-dimensional integral in Clifford algebra.

24.2.2 Three Dimensions

In three-dimensional space, in a region with uniform and constant material properties μ and ϵ, the electromagnetic field can be represented (see Chapter 10) by a Clifford-valued monogenic function f in four dimensions:

$$f = F = \sqrt{\mu} H\sigma - i\sqrt{\epsilon} E e_0 \tag{24.2}$$

where $H = H_x e_1 + H_y e_2 + H_z e_3$ and $E = E_x e_1 + E_y e_2 + E_z e_3$ are the magnetic and electric fields[1]. The Clifford units e_1, e_2, e_3 play the role of spatial basis vectors, the extra dimension e_0 serves as the unit of frequency, and $\sigma = -e_1 e_2 e_3$ plays the role of the negatively signed unit of volume. The imaginary unit i behaves as usual with $i^2 = -1$, and the Clifford units follow the rules in Section 5.3.1 with negative signature:

$$e_p e_q = -e_q e_p \quad \text{and} \quad e_p e_p = -1 \tag{24.3}$$

In this case, the Cauchy extension can be obtained when the integral is recast from complex algebra into the context of Clifford algebra (Clifford 1878, McIntosh 1989) as follows:

$$f(\mathbf{r}) = \begin{cases} f^+(\mathbf{r}) = +\int_\Sigma E_k(\mathbf{s}-\mathbf{r})n(\mathbf{s})f(\mathbf{s})\,d\sigma(\mathbf{s}) \\[2mm] f^-(\mathbf{r}) = -\int_\Sigma E_k(\mathbf{s}-\mathbf{r})n(\mathbf{s})f(\mathbf{s})\,d\sigma(\mathbf{s}) \end{cases} \tag{24.4}$$

where E_k is the Cauchy kernel[1], points $\mathbf{s}$, and $\mathbf{r}$ are, respectively, on and away from the boundary Σ, n is the outward unit normal, and $d\sigma$ is[2] the element of boundary measure. The positive sign is taken for points $\mathbf{r}$ inside the boundary, and the negative sign is taken for points $\mathbf{r}$ outside the boundary.

---

1 For notational simplicity, the wavenumber subscript k is left unwritten on the field quantities H, E, and I. So here E without the subscript is the electric field, and E_k with the subscript k is the Cauchy kernel.
2 The element of boundary measure $d\sigma$ should not be confused with the negatively signed unit of volume $\sigma = -e_1 e_2 e_3$.

Figure 24.1 Cuboctahedron with Cauchy extension of the electromagnetic field from points **s** on the surface Σ to points **r** outside.

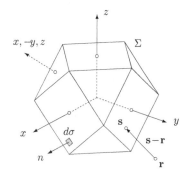

Figure 24.1 shows the case for extending the field from the surface Σ of a cuboctahedron with edges of length 1 m to a point **r** located in the region outside. In this case, the field extension is calculated using the second line of Eq. (24.4).

24.2.3 Singularity

The Cauchy kernel, in the Clifford formalism employed here, takes values in one to three dimensions:

$$
E_k(\mathbf{r}) = \begin{cases}
1D: & -\frac{1}{2}(\hat{\mathbf{r}} - ie_0)e^{-ikr} \\
2D: & ikr\left(\hat{\mathbf{r}}H_1^{(2)}(kr) - e_0 H_0^{(2)}(kr)\right)\frac{1}{4r} \\
3D: & -(\hat{\mathbf{r}} + ikr(\hat{\mathbf{r}} - ie_0))\frac{e^{-ikr}}{4\pi r^2}
\end{cases}
\tag{24.5}
$$

where the $H^{(2)}$ are the Hankel functions of second kind with orders 0 and 1 (Abramowitz & Stegun 1972), and[3] $k = \omega\sqrt{\mu}\sqrt{\epsilon}$ is the wavenumber.

The complex Cauchy kernel E_* and the Clifford Cauchy kernels E_k in two and higher dimensions become singular when evaluated for points u or **r** progressively closer to the boundary, all dividing by zero when reaching the point $u = z$ or $\mathbf{r} = \mathbf{s}$. Evaluating the Cauchy extension as in Eq. (24.1) or Eq. (24.4) without paying any special attention to the singularity usually gives accurate results for points far from the boundary, where either $|\mathbf{s} - \mathbf{r}| \gg d$ or $|z - u| \gg d$ with d as the effective diameter[4] of the boundary. However, the accuracy of the results usually deteriorates closer towards the boundary, even before making contact. Examples calculating the Cauchy extension in a simplistic way showing the large error near the boundary are given in Section 24.3.

24.2.4 The Taming Function

In addition to the (monogenic) electromagnetic field f, it is useful to introduce an auxiliary monogenic function w, subtracting it from the original field f on both sides of Eq. (24.4).

---

3 In the case of meta-materials, for which either the electric permeability μ or the permittivity ϵ may be negative, care must be taken because $\sqrt{\mu}\sqrt{\epsilon} \neq \sqrt{\mu\epsilon}$.
4 Some such measure in N dimensions as the Nth root of the volume.

This produces a somewhat more complicated form of Cauchy extension:

$$f(\mathbf{r}) = w(\mathbf{r}) \pm \int_{\Sigma} E_k(\mathbf{s} - \mathbf{r})n(\mathbf{s})[f(\mathbf{s}) - w(\mathbf{s})] \, d\sigma(\mathbf{s}) \tag{24.6}$$

The addition of w outside the integral cancels the subtraction inside the integral because w has been specifically chosen as a monogenic function, which means it satisfies Eq. (24.4) with w in place of f. In that case the left-hand side of Eq. (24.6) has the same value regardless of whether w is included or not. However, with w in place the integrand on the right-hand side is different. It is now possible to choose w so that the integrand is no longer singular. A particular function w which eliminates the singularity is called a taming function.

The taming function w is constructed in $N + 1$ dimensions so the value of the difference $[f(\mathbf{s}) - w(\mathbf{s})]^{1/N}$ approaches zero at a rate proportional to or higher than the rate at which point $\mathbf{s}$ on the boundary approaches point $\mathbf{r}$, somewhere close to or even on the boundary. In this case, the product $E_k(\mathbf{s} - \mathbf{r})[f(\mathbf{s}) - w(\mathbf{s})]$ takes the form:

$$\lim_{\mathbf{s} \to \mathbf{r}} \frac{f(\mathbf{s}) - w(\mathbf{s})}{(\mathbf{s} - \mathbf{r})^N} \sim \lim_{\mathbf{s} \to \mathbf{r}} \frac{(\mathbf{s} - \mathbf{r})^N}{(\mathbf{s} - \mathbf{r})^N} \tag{24.7}$$

In the limit, the ratio remains finite and approaches some well-behaved fixed value.

In essence, the taming function w plays the role of a monogenic approximation to the field f. If the approximation were known perfectly, i.e. $w(\mathbf{s}) = f(\mathbf{s})$, there would be no need to calculate the integral numerically. When the approximation is imperfect, the integral is carried only on the difference between the approximation w and the field f. Any numerical error is on the difference, not the whole field.

24.2.5 Construction

The method for constructing the taming function here, for electromagnetic fields of four Clifford dimensions, is based on the method described by Lee, Liu, Hong & Chen (2016) in three dimensions for quaternions (Chapter 7).

The taming function w may be constructed in the form:

$$w(\mathbf{r}) = \left[p(\mathbf{r})p(\mathbf{x})^{-1} \right] f(\mathbf{x}) \tag{24.8}$$

from any source T (⁂, known as the taming source), located at a point $\mathbf{q}$ (•) away from the boundary, which generates a monogenic field $p(\mathbf{r})$.

The construction is only effective when the point $\mathbf{r}$ (at which the extension is calculated) and the point $\mathbf{x}$ are located either on the boundary or somewhere in the region on the opposite side of the boundary from the source T.

Figure 24.2 shows the case with the monogenic field $p(\mathbf{r})$ generated by the source T at point $\mathbf{q}$ inside the boundary, and with points $\mathbf{x}$ and $\mathbf{r}$ approaching the boundary from the outside. The field $p(\mathbf{r})$ is monogenic since, by construction, the region occupied by the point $\mathbf{r}$ contains no sources. The function $w(\mathbf{r})$ inherits its monogenic nature from $p(\mathbf{r})$. The factors $p(\mathbf{x})^{-1}$ and $f(\mathbf{x})$ in Eq. (24.8) do not interfere with that inheritance because the point $\mathbf{x}$, although freely chosen, is a fixed point under differentiation with respect to $\mathbf{r}$, meaning that $p(\mathbf{x})^{-1}$ and $f(\mathbf{x})$ behave as constants.

The value of the monogenic field p can be calculated at any location, but the value of the field f is initially known only on the boundary. So in principle, although the point $\mathbf{x}$ can be either on or off the boundary, in practice it is restricted to being on the boundary.

Figure 24.2 Variable of integration **s** passes point of evaluation **r** close to boundary Σ.

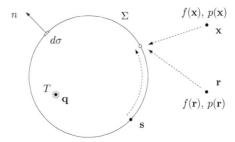

When a Hertzian dipole[5] is chosen for the taming source T, the monogenic field is

$$p(\mathbf{r}) = \mathcal{F}_{(\mathrm{Hz})}(\mathbf{r} - \mathbf{q}, Ih) \tag{24.9}$$

and the term in square brackets of Eq. (24.8) is a ratio of two bivectors. This ratio can be calculated as the product of the numerator with the reciprocal of the denominator. The reciprocal of a Clifford bivector p written as an electromagnetic field:

$$p = \sqrt{\mu}H\sigma - i\sqrt{\epsilon}Ee_0 = a\sigma + be_0 \tag{24.10}$$

where a and b are Clifford vectors, is given (as in Eq. (10.29)) by

$$\frac{1}{a\sigma + be_0} = \frac{1}{p} = p^{-1} = (a - ba^{-1}b)^{-1}\sigma + (b - ab^{-1}a)^{-1}e_0 = c^{-1}\sigma + d^{-1}e_0 = x\sigma + ye_0 \tag{24.11}$$

The effectiveness of the construction of the taming function w when using the Hertzian dipole as the taming source is twofold. First, if position, strength, and orientation of the dipole source are chosen so that the monogenic field p well approximates the field f at least at specific points **s** and **x**, then the term in square brackets in the integrand:

$$\left[f(\mathbf{s}) - \frac{p(\mathbf{s})}{p(\mathbf{x})}f(\mathbf{x}) \right] \xrightarrow{p \to f} \left[f(\mathbf{s}) - \frac{f(\mathbf{s})}{f(\mathbf{x})}f(\mathbf{x}) \right] \longrightarrow \left[f(\mathbf{s}) - f(\mathbf{s}) \right] \longrightarrow 0 \tag{24.12}$$

reduces to zero without any particular constraint on the points **x** and **r**. Second, if the point **x** is chosen to coincide with the point **r** (although that is not possible if **r** is away from the boundary), then:

$$\left[f(\mathbf{s}) - \frac{p(\mathbf{s})}{p(\mathbf{x})}f(\mathbf{x}) \right] \xrightarrow{\mathbf{x} \to \mathbf{r}} \left[f(\mathbf{s}) - \frac{p(\mathbf{s})}{p(\mathbf{r})}f(\mathbf{r}) \right] \xrightarrow{\mathbf{s} \to \mathbf{r}} \left[f(\mathbf{r}) - f(\mathbf{r}) \right] \longrightarrow 0 \tag{24.13}$$

which reduces to zero as **s** approaches **r**.

Together Eqs. (24.5), (24.6), (24.8), and (24.9), with Eq. (24.11), provide all that is required to extend the field away from the boundary without undue problems. Better results may be expected if both conditions (24.12) and (24.13) are met.

If the first condition is not met, either because the Hertzian taming source is poorly positioned and therefore p does not well approximate the field f, or because some other monogenic function is chosen, the second condition is still met. In this case, a simpler alternative taming source is the Dirac delta function, δ. The monogenic field generated by a delta function is

$$p(\mathbf{r}) = F_k(\mathbf{r} - \mathbf{q}) = -E_k(\mathbf{q} - \mathbf{r}) \tag{24.14}$$

---

5 See Chapter 22.

where F_k is the (fundamental) solution of the equation $D_k F_k = \delta$, in which $D_k = k e_0 + \frac{\partial}{\partial x} e_1 + \frac{\partial}{\partial y} e_2 + \frac{\partial}{\partial z} e_3$ is the k-Dirac operator, and where E_k is the Cauchy kernel in Eq. (24.5). In this case, the monogenic field p is a vector field rather than a bivector (electromagnetic) field.

Results for taming the singularity using both a Hertzian dipole and a Dirac delta function are given in Section 24.3.

24.3 Practice

The Cauchy extension has been calculated along the radial line $x = -y = z$ (as shown in Figure 24.1), starting on the surface at the centre of one of the triangular faces of the cuboctahedron (0.4714 m from the origin) and continuing outwards to 10 times that distance, using the field at the surface as generated by a Hertzian dipole standing at the origin aligned vertically, as in Figure 22.1.

The calculation has been repeated at coarse and finer resolutions. The most coarse resolution (1/1) uses one tile for each face, and the finest resolution (1/30) equally subdivides the faces into 30×30 = 900 tiles. A polynomial function of two dimensions is used to represent each component of the electromagnetic field on the surface of each tile as follows:

$$\begin{aligned}
f(r, \theta) &= a_0 + a_1 x + a_2 y + a_3 (x^2 + y^2) + a_4 (x^2 - y^2) + a_5 2xy + a_6 (x^2 + y^2)x \\
&\quad + a_7 (x^2 + y^2)y \\
&= a_0 + a_1 r \cos\theta + a_2 r \sin\theta + a_3 r^2 + a_4 r^2 \cos 2\theta \\
&\quad + a_5 r^2 \sin 2\theta + a_6 r^2 \cos\theta + a_7 r^2 \sin\theta
\end{aligned} \tag{24.15}$$

The polynomial includes all terms to the third order except for $x(x^2 - 3y^2) = r^3 \cos 3\theta$ and $y(3x^2 - y^2) = r^3 \sin 3\theta$.

The extension has been calculated both without and with the two taming functions, generated using delta function (Eq. (24.14)) and Hertzian dipole (Eq. (24.9)) as taming sources, with the integrals (24.6) and (24.4).

24.3.1 Example Code

The code listed in Example 24.1a and 24.1b provides the two monogenic functions $p(\mathbf{r})$, for taming sources of Hertzian dipole (Eq. (24.9)) and delta function (Eq. (24.14)). In Example 24.1a, the values of the material properties are both set to unity. Their effects cancel when the ratio of monogenic functions is calculated in Eq. (24.8), so any non-zero values will suffice.

Example 24.1(a)

```
1:   /* Hertzian dipole */
2:   number mono_function1(k,r,q)
3:       cliff_comp *k;
```

```
 4:          number r,q;
 5:     {
 6:        cliff_comp one;
 7:        number p,rmq;
 8:
 9:        pack_comp(&one,1.0,0.0,0);
10:        rmq=sub_number(r,q,0);
11:        p=hertzian_dipole(&one,&one,k,rmq);
12:        recycle_number(rmq);
13:        return(p);
14:     }
```

Example 24.1(b)

```
 1:    /* fundamental solution for k-Dirac operator */
 2:    number mono_function2(k,r,q)
 3:          cliff_comp *k;
 4:          number r,q;
 5:     {
 6:        number p,rmq;
 7:
 8:        rmq=sub_number(r,q,0);
 9:        p=fundamental_solution3D(k,rmq);
10:        recycle_number(rmq);
11:        return(p);
12:     }
```

The taming function $w(\mathbf{r})$ in Eq. (24.8) is calculated using the sub-routine in Example 24.1c, using either taming source at line 31. The ratio in Eq. (24.8) is calculated either on line 34 or 35 depending on the nature of the entity produced by the source, being either bivector or vector. The nature of the entity is examined using the tests is_vector() and is_bivector().

Example 24.1(c)

```
 1:    number taming(k,qinside,emfield,itile,isample)
 2:       cliff_comp *k;
 3:       number qinside;
 4:       surface_field *emfield;
 5:       int itile,isample;
 6:    {
 7:       boundary *solid;
 8:       tile *tile_array,*thetile;
 9:       tile_field *field_array;
10:       number *tile_position,*points;
11:       number origin,x,fx,px,dx,dy;
12:       double z0[2];
```

(Continued)

Example 24.1(c) (Continued)

```
13:
14:    field_array=emfield->field;              /* array[n] of tile
                                                    fields */
15:    fx=field_array[itile].field[isample];  /* field of sample */
16:
17:    solid=emfield->surface;                  /* geometry of all tiles */
18:    tile_array=solid->faces;                 /* array[n] of tile
                                                    faces */
19:    tile_position=solid->places;             /* array[n] of locations
                                                    of tiles */
20:    origin=tile_position[itile];             /* corner of tile */
21:    thetile=(&tile_array[itile]);            /* structure with tile
                                                    shape */
22:    dx=thetile->dx;                          /* first side of tile */
23:    dy=thetile->dy;                          /* second side of tile */
24:    points=thetile->points;                  /* includes origin */
25:    if(isample==0) { z0[0]=0.0;      z0[1]=0.0;        /* for centre
                                                             point */
26:                     x=point_on_tile(z0,points[0],dx,dy); }
27:    else            { get_coef(points[0],0,z0,0);      /* for other
                                                             points */
28:                     x=point_on_tile(z0,points[isample],dx,dy); }
29:    x=addto_number(x,origin);                /* locate in space */
30:
31:    px=mono_function(k,x,qinside);
32:    recycle_number(x);
33:
34:    if(is_bivector(px))       { x=central_mul(inverseof_emfield(px,1),
                                                 fx,1); }
35:    else { if(is_vector(px))  { x=central_mul(inverseof_vector(px,1),
                                                 fx,1);   }
36:          else { printf("entity neither bivector or vector in
                            routine: taming()\n");
37:                 brexit(0); }}
38:    return(x);
39: }
```

The code listed in Example 24.1d calculates the Cauchy extension at one point 'r' using a taming source located at point 'q', as in Eq. (24.6). The point 'r' is either on the boundary or in the region on the side of the boundary indicated by the value of the flag 'side'. The point 'q' is on the other side. No checks are made to verify that the two points are consistent with the flag provided, that is the responsibility of the calling routine.

Example 24.1(d)

```
1:   /* cauchy extension at one point */
2:   number cauchy_extension_x(k,r,side,q,tam,emfield,weights)
3:       cliff_comp *k;     /* wavenumber */
```

```
 4:        number r;          /* location of point */
 5:        int side;          /* +1 inside, -1 outside */
 6:        number q;          /* location of taming source */
 7:        number tam;        /* taming factor */
 8:        surface_field *emfield;
 9:        matrix weights[2]; /* weights for tile integration */
10:   {
11:   double z0[2];
12:   boundary *solid;
13:   number *tile_position,*field,*points;
14:   tile *tile_array,*thetile;
15:   tile_field *field_array, *thefield;
16:   int i,j,ntiles,npoints;
17:   number dx,dy,origin,nm,x,s,sr,Ek,Ekn,Eknf,sum;
18:   number pr,ps,wr,ws,fs;
19:   tile_field integrand;
20:
21:   ntiles=emfield->ntiles;       /* number of tiles with field */
22:   solid=emfield->surface; /* geometry of all tiles */
23:   if(ntiles!=solid->ntiles)    /* must match or something
                                       missing */
24:     {
25:       printf("field tiles (%d) fails to match surface tiles",
                  ntiles);
26:       printf(" (%d) in routine: cauchy_extension_x()\n",
                  solid->ntiles);
27:       brexit(0);
28:     }
29:   field_array=emfield->field;   /* array[n] of tile fields */
30:   tile_array=solid->faces;      /* array[n] of tile faces */
31:   tile_position=solid->places;  /* array[n] of locations of tiles */
32:   sum=fresh_number();
33:   for(i=0;i<ntiles;i++)          /* loop and sum over tiles */
34:     {
35:       thefield=(&field_array[i]);    /* field quantities */
36:       npoints=thefield->npoints;     /* number 10or9 of field
                                            points */
37:       field=thefield->field;         /* array[10or9] of field
                                            values */
38:       origin=tile_position[i];       /* corner of tile */
39:       thetile=(&tile_array[i]);      /* structure with tile
                                            shape */
40:       dx=thetile->dx;                /* direction of first side */
41:       dy=thetile->dy;                /* direction of second
                                            side */
42:       nm=thetile->normal;            /* direction of unit
                                            normal */
43:       points=thetile->points;        /* array[10or9] of local
                                            points on tile */
44:       if(thetile->npoints!=npoints)  /* points on tile includes
                                            centre */
45:         {
```

(Continued)

Example 24.1(d) (Continued)

```
46:              printf("tile points (%d) fails to match field points",
                      thetile->npoints);
47:              printf(" (%d) in routine: cauchy_extension_x()\n",
                      npoints);
48:              brexit(0);
49:            }
50:        integrand.npoints=npoints;        /* 10 or 9 field points
                                                integrated */
51:        integrand.field=integrand.field10; /* maximum of 10 or 9 */
52:        for(j=0;j<npoints;j++)            /* all points */
53:          {
54:            if(j==0) { z0[0]=0.0;    z0[1]=0.0;
                                        /* for centre point */
55:                    x=point_on_tile(z0,points[0],dx,dy);
                                        /* location on tile */
56:                    get_coef(points[0],0,z0,0); }
                                        /* for other points */
57:            else    { x=point_on_tile(z0,points[j],dx,dy); }
                                        /* location on tile */
58:            s=addto_number(x,origin);    /* location of point in 3D
                                                space */
59:            ps=mono_function(k,s,q);
60:            ws=central_mul(ps,tam,1);
61:            fs=sub_number(field[j],ws,2);
62:            sr=sub_number(s,r,1);        /* s-r */
63:            Ek=cauchy_kernel3D(k,sr);
64:            Ekn=central_mul(Ek,nm,1);    /* need normal */
65:            Eknf=central_mul(Ekn,fs,1);
66:            integrand.field[j]=Eknf;
67:            recycle_number(sr);
68:            recycle_number(fs);
69:          }
70:        sum=add_number(sum,integrate_tile(thetile,&integrand,
                                                weights),3);
71:        recycle_array(npoints,integrand.field);
72:      }
73:    switch(side)              /* determine which side of boundary */
74:      {
75:      case -1: sum=negativeof_number(sum,1); break; /* outside */
76:      case  1:                          break; /* inside   */
77:      default:
78:        printf("side of boundary (%d) not (+1,-1) in routine:
                cauchy_extension_x()\n",side);
79:        brexit(0);
80:        break;
81:      }
82:    pr=mono_function(k,r,q);
83:    wr=central_mul(pr,tam,1);
84:    sum=add_number(sum,wr,3);
85:    sum=keepone_grade(sum,2);
86:    return(sum);
87:  }
```

The Cauchy integral is effected for any single tile by the invocation of routine integrate_tile() on line 70. The loop from line 52 to 69 sets up values of the integrand on each tile from the Cauchy kernel, the value of the field, and the value of the taming function at all points on the tile for which the field is known.

The sum of all integrals is also effected on the same line by invoking the routine add_number() repeatedly within the loop from line 33 to 72. The value of the integral is inverted if the point 'r' is outside the boundary on line 75. Finally, the taming function is added outside the integral on lines 82–84.

The result at this stage should reside entirely within grade 2. However, there are also often some non-zero values in grades 0 and 4 due to imprecise cancellation of numerical calculations. When those values are effectively zero to within machine precision, that is an indication that the integral is probably working well. If these values are significantly greater than machine precision, that is an indication that the integral is probably not working well. That could be because there is an error in the implementation or because the field over the tiles is not well enough represented[6] by the chosen basis functions, here as in Eq. (24.15). In the implementation here no consistency check is made, and any imprecise zeros in grades other than two are removed by the filter on line 85.

24.3.2 Example Output

Figure 24.3 shows the magnitude of the field from the Hertzian dipole at the origin, and the error when using the simplistic untamed version of the Cauchy integral (Eq. (24.4)) to

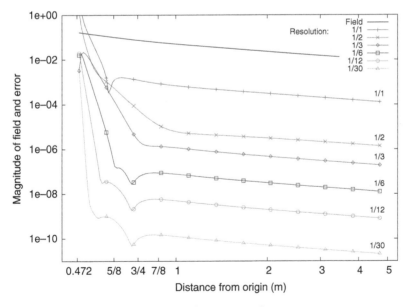

Figure 24.3 Dipole field and error in Cauchy extension.

---

6 In principle, the basis functions need to represent the trace of a three-dimensional monogenic function on the two-dimensional surface.

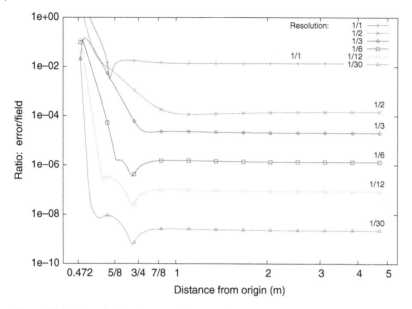

Figure 24.4 Error in Cauchy extension as ratio.

reproduce the field by extension from the surface, for resolutions of 1/1, 1/2, 1/3, 1/6, 1/12, and 1/30.

Figure 24.4 shows the same results with the error plotted as a ratio to the dipole field.

In both figures, a large error near the surface is evident. The accuracy away from the surface is improved when the faces are taken at higher resolution, but the accuracy at the surface is not improved in that way. In the far field (here further than 1 m from the origin) the relative error is effectively constant with values as shown in Figure 24.5.

Figure 24.6 shows the errors in the Cauchy extension calculated in the near field (less than 1 m from the origin) when using the tamed integral formulation of Eq. (24.6) with the Hertzian dipole as taming source T, and Figure 24.7 shows the error when using the Dirac delta function. In both figures, the original (untamed) errors are shown as dashed lines, and the corresponding tamed errors are shown as continuous lines with symbols attached.

For both the taming sources, the error at the surface is reduced by one or more orders of magnitude. However, in neither case is it reduced to the asymptotic values in the far field. Far-field behaviour as more or less straight lines is extended towards the surface to about 0.55 m for the Hertzian taming source but less so, only to around 0.7 m for the Dirac delta taming source. This is consistent with the Hertzian taming source accommodating two mechanisms in the removal of the singularity, whereas the Dirac delta taming source relies only on one. Nevertheless, in this example, as shown by Figures 24.6 and 24.7, the overall effectiveness of the two approaches is not significantly different.

For this experiment, the location for the taming source T was chosen arbitrarily at $q = 0.1(e_1 + e_2 + e_3)$, without any attempt to maximise the improvement in accuracy. A more sophisticated approach in the case of the Hertzian taming source is to choose the location and orientation of the dipole to best fit the field on the surface in the vicinity where the field is to be evaluated. Doing that for the data here would be somewhat dishonest

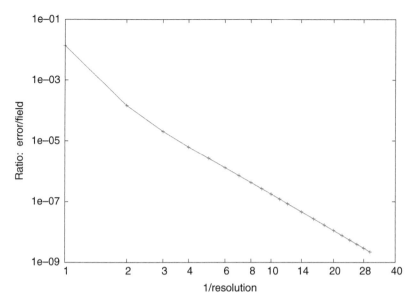

Figure 24.5 Error in Cauchy extension as function of resolution.

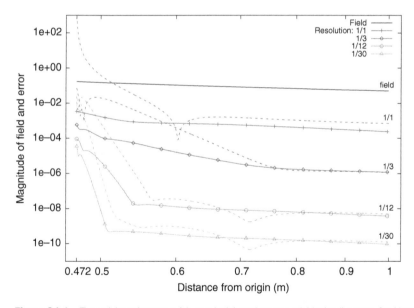

Figure 24.6 Tamed (continuous with symbols) and untamed (dashed) errors for Hertzian taming source.

because the field data itself is taken from another Hertzian dipole and the outcome would be everywhere perfect.

   If the data were generated from some other source, such as two or more Hertzian dipoles or a dipole with non-zero length, the outcome of taming source fitting would not be perfect everywhere at the same time. In this case, in order to achieve the highest accuracy for every

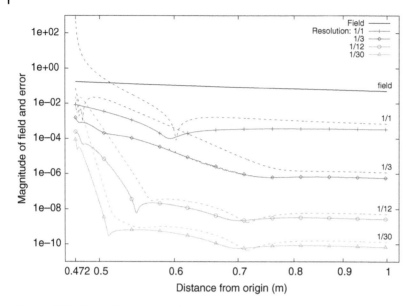

Figure 24.7 Tamed (continuous with symbols) and untamed (dashed) errors for Dirac delta taming source.

point to which the field is extended, it is possible to relocate the taming source for each point. However, that is more computationally intensive.

References

Abramowitz, M. and Stegun, I. (1972), *Handbook of Mathematical Functions*, 9th edn, Dover, New York, US-NY.

Brebbia, C., Telles, J. and Wrobel, L. (1984), *Boundary Element Techniques – Theory and Applications in Engineering*, Springer-Verlag, Berlin.

Cauchy, A. (1841), Résumé d'un mémoire sur la mécanique céleste et sur un nouveau calcul appelé calcul des limites, *in* Bachelier, ed., 'Exercices d'Analyse et de Physique Mathématique', Vol. 2, Bachelier, Paris, France, pp. 41–98.

Clifford, W. K. (1878), 'Applications of Grassmann's extensive algebra', *American Journal of Mathematics* **1**(4), 350–358.

Lee, J., Liu, L., Hong, H. and Chen, J. (2016), 'Applications of the Clifford algebra valued boundary element method to electromagnetic scattering problems', *Engineering Analysis with Boundary Elements* **71**, 140–150.

McIntosh, A. (1989), Clifford algebras and the higher-dimensional Cauchy integral, *in* Z. Ciesielski, ed., 'Approximation and Function Spaces', number 22 *in* 'BCP series', Banach Center Publications, Warsaw, Poland, pp. 253–267.

Shahpari, M. and Seagar, A. (2020), 'Leveraging prior known vector Green's functions in solving perturbed Dirac equation in Clifford algebra', *Advances in Applied Clifford Algebras* **30**(4), 56/1–56/17.

Smithies, F. (1997), *Cauchy and the Creation of Complex Function Theory*, Cambridge University Press, Cambridge, UK.

25

Electromagnetic Scattering

25.1 Background

Electromagnetic scattering problems formulated using the Cauchy integral (see Chapter 24) lead to a system of linear equations. The simplest useful case is perfect reflection, as for a perfect electrical conductor sitting in free (non-conducting) three-dimensional space. The formulation gives a linear system (Seagar 2016):

$$Mf = (I - P^- Q^+)f = P^- g \tag{25.1}$$

where g is a boundary field derived from the incident field, P^- and Q^+ are projection operators, I is the identity operator, and f is the unknown reflected (scattered) field at the boundary[1]. The linear system for general values of material properties is given in Seagar (2016).

In essence, Eq. (25.1) is a linear system where g is the known data vector, f is the unknown vector, and the term in parentheses is a matrix M. The individual elements in the vectors are Clifford variables in the form:

$$u = \sqrt{\mu} H\sigma - i\sqrt{\epsilon} E e_0 = p\sigma + q e_0 \tag{25.2}$$

as in Section 10.1.2.

The linear system can be solved formally by inverting the matrix as follows:

$$f = M^{-1} P^- g = (I - P^- Q^+)^{-1} P^- g \tag{25.3}$$

using whatever direct or indirect (iterative) technique seems most suitable.

Indirect techniques based on a sequence of projections are investigated in Chantaveerod & Seagar (2009), Seagar & Chantaveerod (2015), Seagar & Espinosa (2016), and Seagar (2017, 2018). Those methods seem to work quite well for all materials in one dimension. One dimension has the advantage that the Cauchy integral can be calculated analytically in closed form, and the boundary field can be sampled in full resolution at all of the (two) points on the boundary. Progressing to two or three dimensions means the Cauchy integral must be evaluated numerically and the sampling of the boundary is necessarily finite and incomplete.

---

1 A closed surface when in three dimensions.

Numerical Calculations in Clifford Algebra: A Practical Guide for Engineers and Scientists,
First Edition. Andrew Seagar.
© 2023 John Wiley & Sons Ltd. Published 2023 by John Wiley & Sons Ltd.

In two dimensions, this leads to a sequence of projections which may or may not converge effectively. Whether convergence is achieved or not depends on the particular projections used in any specific case, and those depend on the shape and material properties of the scatterer. Both convincing (good) results (Seagar 2018) and unconvincing (poor) results (Seagar & Espinosa 2016) have been demonstrated. If the projections used are ineffective, other projections derived from the originals may in principle be more effective. However, this approach has not yet been demonstrated in practice.

Direct techniques require calculations taking the form of some kind of matrix inversion, which at the simplest could be one such as Gaussian elimination. In order to do that, it is necessary to calculate the inverse of the individual elements in the matrix. For the scattering problem here, the matrix elements are constructed from the products formed by multiplying the Cauchy kernel with the vector normal to the boundary. The Cauchy kernel (Eq. (24.5)) is a vector in space–time, with all four Clifford units e_0, e_1, e_2, e_3. The normal is a vector in space only with the three spatial units e_1, e_2, e_3. The products fall in both grades 0 and 2. The inverses of those products fall into all even grades in four dimensions, i.e. 0, 2, and 4.

25.2 Theory

In order to solve the linear system in Eq. (25.1) directly within the realms of Clifford algebra, it is necessary to calculate the inverse of four-dimensional Clifford numbers with non-zero coefficients in grades 0, 2, and 4, in the form of the extended electromagnetic field:

$$v = s + p\sigma + qe_0 + r\sigma e_0 \tag{25.4}$$

as in Section 10.3.1. The inverse[2] (which also occupies grades 0, 2, and 4) can then be used in operations to make linear combinations of rows in the matrix, eliminating the off-diagonal elements one at a time.

Table 25.1 provides a comparison of the computational effort for solving field problems in three dimensions using Clifford and equivalent scalar n-tuple approaches. The middle set of three columns applies to electromagnetic fields when solution is effected for both magnetic and electric fields simultaneously. The Clifford entities utilised are extended electromagnetic fields with eight components. This is compared to scalar approaches with either the three Cartesian components (p_x, p_y, p_z) of the magnetic field, and the three Cartesian components (q_x, q_y, q_z) of the electric field as six independent entities bundled together in a 6-tuple, or with eight entities in the form of an 8-tuple if scalar s and pseudo-scalar $r\sigma e_0$ are also included.

The same set of columns also apply to other wavelike phenomena, such as acoustics. In that case, the local deviation in particle velocity plays the role of the magnetic field, and the local deviation in pressure plays the role of the electric field. See Seagar (2018) for details.

The right-hand set of four columns applies to electromagnetic fields when solution is effected separately for just one of the fields, magnetic or electric. That is possible for cases

---

2 In Section 10.3.2.

Table 25.1 Computational effort for alternative methods of matrix solution for electromagnetic field problems.

Row		Simultaneous H and E fields			Separate H or E field			
		Extended EM field	Scalar 8-tuple	Scalar 6-tuple	Quaternion	Pauli matrix	Scalar 4-tuple	Scalar 3-tuple
1	Clifford dimension	4	1	1	3	3	1	1
2	Grades utilised	$0, 2, 4$	0	0	$0, 2$	$0, 2$	0	0
3	Entities/field point	1	8	6	1	1	4	3
4	Components/entity	8	1	1	4	4	1	1
5	Components/field point	8	8	6	4	4	4	3
6	Field points/problem	n	n	n	n	n	n	n
7	Entities/problem	n	$8n$	$6n$	n	n	$4n$	$3n$
8	Operations/problem	n^3	$(8n)^3$	$(6n)^3$	n^3	n^3	$(4n)^3$	$(3n)^3$
9	Multiplications/operation	8^2	1^2	1^2	4^2	4^2	1^2	1^2
10	Multiplications/problem	$8^2 n^3$	$8^3 n^3$	$6^3 n^3$	$4^2 n^3$	$4^2 n^3$	$4^3 n^3$	$3^3 n^3$
11	Relative time (serial)	1	8	$3\frac{3}{8}$	1	1	4	$1\frac{11}{16}$
12	Parallelisation/field point	8^2	1^2	1^2	4^2	4^2	1^2	1^2
13	Relative time (parallel)	$\frac{1}{64}$	8	$3\frac{3}{8}$	$\frac{1}{16}$	$\frac{1}{16}$	4	$1\frac{11}{16}$
14	Vector comparative time	$\frac{1}{216}$	$2\frac{10}{27}$	1	$\frac{1}{27}$	$\frac{1}{27}$	$2\frac{10}{27}$	1

where the fields are static. The same set of four columns applies also to other physical problems in which the mathematical solution can be reduced to a single field in three dimensions.

The inverse of a matrix of size $n \times n$ by Gaussian elimination entails in the order of n^3 multiplication operations. Within Clifford algebra, each multiplication operation between two extended electromagnetic fields involves $8^2 = 64$ complex multiplications, as shown in row 9 of Table 25.1, giving a total of $8^2 n^3$ complex multiplications for the whole matrix inversion, as in row 10. Cast outside of Clifford algebra there are either $(8n)^3$ operations for a scalar 8-tuple, or only $(6n)^3$ if a scalar 6-tuple formulation is used.

Comparing the relative time for the number of operations in row 10 shows in row 11 that the solution within Clifford algebra is in principle between three and four times faster than the fastest of the two scalar solutions. The improvement in speed arises because the savings in reducing the size of the matrix outweigh the extra cost of the Clifford multiplications in comparison to scalar multiplication.

A further improvement in speed can be achieved by executing the 64 individual scalar multiplications internal to each single Clifford multiplication, in parallel, using such additional hardware as a processor with multiple arithmetic cores or a graphics processor unit (GPU). Note that the same acceleration, at the level of multiplying the entity representing the field, cannot be achieved for the solution using 6- or 8-tuple because the multiplications in that case are already at the individual scalar level and can be partitioned no further. Even if that were possible, any acceleration at the base scalar level can be applied to both Clifford and n-tuple approaches equally, not further affecting the speed of the two methods in relation to one another.

Combining the improvement in speed from the Clifford formulation with the improvement from the hardware parallelisation gives the relative times for the alternative methods in row 13 of Table 25.1, listed again in row 14 normalised against the fastest of the two scalar solutions. For the solution of a full electromagnetic field, the Clifford approach using the extended electromagnetic fields is in principle $6^3 = 216$ times faster.

The case for the solution of fields involving three Cartesian components using either quaternions[3] or Pauli matrices, or the equivalent scalar 3- or 4-tuple approaches is similar, although the advantages of the Clifford approach are not as pronounced as for whole electromagnetic fields. For the quaternions and Pauli matrices, the combined improvement for both software and hardware is at best $3^3 = 27$ times, with most of that coming from the hardware.

25.3 Practice

25.3.1 Example Code

Example 25.1 lists the code of a routine which provides for testing the inversion of matrices with elements which are either scalars, bicomplex numbers, quaternions, Pauli matrices, or extended electromagnetic fields. The tests are similar to those in Example 13.3 for bicomplex numbers, with additional provisions to allow different sizes of matrix, for measuring the computational time and also for repeating tests to allow for statistically stronger estimates of accuracy and time.

Example 25.1

```
 1:   void application12(argc,argv)  /* matrix inversion */
 2:          int argc; char** argv;
 3:   {
 4:     int entity,size,rep;
 5:
 6:     entity=0;
 7:     size=10;
 8:     rep=10;
 9:     if(argc==4)
10:       {
11:         printf("%s %s %s %s\n",argv[0],argv[1],argv[2],argv[3]);
12:         sscanf(argv[1],"%d",&entity);
13:         sscanf(argv[2],"%d",&size);
14:         sscanf(argv[3],"%d",&rep);
15:       }
```

---

3 Maxwell (1873a,) used quaternions.

```
16:     if(entity<0||entity>4||size<1||rep<1||argc!=4)
17:       {
18:         printf("usage: application 12 entity size time-limit\n");
19:         printf(" entity=0(scalar), 1(bicomplex), 2(quaternion),");
20:         printf(" 3(Pauli matrix), 4(extended EM field)\n");
21:         printf(" size(>0), time-limit(>0 seconds)\n");
22:         brexit(0);
23:       }
24:     switch(entity)
25:       {
26:       case 0:
27:         printf("matrix of scalars\n");
28:         printf("each element of matrix has 1 Clifford component\n");
29:         test_scalar(size,rep);
30:         break;
31:       case 1:
32:         printf("matrix of bicomplex numbers\n");
33:         printf("each element of matrix has 2 Clifford components\n");
34:         test_bicomplex(size,rep);
35:         break;
36:       case 2:
37:         printf("matrix of quaternions\n");
38:         printf("each element of matrix has 4 Clifford components\n");
39:         test_quaternion(size,rep);
40:         break;
41:       case 3:
42:         printf("matrix of Pauli matrices\n");
43:         printf("each element of matrix has 4 Clifford components\n");
44:         test_pauli(size,rep);
45:         break;
46:       case 4:
47:         printf("matrix of extended electromagnetic fields\n");
48:         printf("each element of matrix has 8 Clifford components\n");
49:         test_xemfield(size,rep);
50:         break;
51:       default:
52:         printf("unrecognised switch value (%d) in routine:
                     application12()\n",entity);
53:         break;
54:       }
55:     fflush(stdout);
56:   }
```

25.3.2 Example Output

Figures 25.1 and 25.2 show, as in Example 13.3b, the error in the matrix elimination and in the product of the matrix with its inverse, here plotted for different sizes of matrix. In the case of the electromagnetic field, the size of the matrix is the same as the number of points at which the solution for the field is determined. In the case of scalars, the size of the matrix for the same number of field points is six or eight times larger because the six components

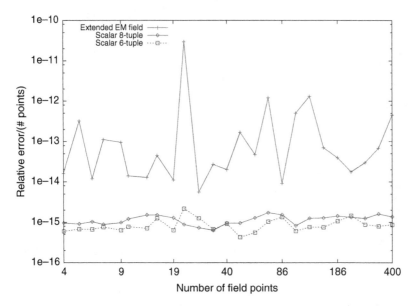

Figure 25.1   Relative error in reduced matrix normalised against the number of field points.

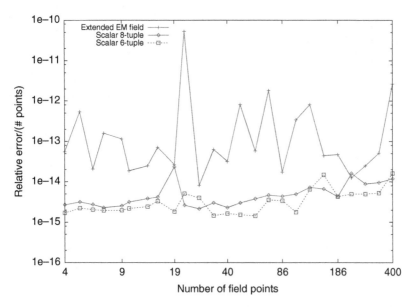

Figure 25.2   Relative error in product of matrix and inverse normalised against the number of field points.

in the electromagnetic field or the eight components in the extended electromagnetic field are treated as separate entities when formulating the linear equations.

If the errors accumulate randomly, an increase proportional to the square root of the number of elements in the matrix would be expected. The number of elements is the square of the size of the matrix, so the error should in principle increase linearly with size. Here the error is plotted after dividing by the size. This compensates for any linear increase so that the overall trend in the curves is largely constant rather than linear.

With the linear trend removed, the error is in the vicinity of machine precision for scalars, although the error for the product is higher because the inverse matrix used in the product already has errors of its own. The relative error using the extended electromagnetic field is consistently higher.

Figure 25.3 compares the execution times for solving the three alternative linear systems. Execution time is anticipated as proportional to the cube of the size the system, so the results are normalised against that value. With the cubic trend removed, the overall behaviour in the curves is more or less constant.

The solution implemented as the extended electromagnetic field is consistently faster than both of the two alternatives. The ratio of the speeds is consistent with the expected values of (from fastest to slowest as ratios) $8 : 3\frac{3}{8} : 1$.

Comparisons involving quaternions and Pauli matrices with scalar 4- and 3-tuples are shown in Figures 25.4–25.6. The results for quaternions and Pauli matrices are almost identical so that their curves in the figures overlie one another. The overall behaviour of the solutions is much the same as for the electromagnetic fields, in this case with ratio of speeds $4 : 1\frac{11}{16} : 1$.

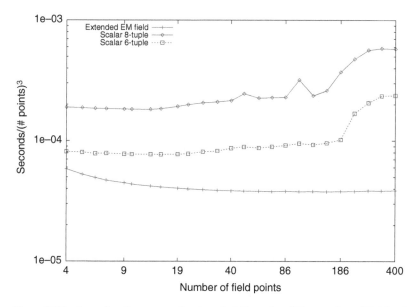

Figure 25.3 Inversion time normalised against the cube of the number of field points.

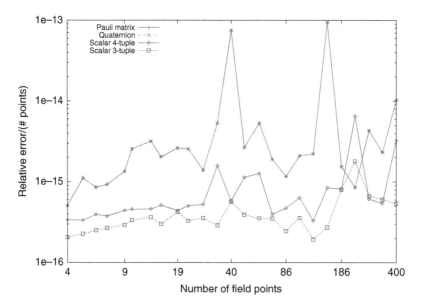

Figure 25.4   Relative error in reduced matrix normalised against the number of field points.

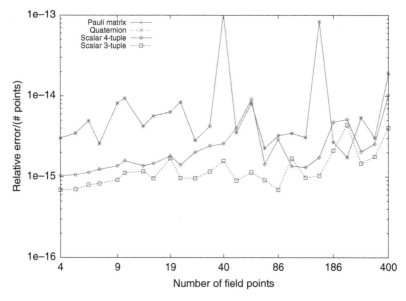

Figure 25.5   Relative error in product of matrix and inverse normalised against the number of field points.

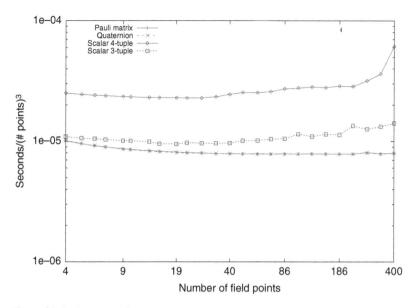

Figure 25.6   Inversion time normalised against the cube of the number of field points.

References

Chantaveerod, A. and Seagar, A. (2009), 'Iterative solutions for electromagnetic fields at perfectly reflective and transmissive interfaces using Clifford algebra and the multi-dimensional Cauchy integral', *IEEE Transactions on Antennas and Propagation* **57**(11), 3489–3499.

Maxwell, J. C. (1873a), *A Treatise on Electricity and Magnetism*, Vol. 1, Clarendon Press, Oxford, UK.

Maxwell, J. C. (1873b), *A Treatise on Electricity and Magnetism*, Vol. 2, Clarendon Press, Oxford, UK.

Seagar, A. (2016), *Application of Geometric Algebra to Electromagnetic Scattering – the Clifford-Cauchy-Dirac Technique*, Springer, Singapore.

Seagar, A. (2017), Accuracy of CCD technique for EM scattering from lossless and lossy dielectrics with different basis functions, *in* 'International Conference on Electromagnetics in Advanced Applications (ICEAA'17)', Verona, Italy, pp. 425–428.

Seagar, A. (2018), 'First order wavelike phenomena with electromagnetic and acoustic applications', *IEEE Transactions on Antennas and Propagation* **57**(7), 3526–3531.

Seagar, A. and Chantaveerod, A. (2015), An EM scattering algorithm for all materials, *in* 'International Conference on Electromagnetics in Advanced Applications (ICEAA'15)', Turin, Italy, pp. 256–259.

Seagar, A. and Espinosa, H. (2016), A numerical comparison between EFIE/MoM and CCD methods for EM scattering in two dimensions, *in* 'International Conference on Electromagnetics in Advanced Applications (ICEAA'16)', Cairns, Australia, pp. 328–330.

Part IV

Programming

26

Interfaces

This chapter contains a list of the application programmer interfaces (APIs) organised in terms of function, with cross references to the pages in Chapter 27 where the full descriptions are given. For short descriptions, see the quick reference Section 27.5.

26.1 Configuration and Observation

These interfaces do not change the values of any numerical entities. They are used to control the effect of and display the result of other operations.

26.1.1 Management

It is necessary to establish the numerical context for calculation before any can occur. This includes reserving some memory, declaring the number of dimensions, and declaring the signature for each dimension.

Before anything else is attempted, the sub-routines `recovery_level()`, `create_cliffmem()`, and `declare_primal()` should be invoked in that order. These do not have default values. The sub-routine `pack_signature()` can be used to make it easier to set up uniform signatures. For mixed signatures, the array 'lambda [n]' needs to be initialised in some other manner. It is also appropriate to establish the printing format using `format_printing()` early, although the printing format does have a default, set from within `create_cliffmem()`.

	Page
`void create_cliffmem(int n);`	337
`void free_cliffmem(int p);`	345
`void get_mem_usage(int n[2][3]);`	349
`int *pack_signature(int n, int *lambda, int sig);`	396
`void declare_primal(int n, int *lambda);`	339
`int access_primal(int **lambda);`	316

Numerical Calculations in Clifford Algebra: A Practical Guide for Engineers and Scientists,
First Edition. Andrew Seagar.
© 2023 John Wiley & Sons Ltd. Published 2023 by John Wiley & Sons Ltd.

	Page
`int check_number(number x);`	329
`void recycle_number(number x);`	440
`void recycle_allnumber(void);`	439
`int sigsetjmp(sigjmp_buf buf, int savesigs);`	451
`void recovery_level(int n, int s, sigjump_buf buf);`	438
`void siglongjmp(sigjmp_buf buf, int val);`	451
`void brexit(int s);`	324
`int format_printing(char *fmt);`	344
`int format_matrix(int k);`	344
`void random_seed(int k);`	432

26.1.2 Printing

Routines of the form `print_xxx()` take the default format for printing, which may be set up earlier by the sub-routine `format_printing()`. Routines of the form `printf_xxx()` with the extra 'f' ignore the format established by format_printing() in favour of the format provided as the first of the list of arguments.

Many printing routines behave as an identity function, returning the same entity as passed to them for printing. This makes it easy to examine the value of a variable as an aid to debugging almost anywhere, as for example embedded in the argument list of another function, without too much disruption to existing code.

	Page
`int format_printing(char *fmt);`	344
`int clone_format(int n, char *fmt);`	331
`void print_mem_usage(char *text);`	411
`void print_signature(void);`	416
`number print_num_info(number x);`	412
`number print_tree(number x);`	417
`number print_nodes(number x);`	412
`number print_bounds(number x);`	404
`void print_fork(number f);`	407
`void print_leaf(number l);`	408
`int print_unit(bitmap u);`	417
`int print_coef(double x, double y);`	405
`int printf_coef(char *fmt, double x, double y);`	419
`int print_comp(cliff_comp *c);`	406
`int printf_comp(char *fmt, cliff_comp *c);`	420
`number print_number(number x, int cmax);`	413
`number printf_number(char *fmt, number x, int cmax);`	422

Page

```
number print_single_coefficient(number x, bitmap u);              416

number printf_single_coefficient(char *fmt, number x, bitmap u);  425

number print_listof_units(number x, int cmax);                    409

number print_listof_coefficients(number x, int cmax);             409

number printf_listof_coefficients(char *fmt, number x, int cmax); 422

number print_gradedlistof_units(number x, int cmax);              408

number print_gradedlistof_coefficients(number x, int cmax);       407

number printf_gradedlistof_coefficients(char *fmt, number x,

                                        int cmax);                521
```
```
number print_vector(number x);                                    418

number printf_vector(char *fmt, number x);                        426

number printf_3vector(char *fmt, number x);                       418

number printf_Nvector(char *fmt, int n, number x);                423
```
```
number print_bicomplex(number x);                                 404

number printf_bicomplex(char *fmt, number x);                     419
```
```
number print_quaternion(number x);                                415

number printf_quaternion(char *fmt, number x);                    425
```
```
number print_pauli(number x);                                     415

number printf_pauli(char *fmt, number x);                         424
```
```
number print_emfield(number x);                                   406

number printf_emfield(char *fmt, number x);                       421

number printfs_emfield(char *fmt, char *sep, number x);           426
```
```
number *print_array(int n, number *a, int cmax);                  403

number *print_onecoefficientof_array(int n, number *a, bitmap u); 413

number *printf_onecoefficientof_array(char *fmt, int n,
                                      number *a, bitmap u);        423

number *print_magnitudeof_array(int n, number *a);                410
```
```
int format_matrix(int k);                                         344
```
```
matrix *print_matrix(matrix *m, int cmax);                        411

matrix *print_onecoefficientof_matrix(matrix *m, bitmap u);       414

matrix *printf_onecoefficientof_matrix(char *fmt, matrix *m,      424
                                       bitmap u);

matrix *print_magnitudeof_matrix(matrix *m);                      410

matrix *print_row(int i, matrix *m, int cmax);                    415

matrix *print_column(int j, matrix *m, int cmax);                 405

matrix *print_element(int i, int j, matrix *m, int cmax);         406

matrix *printf_element(char *fmt, int i, int j,
                       matrix *m, int cmax);                       420
```

26.2 Simple Entities

Simple entities encompass everything that involves single instances of Clifford numbers, either whole or in parts as units or components.

26.2.1 Units

The first four of these routines are probably of more interest than the others. There may not be any need to use any of the other routines directly. Potential indirect uses are within `combine()`, `insert()`, `function()`, `multiply()`, and `test()` functions that you construct yourself in order to elicit special effects from sub-routines `merge_comp()`, `func_number()`, `mul_comp()`, and `filter_number()`. See Chapter 16 for details.

	Page
`bitmap pack_unit(int n, int *primal);`	397
`int unpack_unit(bitmap u, int *primal);`	472
`bitmap spack_unit(char *evolved);`	454
`char *sscan_unit(char *evolved);`	457
`bitmap primal(int n);`	402
`bitmap pseudo_scalar(int n);`	427
`int dimension(bitmap u);`	341
`int grade(bitmap u);`	350
`int r_unit(bitmap u);`	428
`int i_unit(bitmap u);`	353
`int c_unit(bitmap u);`	325
`int s_unit(bitmap u);`	448
`int inv_unit(bitmap u);`	359
`int x_count(bitmap u, bitmap v);`	474
`int u_sign(bitmap u, bitmap v);`	468
`int print_unit(bitmap u);`	417

26.2.2 Components

The vast majority of the routines which operate on components pay proper attention to both the value of the coefficient and the unit. There are a few routines of the form xxx_coef() which ignore the unit take only the coefficient into the calculation. Most routines return a pointer to the result to make nesting of operations easier.

	Page
`cliff_comp *pack_comp(cliff_comp *c, double x, double y,` `                bitmap u);`	390
`void unpack_comp(cliff_comp *c, double *x, double *y, bitmap *u);`	469
`cliff_comp *zero(cliff_comp *c);`	476
`cliff_comp *random_complex(cliff_comp *c);`	429

Page

```
cliff_comp *real_part(cliff_comp *a, cliff_comp *c);                         433
cliff_comp *imaginary_part(cliff_comp *a, cliff_comp *c);                    354
cliff_comp *complex_conjugation(cliff_comp *a, cliff_comp *c);               334
cliff_comp *negation(cliff_comp *a, cliff_comp *c);                          379
cliff_comp *half(cliff_comp *a, cliff_comp *c);                              351
cliff_comp *identity(cliff_comp *a, cliff_comp *c);                          353
cliff_comp *twice(cliff_comp *a, cliff_comp *c);                             467
cliff_comp *reversion(cliff_comp *a, cliff_comp *c);                         443
cliff_comp *involution(cliff_comp *a, cliff_comp *c);                        365
cliff_comp *clifford_conjugation(cliff_comp *a, cliff_comp *c);              329
cliff_comp *supplement(cliff_comp *a, cliff_comp *c);                        465
cliff_comp *inverse(cliff_comp *a, cliff_comp *c);                           360
cliff_comp *cartesian(cliff_comp *a, cliff_comp *c);                         326
cliff_comp *polar(cliff_comp *a, cliff_comp *c);                             398
cliff_comp *degrees(cliff_comp *a, cliff_comp *c);                           339
cliff_comp *radians(cliff_comp *a, cliff_comp *c);                           428
cliff_comp *squareof_magnitude(cliff_comp *a, cliff_comp *c);                456
cliff_comp *squareof_latitude(cliff_comp *a, cliff_comp *c);                 456
cliff_comp *square_root(cliff_comp *a, cliff_comp *c);                       455
cliff_comp *square_root_coef(cliff_comp *a, cliff_comp *c);                  455
cliff_comp *log_coef(cliff_comp *a, cliff_comp *c);                          371
cliff_comp *power_coef(double p, cliff_comp *b, cliff_comp *c);              401
cliff_comp *addition_to(cliff_comp *a, cliff_comp *b);                       320
cliff_comp *subtraction_from(cliff_comp *a, cliff_comp *b);                  464
cliff_comp *addition(cliff_comp *a, cliff_comp *b,
                     cliff_comp *c);                                         319
cliff_comp *subtraction(cliff_comp *a, cliff_comp *b,
                        cliff_comp *c);                                      463
cliff_comp *rscale(double s, cliff_comp *b, cliff_comp *c);                  447
cliff_comp *central(cliff_comp *a, cliff_comp *b, cliff_comp *c);            327
cliff_comp *inner(cliff_comp *a, cliff_comp *b, cliff_comp *c);              355
cliff_comp *outer(cliff_comp *a, cliff_comp *b, cliff_comp *c);              385
cliff_comp *regress(cliff_comp *a, cliff_comp *b, cliff_comp *c);            442
cliff_comp *left(cliff_comp *a, cliff_comp *b, cliff_comp *c);               370
cliff_comp *right(cliff_comp *a, cliff_comp *b, cliff_comp *c);              444
int print_comp(cliff_comp *c);                                              406
int printf_comp(char *fmt, cliff_comp *c);                                  420
```

Functions which apply a test to a component and return an integer value of 1 (true) or 0 (false) are useful in constructing filters and in performing conditional execution.

	Page
`int no_real(cliff_comp *c);`	381
`int no_imaginary(cliff_comp *c);`	381
`int grade_zero(cliff_comp *c);`	351
`int grade_one(cliff_comp *c);`	350
`int grade_two(cliff_comp *c);`	351
`int non_scalar(cliff_comp *c);`	382
`int not_grade_one(cliff_comp *c);`	383
`int not_grade_two(cliff_comp *c);`	384
`int spatial_part(cliff_comp *c);`	454
`int temporal_part(cliff_comp *c);`	466
`int null_signature(cliff_comp *c);`	384
`int positive_signature(cliff_comp *c);`	400
`int negative_signature(cliff_comp *c);`	380

26.2.3 Numbers

Many routines which use numbers as arguments include an extra argument, the disposal flag, which can be used to dispose of the numbers in the other arguments after they have been used. This makes nesting of operations neater.

26.2.3.1 Establishing and Recovering Values
There are no especially good ways to get numerical values into and out of the internal 'number' format. Generally one of the following should get you started, although in the longer term, it may be better to construct your own interface to meet your particular special needs.

	Page
`number pack_number(int n, cliff_comp *c);`	392
`number pack_components(int n, double (*y_complex)[2], bitmap *y_unit);`	390
`int count_components(number x);`	336
`int unpack_number(number x, cliff_comp *c);`	471
`int unpack_components(number x, double (*y_complex)[2], bitmap *y_unit);`	469
`cliff_comp *find_unit(number x, bitmap u);`	344
`cliff_comp *find_comp(number x, cliff_comp *a);`	343
`number merge_comp(number x, cliff_comp *y_comp, cliff_comp *combine(cliff_comp *a, cliff_comp *b), cliff_comp *insert(cliff_comp *a, cliff_comp *c));`	377
`double *get_coef(number x, bitmap u, double c[2], int d);`	348
`double get_real(number x, bitmap u, int d);`	349
`double get_imag(number x, bitmap u, int d);`	348

26.2.3.2 Functions

Most of these routines apply a particular function to components independently and in isolation to one another, often leaving the unit unchanged.

	Page
`number magnitudeof_number(number x, int d);`	374
`number latitudeof_number(number x, int d);`	370
`number func_number(number x, cliff_comp *function(cliff_comp *a,` `cliff_comp *c));`	347
`number clonefuncof_number(number x, cliff_comp` `*function(cliff_comp *a, cliff_comp *c));`	332

Functions that apply a test to a number and return an integer value of 1 (true) or 0 (false) are useful in controlling loops and in performing conditional execution. You will need to write your own function if the tests listed here do not meet your needs.

	Page
`int is_single(number x);`	368
`int is_multi(number x);`	366
`int is_zero(number x);`	369
`int is_nonzero(number x);`	367
`int is_scalar(number x);`	367
`int is_vector(number x);`	368
`int is_bivector(number x);`	366

26.2.3.3 Addition and Subtraction

Addition to and subtraction from an existing number (overwriting the original in the process) is more efficient than producing an entirely new number as the result. When overwriting it is not necessary to construct the tree for a new number, only to modify an existing one.

	Page
`number addto_number(number x, number y);`	322
`number subfrom_number(number x, number y);`	460
`number add_number(number x, number y, int d);`	318
`number sub_number(number x, number y, int d);`	458
`number addfuncto_number(number x, number y, cliff_comp` `*function(cliff_comp *a, cliff_comp *c));`	319
`number subfuncfrom_number(number x, number y, cliff_comp` `*function(cliff_comp *a, cliff_comp *c));`	461

26.2.3.4 Multiplication

Most of the multiplication routines use `mul_comp()` internally. The sub-routines `scalar_mul()` and `inner_mul()` produce the same numeric values as different kinds of entities.

26.2.3.5 Geometric

The geometric operations or reflection, projection and rotation are designed to work consistently with numbers of either positive or negative but not mixed signature.

26.2.3.6 Filtering

Filters are useful when you want to ignore some of the components. This is often the case in applications where certain components are known in principle to be zero; however, the numerical calculations leave some small non-zero residue. You can construct your own filters using `keep_filter()`, `clone_filter()`, and `filter_number()`. You will need to write a function which applies a test to a component and returns an integer value of 1 (true) or 0 (false) if the tests already provided (towards the end of Section 26.2.2) do not meet your needs.

26.3 Higher Entities

26.3.1 Vectors

Some special routines are provided for vectors in three dimensions, and for three-dimensional spatial vectors embedded in four-dimensional space–time. The Cartesian vector units $\hat{x}, \hat{y}, \hat{z}$ for vectors without a time dimension are identified with the primal units e_0, e_1, e_2. For space–time, the primal unit e_0 is used for time (or frequency) and the Cartesian vector units $\hat{x}, \hat{y}, \hat{z}$ are identified with the primal units e_1, e_2, e_3.

The geometric operations are designed to work consistently with vectors of either positive or negative but not mixed signature. When applied to entities occupying higher grades, the effect produced by positive and negative signatures may differ. Which signature is most suitable depends on the effects required by your application. For example when electromagnetic fields are represented as bivectors as in Chapter 10, the vector reflection operation with negative signature behaves like a physical mirror and reflects an electromagnetic wave as one would expect.

26.3.2 Bicomplex Numbers

26.3.3 Quaternions

26.3.4 Pauli Matrices

26.3.5 Electromagnetic Fields

Electromagnetic fields reside in grade 2 of a four-dimensional Clifford number. The difference between the two sub-routines pack_spacetime() and pack_emfield() is that the latter scales by the material properties $\sqrt{\mu}$ and $\sqrt{\epsilon}$, whereas the former does not.

26.4 Multiple Entities

26.4.1 Arrays

Arrays of Clifford numbers are effectively arrays of integer valued tokens. Creation of these arrays is achieved using the normal mechanisms within the 'c' programming language (King 2008). Operations on the arrays can be achieved by putting them in a loop and operating on each element one at a time.

	Page
`number *fresh_array(int n, number *a);`	346
`number *recycle_array(int n, number *a);`	439
`number *copy_array(int n, number *a, number *b);`	335
`number *clone_array(int n, number *a, number *b);`	330
`number *postscale_array(int n, number *a, number s, number *b);`	400
`number *prescale_array(number s, int n, number *a, number *b);`	401
`number *addto_array(int n, number *a, number *b);`	321
`number *subfrom_array(int n, number *a, number *b);`	459
`number *add_array(int n, number *a, number *b, number *c);`	317
`number *subtract_array(int n, number *a, number *b, number *c);`	462
`number peakof_array(int n, number *a);`	398
`number meanabsof_array(int n, number *a);`	377
`number rmsof_array(int n, number *a);`	445
`number sumof_array(int n, number *a);`	464
`number weightedsumof_array(int n, number *a, cliff_comp *w);`	474
`number *print_array(int n, number *a, int cmax);`	403
`number *print_onecoefficientof_array(int n, number *a, bitmap u);`	413
`number *printf_onecoefficientof_array(char *fmt, int n,` `number *a, bitmap u);`	423
`number *print_magnitudeof_array(int n, number *a);`	410

26.4.2 Fast Fourier Transforms

These transforms only operate on arrays with length equalling a power of 2.

	Page
`number *transform(int n, number *a, number *b);`	466
`number *inverse_transform(int n, number *a, number *b);`	361

26.4.3 Series

Power series can be used to implement polynomials and other functions. The routine `poly_number()` applies an array 'a[n]' of coefficients to successive integer powers of the Clifford number 'x'. The two routines `binomial_series()` and `bessel_clifford_series()` generate arrays of coefficients for those particular series. It is expected that the user will write some such similar sub-routines to construct specific series for their own purposes.

	Page
`void binomial_series(int n, number *a, int denom);`	324
`void bessel_clifford_series(int n, number *a, int order);`	323
`number poly_number(int n, number *a, number x, number c, int d);`	399
`void make_series(int p, int n);`	376
`number series(label f, number x, int p);`	450
`number x_series(label f, number x, int p);`	475
`number xx_series(label f, number x, int p);`	475

The routine `make_series()` generates and stores internally arrays of coefficients for the power series representations of the common functions 'f' = COS, SIN, COSH, SINH, EXP, LOG, CM_INV (central multiplicative inverse, or reciprocal), SQRT and (nth) ROOT.

The values of the internally stored functions are obtained by invoking either of the sub-routines `series()`, `x_series()`, or `xx_series()`. The versions `x_series()` and `xx_series()` provide extended ranges of convergence, but only for some of the functions.

26.4.4 Matrices

Matrices of Clifford numbers are supported for the basic operations of addition and subtraction and also multiplication and inversion with the central product.

	Page
`matrix *declare_matrix(int nrows, int ncols, matrix *m,`	
`                       number *data);`	338
`matrix *refresh_matrix(matrix *m);`	442
`matrix *recycle_matrix(matrix *m);`	440
`matrix *random_matrix(matrix *m, int n, int *grades);`	430
`matrix *identity_matrix(matrix *m);`	354
`int check_matrix(matrix *m);`	328
`double check_identity(matrix *m);`	327
`matrix *copy_matrix(matrix *a, matrix *b);`	336
`matrix *clone_matrix(matrix *a, matrix *b);`	331
`matrix *addto_matrix(matrix *a, matrix *b);`	321
`matrix *subfrom_matrix(matrix *a, matrix *b);`	459
`matrix *add_matrix(matrix *a, matrix *b, matrix *c, int d);`	318
`matrix *subtract_matrix(matrix *a, matrix *b, matrix *c, int d);`	463
`matrix *multiply_matrix(matrix *a, matrix *b, matrix *c, int d);`	379

Reference

King, K. N. (2008), *C Programming: A Modern Approach*, 2nd edn, W. W. Norton, New York, US-NY.

27

Descriptions

This chapter contains detailed descriptions of the application programmer interfaces (APIs) in alphabetical order. For a list of the interfaces organised in terms of function with cross references to these pages, see Chapter 26. For a quick reference, see Section 27.5.

27.1 Arguments

The functional type of the arguments specified in the interfaces to the individual sub-routines is identified in the descriptions by single letter codes as listed in Table 27.1.

Arguments of type 'i' (input) are set by the user to some value which is used to control the behaviour of the sub-routine. Arguments of type 'o' (output) are used by the sub-routine to pass some result back to the user. Some arguments have both types input and output. In this case, the values set up by the user are modified by the sub-routine.

Arguments of type 'r' (recycled) are numbers which are in some way consumed by the sub-routine and no longer valid on exit from the sub-routine. An example is the sub-routine addto_number() which uses the memory associated with the first argument as the basis on which to construct the result, since many of the components are already in place. This is faster than constructing a whole new number.

Arguments of type 'd' (disposed) are either recycled or not, according to the value of the flag set by the user in another argument. This extra argument, the disposal flag 'd', is found as the last in the list of arguments. It indicates which of the previous arguments should be recycled and which should be not recycled. Each bit in the flag indicates one argument. The first (rightmost or least significant) bit is for the first argument. The second bit is for the second argument. Multiple bits may be set to indicate the disposal of multiple arguments. Not all arguments can be recycled, only those identified as type 'd'.

27.2 Data types

Some new data types are defined for the storage of particular entities, as listed in Table 27.2.

The data type 'number' provides the user with a means to manipulate Clifford numbers. As implemented, the value of a variable of data type 'number' is a signed integer which serves as a token to identify the particular number assigned to that variable. The token is

Numerical Calculations in Clifford Algebra: A Practical Guide for Engineers and Scientists,
First Edition. Andrew Seagar.
© 2023 John Wiley & Sons Ltd. Published 2023 by John Wiley & Sons Ltd.

Table 27.1 Argument types.

Code	Type	Usage
i	Input	Value of argument is used to transfer data into routine
o	Output	Value of argument is used to transfer results outside routine
r	Recycled	Memory associated with value of argument is recycled before exit
d	Disposed	Memory associated with value of argument is disposed of as requested

Table 27.2 New data types.

New type	Definition	Use
`number`	`typedef int number`	Holds value of token via which memory is referenced
`bitmap`	`typedef unsigned int bitmap`	Holds evolved unit as bit pattern, one bit for each primal unit

`cliff_comp`	`typedef struct{`
	
	`    bitmap unit;`
	`    double coef[2];`
	`} cliff_comp;`
		Holds single Clifford component, complex value plus unit
		`/* evolved unit */`
		`/* real and imaginary parts */`

`label`	`typedef enum{`
	`    COS,SIN,COSH,SINH,EXP,`
	`    LOG,CM_INV,SQRT,ROOT`
	`} label;`
		Labels to identify pre-defined power series of functions

`matrix`	`typedef struct{`
	`    int nrows;`
	`    int ncols;`
	`    int stride;`
	
	`    number *data;`
	`} matrix;`
		For matrix operations
		`/* number of rows */`
		`/* number of columns */`
		`/* offset between 1st elements in adjacent columns */`
		`/* first element of data`

a unique value used to index into tables containing the data structures where the actual numerical data are stored.

Using the values of variables containing tokens which have become invalid due to recycling the underlying structures, either explicitly by invoking a sub-routine such as `recycle_number()` or implicitly as with the first argument of the sub-routine `addto_number()`, usually leads to some kind of irrevocable error after accessing fragments of underlying structures which are no longer properly connected into any kind of consistent construct. If this happens an error message suggesting some kind of inappropriate or corrupted construct is usually generated. When this occurs, look for variables which have been reused badly.

It is possible to re-use variables of the data type 'number'. If the underlying storage has been recycled, either implicitly or explicitly, the variable can be used to hold the value of a new token as generated by some operation which returns a value in the data type 'number'.

The data type 'bitmap' provides a mechanism for storing the unit of components. As implemented, the value of a variable of data type 'bitmap' is an unsigned integer which serves as a bit pattern in which each bit represents one primal unit. The first (rightmost or least significant) bit represents the primal unit e{0}. The second bit is for the second primal unit, e{1}. Multiple bits may be set to indicate an evolved unit containing multiple primal units. An empty bit pattern, the integer zero, represents the absolute unit e{} of grade 0 and scalars.

The data type 'cliff_comp' provides a structure in which to store individual components of Clifford numbers. All components are stored as complex values, with precision of the real and imaginary parts according to the compiler's data type 'double'. That is often around 15 or 16 decimal places. There are no structures for storing integers or separate real or imaginary values, or variables of single 'float' precision in ways which would be more efficient in the consumption of memory. Utmost memory efficiency is sacrificed for simplicity of implementation.

The data type 'label' provides a list of keywords to be used when calculating simple mathematical functions defined as power series. The labels each identify one of a set of pre-defined arrays of coefficients, which are used to produce the corresponding function. It is usually not necessary to declare variables of this data type, being sufficient to use the literal names for the labels rather than a variable containing them.

The data type 'matrix' provides a structure in which to store the information required to properly access the elements in matrices, or sub-matrices within other matrices. Matrices are stored in memory in column major format, with consecutive elements in each column occupying consecutive locations in memory. When a new matrix is initialised whole columns follow in memory one from the previous, with the first element in each column separated from the first element in the previous column by the number of rows. The distance between the first element in one column and the next is called the 'stride'.

27.3 Formats

The formats in which the numerical components of the data types 'number' or 'cliff_comp' are printed conform to the formats in the 'c' computer programming language (King, 2008) for variables of data type 'double', as in the left-hand column of Table 27.3. The default printing format can be set by invoking the sub-routine printing_format() with the format string as the argument. The string is eventually passed into the standard input–output library routine printf(), so any format supported by printf() for a single variable of data type 'double' should work as intended.

The formats supported by the standard input–output library routine printf() are optionally extended as in the right-hand column of Table 27.3 by appending (exactly) two extra characters to the normal format string. The first character (#, +, 0, −) specifies rounding to one of the nearest integer values. The second character (c, r, i, m, p) specifies how the real and imaginary parts should be transformed before printing. The last four options print only one component. To get both magnitude and phase, you need to print twice.

The default printing format if not set by the user is "%.5e". This prints, without rounding to an integer, both the real and imaginary parts with five decimal places following the decimal point and an exponent for radix 10.

Table 27.3 Printing formats.

Standard			Extended	
% .	Precision	Format	Rounding	Transformation
% .	0–15	e	# ⇒ none	c ⇒ complex
		f	+ ⇒ up	r ⇒ real
			0 ⇒ nearest	i ⇒ imaginary
			− ⇒ down	m ⇒ magnitude
				p ⇒ phase in radians

27.4 Manual Pages

27.4.1 A–E

1	ACCESS_ELEMENT	A

NAME

    access_element :– get pointer to element in array

SYNOPSIS

    `number *access_element(int i, int j, matrix *m);`

ARGUMENTS

    1: i ⋄ `int i`   ~ row
    2: i ⋄ `int j`   ~ column
    3: i ⋄ `matrix *m` ~ matrix

DESCRIPTION

    Get a pointer to the number located at matrix element $m\{i,j\}$ in row i and column j. If unknown, the total number of rows and columns can be obtained using the routines `rows()` and `columns()`. The indexes i, j for the rows and columns start from zero and go up to one less than the totals. No checks are made to determine if the indexes are out of bounds.

RETURN VALUE

    The return value is a pointer to the element in the matrix.

RECYCLING

    ~

ERROR RECOVERY

    It is an error to use indexes which are out of bounds.

SEE ALSO

    `columns()`, `find_comp()`, `find_unit()`, `rows()`

2	ACCESS_PRIMAL	A

NAME

    access_primal :– access primal unit signatures

SYNOPSIS

```
int access_primal(int **lambda);
```

ARGUMENTS

1: o ◊ int **lambda ~ address of pointer to array of signatures

DESCRIPTION

Get a pointer to the signature array currently linked to the mathematical environment.

RETURN VALUE

The return value is the size of the signature array. That is the same as the number of dimensions.

RECYCLING

~

ERROR RECOVERY

It is an error to get a pointer before one has been declared. Changing any of the values in the array changes the signatures for the corresponding primal units. However, this by itself is not sufficient. It is then necessary to invoke declare_primal(), because this also updates internal flags which make some operations which depend on the signature faster. If the flags are not updated those operations will not behave properly.

SEE ALSO

```
declare_primal(), print_signature()
```

3	ADD_ARRAY	A

NAME

add_array :– sum of two arrays

SYNOPSIS

```
number *add_array(int n, number *a, number *b,
                  number *c);
```

ARGUMENTS

1: i ◊ int n   ~ number of elements in arrays
2: i ◊ number *a ~ augend, pointer to first array of numbers
3: i ◊ number *b ~ addend, pointer to second array of numbers
4: o ◊ number *c ~ sum, pointer to array in which to save result

DESCRIPTION

Calculate the sum of two arrays. The second array is added to the first array, and the result is placed in the third array. Initially, the third array must be empty.

RETURN VALUE

The return value is a pointer to the array containing the results.

RECYCLING

~

ERROR RECOVERY

It is an error to provide an array which does not have the number of elements specified. It is an error if the third array is not empty initially.

SEE ALSO

```
add_matrix(), add_number(), addto_array(), subtract_array()
```

4	ADD_MATRIX	A

NAME

add_matrix :– calculate sum of two matrices

SYNOPSIS

```
matrix *add_matrix(matrix *a, matrix *b, matrix *c,
                   int d);
```

ARGUMENTS

1: id ◇ matrix *a ~ augend, matrix

2: id ◇ matrix *b ~ addend, matrix

3: o ◇ matrix *c ~ sum, empty matrix

4: i ◇ int d   ~ disposal flag

DESCRIPTION

Add the second matrix (addend) to the first matrix (augend) and save the result in the third matrix.

RETURN VALUE

The return value is the matrix containing the sum.

RECYCLING

~

ERROR RECOVERY

It is an error if the matrices have different sizes. On error, the recovery strategy brexit(0) is invoked to initiate a user re-start. It is an error if the third matrix is not empty initially. No check is made.

SEE ALSO

```
add_array(), add_number(), addto_matrix(),
subtract_matrix()
```

5	ADD_NUMBER	A

NAME

add_number :– produce sum of two numbers

SYNOPSIS

```
number add_number(number x, number y, int d);
```

ARGUMENTS

1: id ◇ number x ~ augend

2: id ◇ number y ~ addend

3: i ◇ int d   ~ disposal flag

DESCRIPTION

Calculate the sum of two numbers.

RETURN VALUE

The return value is the sum.

RECYCLING

The numbers x and y are recycled according to the value of the disposal flag.

ERROR RECOVERY

~

SEE ALSO

```
add_array(), add_matrix(), addition(), addto_number(),
sub_number()
```

6	ADDFUNCTO_NUMBER	A

NAME

addfuncto_number :– add function of number to another

SYNOPSIS

```
number addfuncto_number(number x, number y,
                        cliff_comp *function(cliff_comp
                        *a, cliff_comp *c));
```

ARGUMENTS

1: ir ◇ number x   ~ augend

2: i ◇ number y   ~ number, argument of function

3: i ◇ cliff_comp *function ~ pointer to function
      (cliff_comp *a, cliff_comp *c)

DESCRIPTION

Apply the function to every component in the second number and add the result onto the first number.

RETURN VALUE

The return value contains the result.

RECYCLING

The augend is recycled. It is overwritten *in situ* and returned as the result.

ERROR RECOVERY

It is an error to add the function of a number to itself. On error, the recovery strategy brexit(6) is invoked, terminating the execution.

SEE ALSO

addto_number(), subfuncfrom_number()

7	ADDITION	A

NAME

addition :– sum of two components

SYNOPSIS

```
cliff_comp *addition(cliff_comp *a, cliff_comp *b,
                     cliff_comp *c);
```

ARGUMENTS

1: i ◇ cliff_comp *a ~ augend, pointer to structure containing component

2: i ◇ cliff_comp *b ~ addend, pointer to structure containing component

3: o ◇ cliff_comp *c ~ pointer to structure in which to save sum

DESCRIPTION

Calculate the sum of two components. Setting the pointers of either input component to the same value as the output component overwrites the input with the output.

RETURN VALUE

The return value is a pointer to the component containing the sum.

RECYCLING

~

ERROR RECOVERY

It is an error if the two input components do not have the same unit. On error, the recovery strategy brexit(6) is invoked, terminating the execution.

SEE ALSO

    add_number(), addition_to(), subtraction()

| 8 | ADDITION_TO | A |

NAME

addition_to :– add component onto another

SYNOPSIS

    cliff_comp *addition_to(cliff_comp *a, cliff_comp *b);

ARGUMENTS

    1: io ◇ cliff_comp *a ~ augend, pointer to structure containing component
    2: i ◇ cliff_comp *b ~ addend, pointer to structure containing component

DESCRIPTION

Adds the second component onto the first component.

RETURN VALUE

The return value is a pointer to the first component, which contains the sum. Setting the pointers of both components to the same value overwrites the input with the output.

RECYCLING

    ~

ERROR RECOVERY

It is an error if the two components do not have the same unit. On error, the recovery strategy brexit(6) is invoked, terminating the execution.

SEE ALSO

    addition(), addto_number(), subtraction_from()

| 9 | ADDSCALEDTO_ROW | A |

NAME

addscaledto_row :– scale one row of matrix and add to another

SYNOPSIS

    matrix *addscaledto_row(int i, number s, int j,
                            matrix *m);

ARGUMENTS

    1: i ◇ int i   ~ augend, row to update by addition
    2: i ◇ number s   ~ scale factor
    3: i ◇ int j   ~ row to scale and add
    4: io ◇ matrix *m ~ matrix

DESCRIPTION

Pre-scale (multiply from the left) the second row (j) and add to the first row (i). The first of the two rows is overwritten with the result. The second row is not modified. If unknown, the total number of rows can be obtained using the routine rows(). The indexes i, j for the rows start from zero and go up to one less than the total. No checks are made to determine if the indexes are out of bounds.

RETURN VALUE

The return value is the original matrix with modified contents.

RECYCLING

    ~

ERROR RECOVERY

    It is an error to use indexes which are out of bounds.

SEE ALSO

    `addto_row(), rows(), subscaledfrom_row()`

10	ADDTO_ARRAY	A

NAME

    addto_array :– add one array to another

SYNOPSIS

    `number *addto_array(int n, number *a, number *b);`

ARGUMENTS

    1: i ◇ `int n`   ~ number of elements in arrays

    2: io ◇ `number *a` ~ augend, pointer to first array of numbers

    3: i ◇ `number *b` ~ addend, pointer to second array of numbers

DESCRIPTION

    Adds the second array onto the first array. The numbers in the first array are overwritten with the results.

RETURN VALUE

    The return value is a pointer to the array containing the results.

RECYCLING

    The numbers in the first array are recycled before they are overwritten.

ERROR RECOVERY

    It is an error to provide an array which does not have the number of elements specified.

SEE ALSO

    `add_array(), addto_matrix(), addto_number(),`
    `subfrom_array()`

11	ADDTO_MATRIX	A

NAME

    addto_matrix :– add one matrix onto another

SYNOPSIS

    `matrix *addto_matrix(matrix *a, matrix *b);`

ARGUMENTS

    1: io ◇ `matrix *a` ~ augend, matrix

    2: i ◇ `matrix *b` ~ addend, matrix

DESCRIPTION

    Add the second matrix (addend) onto the first matrix (augend) and save the result in the first matrix.

RETURN VALUE

    The return value is the augend matrix with modified values.

RECYCLING

    ~

ERROR RECOVERY

It is an error if the matrices have different sizes. On error, the recovery strategy `brexit(0)` is invoked to initiate a user re-start.

SEE ALSO

`add_matrix()`, `addto_array()`, `addto_number()`, `addto_row()`, `subfrom_matrix()`

12	ADDTO_NUMBER	A

NAME

addto_number :– add number to another

SYNOPSIS

`number addto_number(number x, number y);`

ARGUMENTS

1: ir ◇ `number x` ~ augend
2: i ◇ `number y` ~ addend

DESCRIPTION

Adds the second number onto the first number.

RETURN VALUE

The return value is the sum.

RECYCLING

The original number is recycled. It is overwritten *in situ* and returned in place of the new one.

ERROR RECOVERY

It is an error to add a number to itself. On error, the recovery strategy `brexit(6)` is invoked, terminating the execution.

SEE ALSO

`add_number()`, `addfuncto_number()`, `addition_to()`, `addto_array()`, `addto_matrix()`, `subfrom_number()`

13	ADDTO_ROW	A

NAME

addto_row :– add one row of a matrix onto another

SYNOPSIS

`matrix *addto_row(int i, int j, matrix *m);`

ARGUMENTS

1: i ◇ `int i` ~ augend, row to update by addition
2: i ◇ `int j` ~ addend, row to add
3: io ◇ `matrix *m` ~ matrix

DESCRIPTION

Add the second row (`j`, addend) to the first row (`i`, augend). The first of the two rows is overwritten with the result. If unknown, the total number of rows can be obtained using the routine `rows()`. The indexes `i`, `j` for the rows start from zero and go up to one less than the total. No checks are made to determine if the indexes are out of bounds.

RETURN VALUE

> The return value is the original matrix with modified values.

RECYCLING

> ~

ERROR RECOVERY

> It is an error to use indexes which are out of bounds.

SEE ALSO

> `addto_array()`, `addto_matrix()`, `addto_number()`, `rows()`,
> `subfrom_row()`

14	BESSEL_CLIFFORD_SERIES	B

NAME

> bessel_clifford_series :– construct coefficients for Bessel–Clifford power series

SYNOPSIS

> `number *bessel_clifford_series(int n, number *a,`
> `                              int order);`

ARGUMENTS

> 1: i ◇ `int n`   ~ size of array
> 2: o ◇ `number *a` ~ pointer to array to in which to save coefficients
> 3: i ◇ `int order` ~ order of function

DESCRIPTION

> Calculates the first n coefficients (corresponding to powers from zero to n-1) for
> a Bessel–Clifford series of the specified order. The coefficients for the power series
> are saved in the array `a[n]`. Initially, the array must be empty.

RETURN VALUE

> The return value is a pointer to the array containing the results.

RECYCLING

> ~

ERROR RECOVERY

> It is an error to provide an array which does not have the number of elements
> specified.

SEE ALSO

> `binomial_series()`, `make_series()`, `poly_number()`, `series()`

15	BESSEL_POTENTIAL3D	B

NAME

> bessel_potential3D :– calculate Bessel potential in three dimensions

SYNOPSIS

> `number bessel_potential3D(cliff_comp *k, double r);`

ARGUMENTS

> 1: i ◇ `cliff_comp *k` ~ complex wavenumber, pointer to structure containing
> component
> 2: i ◇ `double r`   ~ field location as Cartesian vector

DESCRIPTION

Calculate the Bessel potential in three dimensions at the distance r from the origin. A negative value of k should be used when negative wavenumbers are required.

RETURN VALUE

The return value is the Bessel potential.

RECYCLING

~

ERROR RECOVERY

It is an error to calculate the potential at the origin, where the value is unbounded. On error, the recovery strategy brexit(0) is invoked to initiate a user re-start.

SEE ALSO

hertzian_dipole()

16	BINOMIAL_SERIES	B

NAME

binomial_series :– construct coefficients for binomial power series

SYNOPSIS

number *binomial_series(int n, number *a, int denom);

ARGUMENTS

1: i ◇ int n   ~ size of array
2: o ◇ number *a ~ pointer to array in which to save coefficients
3: i ◇ int denom ~ denominator of fractional exponent

DESCRIPTION

Calculates the first n coefficients (corresponding to powers from 0 to n-1) for a binomial series with a fractional exponent 1/denom. If the value of denom is zero, it is treated as one, and the coefficients are those for the identity function. Common values for denom are -1 for the reciprocal function, 2 for the square root, and 3 for the cube root. The coefficients for the power series are saved in the array a[n]. Initially, the array must be empty.

RETURN VALUE

The return value is a pointer to the array containing the results.

RECYCLING

~

ERROR RECOVERY

It is an error to provide an array which does not have the number of elements specified.

SEE ALSO

bessel_clifford_series(), make_series(), poly_number(), series()

17	BREXIT	B

NAME

brexit :– break and recover or exit

SYNOPSIS

```
void brexit(int s);
```

ARGUMENTS

1: i ◇ int s ~ strategy for recovery

DESCRIPTION

Break execution and either attempt recovery or exit. The action to be taken for strategy s=0 is the one specified by `recovery_level()`. Other actions ignore the value specified by `recovery_level()`.

s=0	Follow strategy in `recovery_level()`
s>0	Do not restart, exit
s<0	Attempt restart
s odd	Prompt on restart
|s|={1,2}	Do not free memory
|s|={3,4}	Recycle all numbers
|s|={5,6}	Free all memory
|s|={7,8}	Free all memory and print statistics

RETURN VALUE

The routine does not return. Execution is either terminated or returned to the location supplied to the c-library routine `sigsetjmp()`.

RECYCLING

~

ERROR RECOVERY

It is an error to initiate a strategy which specifies recovery before establishing the location for recovery using `sigsetjmp()`.

SEE ALSO

```
recovery_level(), siglongjmp(), sigsetjmp()
```

18	C_UNIT	C

NAME

c_unit :– Clifford conjugate of unit

SYNOPSIS

```
int c_unit(bitmap u);
```

ARGUMENTS

1: i ◇ bitmap u ~ unit

DESCRIPTION

Calculate the effect of reversing both the order and the sign of all of the primal units in the unit specified.

RETURN VALUE

The return value is the sign induced by Clifford conjugation of the unit.

RECYCLING

~

ERROR RECOVERY

~

SEE ALSO

    i_unit(), r_unit()

| 19 | CARTESIAN | C |

NAME

    cartesian :– convert component from polar to Cartesian

SYNOPSIS

    cliff_comp *cartesian(cliff_comp *a, cliff_comp *c);

ARGUMENTS

    1: i ◇ cliff_comp *a ~ pointer to structure containing polar component
    2: o ◇ cliff_comp *c ~ pointer to structure in which to save Cartesian
                    component

DESCRIPTION

    Convert the component from polar to Cartesian. The radial part is taken from the real part, and the angular part from the imaginary part. The angle is in radians. Setting the pointers of both components to the same value overwrites the input with the output.

RETURN VALUE

    The return value is a pointer to the component containing the Cartesian form.

RECYCLING

    ~

ERROR RECOVERY

    ~

SEE ALSO

    degrees(), polar(), radians()

| 20 | CAUCHY_KERNEL3D | C |

NAME

    cauchy_kernel3D :– Cauchy kernel in three dimensions

SYNOPSIS

    number cauchy_kernel3D(cliff_comp *k, number r);

ARGUMENTS

    1: i ◇ cliff_comp *k ~ wavenumber, pointer to structure containing
                    component
    2: i ◇ number r   ~ radial vector from origin

DESCRIPTION

    Calculate the Cauchy kernel in three dimensions.

RETURN VALUE

    The return value is a number containing the kernel as a bivector.

RECYCLING

    ~

ERROR RECOVERY

    It is an error to calculate the kernel at the origin, where the value is unbounded.

SEE ALSO

    fundamental_solution3D()

21	Central	C

NAME

central :– central product of components

SYNOPSIS

```
cliff_comp *central(cliff_comp *a, cliff_comp *b,
                    cliff_comp *c);
```

ARGUMENTS

1: i ⋄ cliff_comp *a ~ multiplier, pointer to structure containing component

2: i ⋄ cliff_comp *b ~ multiplicand, pointer to structure containing component

3: o ⋄ cliff_comp *c ~ pointer to structure in which to save product

DESCRIPTION

Calculate the central (Clifford) product of two components. Setting the pointers of either input component to the same value as the output component overwrites the input with the output.

RETURN VALUE

The return value is a pointer to the component containing the central product.

RECYCLING

~

ERROR RECOVERY

~

SEE ALSO

```
central_mul(), inner(), outer()
```

22	Central_Mul	C

NAME

central_mul :– calculate central product of numbers

SYNOPSIS

```
number central_mul(number x, number y, int d);
```

ARGUMENTS

1: id ⋄ number x ~ multiplier

2: id ⋄ number y ~ multiplicand

3: i ⋄ int d   ~ disposal flag

DESCRIPTION

Multiply two numbers using central (Clifford) multiplication.

RETURN VALUE

The return value is the central product.

RECYCLING

The numbers x and y are recycled according to the value of the disposal flag.

ERROR RECOVERY

~

SEE ALSO

```
central(), inner_mul(), multiply_matrix(), outer_mul()
```

23	CHECK_IDENTITY	C

NAME

check_identity :– calculate difference between matrix and identity matrix

SYNOPSIS

```
double check_identity(matrix *m);
```

ARGUMENTS

1: i ◇ matrix *m ~ matrix to compare

DESCRIPTION

Subtract the identity matrix and calculate the root mean square value from all of the coefficients, ignoring the imaginary unit and all Clifford units, giving the maximum RMS difference.

RETURN VALUE

The return value is the root mean square (RMS) value of the difference between the matrix and the identity matrix.

RECYCLING

~

ERROR RECOVERY

It is an error to check a matrix which is not square. On error, the recovery strategy brexit(0) is invoked to initiate a user re-start.

SEE ALSO

```
check_matrix(), identity_matrix()
```

24	CHECK_MATRIX	C

NAME

check_matrix :– check structural integrity of memory trees for numbers in matrix

SYNOPSIS

```
int check_matrix(matrix *m);
```

ARGUMENTS

1: i ◇ matrix *m ~ matrix to check contents

DESCRIPTION

Check value of all integer tokens to see if they each appear to access a structurally consistent memory tree as would be expected for a number construct.

RETURN VALUE

The return value is zero if the matrix structure appears consistent, or −1 if there appear to be inconsistencies. Diagnostic messages are printed in either case, giving as much information as possible.

RECYCLING

~

ERROR RECOVERY

~

SEE ALSO

```
check_identity(), check_number()
```

25	CHECK_NUMBER	C

NAME

    check_number :– check structural integrity of memory tree for number

SYNOPSIS

```
int check_number(number x);
```

ARGUMENTS

    1: i ◇ number x ~ number to check

DESCRIPTION

    Check value of integer token to see if it appears to access a structurally consistent memory tree as would be expected for a number construct.

RETURN VALUE

    The return value is zero if the number structure appears consistent, or −1 if there appear to be inconsistencies. A diagnostic message is printed in either case, giving as much information as possible.

RECYCLING

    ~

ERROR RECOVERY

    ~

SEE ALSO

```
check_matrix()
```

26	CLIFFORD_CONJUGATION	C

NAME

    clifford_conjugation :– Clifford conjugate of component

SYNOPSIS

```
cliff_comp *clifford_conjugation(cliff_comp *a,
                                 cliff_comp *c);
```

ARGUMENTS

    1: i ◇ cliff_comp *a ~ pointer to structure containing component
    2: o ◇ cliff_comp *c ~ pointer to structure in which to save Clifford
                 conjugate

DESCRIPTION

    Calculate the Clifford conjugate of a component. Setting the pointers of both components to the same value overwrites the input with the output.

RETURN VALUE

    The return value is a pointer to the component containing the Clifford conjugate.

RECYCLING

    ~

ERROR RECOVERY

    ~

SEE ALSO

```
cliffordconjugateof_number(), complex_conjugation(),
involution(), reversion()
```

27	CLIFFORDCONJUGATEOF_NUMBER	C

NAME

cliffordconjugateof_number :– produce Clifford conjugate of number

SYNOPSIS

```
number cliffordconjugateof_number(number x, int d);
```

ARGUMENTS

1: id ◊ number x ~ number

2: i ◊ int d   ~ disposal flag

DESCRIPTION

Construct a number containing the Clifford conjugate of the number specified.

RETURN VALUE

The return value is a number containing the Clifford conjugate.

RECYCLING

The number x is recycled according to the value of the disposal flag d. If d=1, the original number is overwritten *in situ* and returned in place of the new one.

ERROR RECOVERY

~

SEE ALSO

```
involutionof_number(), clifford_conjugation(),
reversionof_number()
```

28	CLONE_ARRAY	C

NAME

clone_array :– clone array of numbers

SYNOPSIS

```
number *clone_array(int n, number *a, number *b);
```

ARGUMENTS

1: i ◊ int n   ~ number of elements in arrays

2: i ◊ number *a ~ pointer to first array of numbers

3: o ◊ number *b ~ pointer to second array, in which to save clones

DESCRIPTION

Clone the data from the first array to the second array. This creates new data elements for the second array which take the same values but are independent from the those in the first array. Initially, the second array must be empty.

RETURN VALUE

The return value is a pointer to the array containing the results.

RECYCLING

~

ERROR RECOVERY

It is an error to provide an array which does not have the number of elements specified.

SEE ALSO

```
clone_matrix(), cloneof_number(), copy_array(),
fresh_array()
```

29	C<small>LONE</small>_F<small>ILTER</small>	C

N<small>AME</small>

clone_filter :– clone components passing through filter

S<small>YNOPSIS</small>

```
number clone_filter(number x, int test(cliff_comp *c));
```

A<small>RGUMENTS</small>

1: i ◇ number x   ~ number

2: i ◇ int test(cliff_comp *c) ~ pointer to filter test function

D<small>ESCRIPTION</small>

Construct a clone of the number containing all components which pass the test. The test function returns a value of 1 (true) for components which pass the test, and 0 (false) for components which fail the test.

R<small>ETURN</small> V<small>ALUE</small>

The return value is a number containing all components passing the test.

R<small>ECYCLING</small>

~

E<small>RROR</small> R<small>ECOVERY</small>

~

S<small>EE</small> A<small>LSO</small>

```
filter_number(), keep_filter()
```

30	C<small>LONE</small>_F<small>ORMAT</small>	C

N<small>AME</small>

clone_format :– take a clone of the internal printing format

S<small>YNOPSIS</small>

```
int clone_format(int n, char *fmt);
```

A<small>RGUMENTS</small>

1: i ◇ int n   ~ length of string in which to copy format

2: o ◇ char *fmt ~ pointer to string in which to copy the format

D<small>ESCRIPTION</small>

Obtain a copy of the current printing format string. If the size of array fmt [n] is too small to contain the whole of the internal format string, as much as possible is copied. If the size of array fmt [n] is zero or less, nothing is copied, but the size of the internal format string is still returned.

R<small>ETURN</small> V<small>ALUE</small>

The return value is the length of the internal format string, excluding the null at the end. The string copied is terminated with a null.

R<small>ECYCLING</small>

~

E<small>RROR</small> R<small>ECOVERY</small>

It is an error to provide an array which does not have the number of elements specified.

S<small>EE</small> A<small>LSO</small>

```
format_printing()
```

31	CLONE_MATRIX	C

NAME

clone_matrix :– clone matrix of numbers

SYNOPSIS

```
matrix *clone_matrix(matrix *a, matrix *b);
```

ARGUMENTS

1: i ◇ matrix *a ~ source, matrix

2: o ◇ matrix *b ~ destination, matrix

DESCRIPTION

Clone the data from the first matrix to the second matrix. This creates new data elements for the second matrix which take the same values but are independent from the those in the first matrix. Initially, the second matrix must be empty.

RETURN VALUE

The return value is the cloned matrix.

RECYCLING

~

ERROR RECOVERY

It is an error if the matrices have different sizes, or if the source matrix is a sub-matrix. On error, the recovery strategy brexit(0) is invoked to initiate a user re-start. It is an error if the second matrix is not empty initially. No check is made.

SEE ALSO

```
clone_array(), clone_number(), copy_matrix()
```

32	CLONEFUNCOF_NUMBER	C

NAME

clonefuncof_number :– construct clone of function of number

SYNOPSIS

```
number clonefuncof_number(number x,
                          cliff_comp *function
                          (cliff_comp *a, cliff_comp *c));
```

ARGUMENTS

1: i ◇ number x   ~ number

2: i ◇ cliff_comp *function   ~ pointer to function
       (cliff_comp *a, cliff_comp *c)

DESCRIPTION

Applying the function to every component in the number and construct a new number containing the result.

RETURN VALUE

The return value is a clone of the original number with the function applied.

RECYCLING

~

ERROR RECOVERY

~

SEE ALSO

```
cloneof_number(), func_number()
```

33	CLONEOF_NUMBER	C

NAME

cloneof_number :– construct a clone of a number

SYNOPSIS

```
number cloneof_number(number x);
```

ARGUMENTS

1: i ◇ number x ~ number

DESCRIPTION

Construct a clone of a number. The clone is a new number which is independent of the original number, set to the same value.

RETURN VALUE

The return value is the clone of the original number.

RECYCLING

~

ERROR RECOVERY

~

SEE ALSO

```
clonefuncof_number(), identity(), negativeof_number()
```

34	CLONEONE_GRADE	C

NAME

cloneone_grade :– clone one grade of number

SYNOPSIS

```
number cloneone_grade(number x, int n);
```

ARGUMENTS

1: i ◇ number x ~ number
2: i ◇ int n   ~ grade

DESCRIPTION

Construct a clone of the number containing only the specified grade.

RETURN VALUE

The return value is a number containing only the specified grade.

RECYCLING

~

ERROR RECOVERY

~

SEE ALSO

```
grade(), keepone_grade(), onegradeof_number()
```

35	COLUMNS	C

NAME

columns :– get number of columns for a matrix

SYNOPSIS

```
int columns(matrix *m);
```

ARGUMENTS

  1: i ◇ matrix *m ~ matrix

DESCRIPTION

  Get the number of columns for a matrix. The indexes for the columns in the matrix start at zero and go up to one less than this number.

RETURN VALUE

  The return value is the number of columns in the matrix.

RECYCLING

  ~

ERROR RECOVERY

  ~

SEE ALSO

  access_element(), rows(), stride()

36	COMP_MUL	C

NAME

  comp_mul :– multiply component with number

SYNOPSIS

```
number comp_mul(cliff_comp *x_comp,
                cliff_comp *multiply(cliff_comp *a,
                cliff_comp *b, cliff_comp *c),
                number y);
```

ARGUMENTS

  1: i ◇ cliff_comp *x_comp   ~ multiplier, pointer to structure containing component

  2: i ◇ cliff_comp *multiply(
      cliff_comp *a,   ~ pointer to multiply function
   cliff_comp *b, cliff_comp *c)
  3: i ◇ number y   ~ multiplicand

DESCRIPTION

  Multiply component onto number using the multiply function specified.

RETURN VALUE

  The return value is the product generated using the multiply function specified.

RECYCLING

  ~

ERROR RECOVERY

  ~

SEE ALSO

  mul_comp(), mul_number(), prescale_array(),

37	COMPLEX_CONJUGATION	C

NAME

  complex_conjugation :– complex conjugate of component

SYNOPSIS

```
cliff_comp *complex_conjugation(cliff_comp *a,
                                cliff_comp *c);
```

ARGUMENTS

    1: i ◇ `cliff_comp` `*a` ~ pointer to structure containing component

    2: o ◇ `cliff_comp` `*c` ~ pointer to structure in which to save complex conjugate

DESCRIPTION

    Calculate the complex conjugate of a component. Setting the pointers of both components to the same value overwrites the input with the output.

RETURN VALUE

    The return value is a pointer to the component containing the complex conjugate.

RECYCLING

    ~

ERROR RECOVERY

    ~

SEE ALSO

    `clifford_conjugation()`, `complex_conjugateof_number()`, `imaginary_part()`, `real_part()`

38	COMPLEXCONJUGATEOF_NUMBER	C

NAME

    complexconjugateof_number :– produce complex conjugate of number

SYNOPSIS

```
number complexconjugateof_number(number x, int d);
```

ARGUMENTS

    1: id ◇ `number` `x` ~ number

    2: i ◇ `int d` ~ disposal flag

DESCRIPTION

    Construct a number containing the complex conjugate of the number specified.

RETURN VALUE

    The return value is a number containing the complex conjugate.

RECYCLING

    The number x is recycled according to the value of the disposal flag d. If d=1, the original number is overwritten *in situ* and returned in place of the new one.

ERROR RECOVERY

    ~

SEE ALSO

    `cliffordconjugateof_number()`, `complex_conjugation()`

39	COPY_ARRAY	C

NAME

    copy_array :– copy array of tokens of numbers

SYNOPSIS

```
number *copy_array(int n, number *a, number *b);
```

ARGUMENTS

    1: i ◊ `int n`   ~ number of elements in arrays

    2: i ◊ `number *a` ~ pointer to first array of numbers

    3: o ◊ `number *b` ~ pointer to second array, in which to save numbers

DESCRIPTION

    Copy the tokens from the first array to the second array. This does not create new data values. It simply provides two separate means through which to access one set of data. Initially, the second array must be empty.

RETURN VALUE

    The return value is a pointer to the array containing the results.

RECYCLING

    ~

ERROR RECOVERY

    It is an error to provide an array which does not have the number of elements specified. It is usually an error to access data via one means of access, which has been modified through the other means of access.

SEE ALSO

    `clone_array(), copy_matrix(), fresh_array()`

40	COPY_MATRIX	C

NAME

    copy_matrix :– copy matrix of tokens of numbers

SYNOPSIS

    `matrix *copy_matrix(matrix *a, matrix *b);`

ARGUMENTS

    1: i ◊ `matrix *a` ~ source, matrix

    2: o ◊ `matrix *b` ~ destination, empty matrix

DESCRIPTION

    Copy the tokens from the first matrix to the second matrix. This does not create new data values. It simply provides two separate means through which to access one set of data. Initially, the second matrix must be empty.

RETURN VALUE

    The return value is the matrix with the copies.

RECYCLING

    ~

ERROR RECOVERY

    It is an error if the matrices have different sizes. On error, the recovery strategy `brexit(0)` is invoked to initiate a user re-start. It is an error if the second matrix is not empty initially. No check is made.

SEE ALSO

    `clone_matrix(), copy_array()`

41	COUNT_COMPONENTS	C

NAME

    count_components :– count number of components

SYNOPSIS

```
int count_components(number x);
```

ARGUMENTS

1: i ⋄ number x ~ number

DESCRIPTION

Counts the number of non-zero components in a number.

RETURN VALUE

The return value is the number of non-zero components unless the number is exactly zero. In that case, there is one component of zero value.

RECYCLING

~

ERROR RECOVERY

~

SEE ALSO

```
unpack_components(), unpack_number()
```

42	CREATE_CLIFFMEM	C

NAME

create_cliffmem :– create Clifford memory

SYNOPSIS

```
void create_cliffmem(int n);
```

ARGUMENTS

1: i ⋄ int n ~ number of components

DESCRIPTION

Allocate memory for data nodes to contain n components and an equal number of structural nodes. If the memory allocated initially is later found during some operation to be insufficient, it is expanded transparently by 19%, as many times as necessary without affecting any data values accessed through the data type `number`.

RETURN VALUE

~

RECYCLING

~

ERROR RECOVERY

It is an error to request a non-positive amount of memory.

SEE ALSO

```
free_cliffmem()
```

43	DECLARE_INVERSE	D

NAME

declare_inverse :– define function used in matrix inversion to invert elements

SYNOPSIS

```
number (*declare_inverse(number (*new_inverse)
        (number xn)))(number xo);
```

ARGUMENTS

1: i ◇ number (*new_inverse) () ~ pointer to new function which returns a number

DESCRIPTION

The function declare_inverse() takes one argument 'new_inverse'. The argument 'new_inverse' is a pointer to a function (the new inverse function) which takes one argument, the number 'xn' (x *new*), and produces one number as a return value.

The function declare_inverse() produces as a return value a pointer to an unnamed function (the old inverse function) which itself takes one argument, the number 'xo' (x *old*), and produces one number as a return value.The purpose of the function declare_inverse() is to exchange the previously declared (old) inverse function with a new inverse function. The old inverse function may be saved and restored later if required.The new function will be used in subsequent matrix operations whenever it is necessary to calculate the inverse of elements in the matrix, such as in row elimination operations. The new inverse function should be chosen to produce a correct multiplicative inverse under central multiplication for the numerical entities represented by the elements in the matrix.

RETURN VALUE

The return value is a pointer to the old inverse function.

RECYCLING

~

ERROR RECOVERY

It is an error to provide an inappropriate inverse function.

SEE ALSO

declare_matrix(), inv_scalar(), inv_bicomplex(),
inv_quaternion(), inv_pauli(), inv_xemfield()

44	DECLARE_MATRIX	D

NAME

declare_matrix :– define matrix size and memory space

SYNOPSIS

```
matrix *declare_matrix(int nrows,  int ncols,
                         matrix *m, number *data);
```

ARGUMENTS

1: i ◇ int nrows   ~ number of rows
2: i ◇ int ncols   ~ number of columns
3: o ◇ matrix *m   ~ pointer to structure for matrix
4: i ◇ number *data ~ pointer to memory allocated to numbers in matrix

DESCRIPTION

Define a matrix by initialising its structure m with the number of rows and columns and a memory space of size nrows*ncols*sizeof(number). The memory space can be pre-allocated globally, locally on the stack, or dynamically using the system memory allocation routine malloc().

RETURN VALUE

The return value is the pointer to the structure of the matrix.

RECYCLING

~

ERROR RECOVERY

It is an error to provide inconsistent values. If the size of the matrix is not positive, the recovery strategy `brexit(0)` is invoked to initiate a user re-start.

SEE ALSO

`access_element()`, `columns()`, `declare_inverse()`, `rows()`, `stride()`

| 45 | DECLARE_PRIMAL | D |

NAME

declare_primal :– declare signatures of primal units

SYNOPSIS

```
void declare_primal(int n, int *lambda);
```

ARGUMENTS

1: i ◊ `int n`   ~ number of dimensions
2: i ◊ `int *lambda` ~ pointer to array of signatures

DESCRIPTION

Establish the number of dimensions as n and link the array `lambda` into the mathematical environment. The array is not copied so that changing any of its values after invoking `declare_primal()` changes the signatures for the corresponding primal units. However, this by itself is not sufficient. It is necessary to invoke `declare_primal()` a second time because this also updates internal flags which make some operations which depend on the signature faster. If the flags are not updated, those operations will not behave properly.

RETURN VALUE

~

RECYCLING

~

ERROR RECOVERY

It is an error to provide an array which does not have the number of elements specified. It is an error for the dimension to be not greater than zero, or for the value of the signatures to be other than -1, 0, or $+1$. On error, the recovery strategy `brexit(0)` is invoked to initiate a user re-start.

SEE ALSO

`access_primal()`, `pack_signature()`

| 46 | DEGREES | D |

NAME

degrees :– convert component in polar form from radians to degrees

SYNOPSIS

```
cliff_comp *degrees(cliff_comp *a, cliff_comp *c);
```

ARGUMENTS

> 1: i ◇ `cliff_comp *a` ~ pointer to structure containing polar component in radians
>
> 2: o ◇ `cliff_comp *c` ~ pointer to structure in which to save polar component in degrees

DESCRIPTION

> Convert the component from radians to degrees. The radial part is taken from the real part, and the angular part from the imaginary part. The angle is initially in radians. Setting the pointers of both components to the same value overwrites the input with the output.

RETURN VALUE

> The return value is a pointer to the component containing the polar form with the angle in degrees.

RECYCLING

> ~

ERROR RECOVERY

> ~

SEE ALSO

> `polar(), radians()`

47	DENOM_XEMFIELD	D

NAME

> denom_xemfield :– produce denominator used in inverse of extended electromagnetic field

SYNOPSIS

> `number denom_xemfield(number x, int d);`

ARGUMENTS

> 1: id ◇ `number x` ~ extended electromagnetic field
>
> 2: i ◇ `int d` ~ disposal flag.

DESCRIPTION

> Construct the denominator of the ratio used when calculating the inverse of the extended electromagnetic field specified. The denominator is guaranteed to be a scalar. If there is no finite inverse, the denominator is zero.

RETURN VALUE

> The return value contains the denominator of the ratio.

RECYCLING

> The extended electromagnetic field x is recycled according to the value of the disposal flag.

ERROR RECOVERY

> It is assumed that the signature is all negative; however, no checks are performed.

SEE ALSO

> `inv_xemfield(), inverseof_xemfield(),`
> `reciprocalof_emfield()`

48	DIMENSION	D

NAME

> dimension :– dimension of unit

SYNOPSIS

> ```
> int dimension(bitmap u);
> ```

ARGUMENTS

> 1: i ◇ bitmap u ~ unit

DESCRIPTION

> Calculate the number of dimensions required to accommodate the unit.

RETURN VALUE

> The return value is the dimension of the specified unit.

RECYCLING

> ~

ERROR RECOVERY

> ~

SEE ALSO

> ```
> grade(), pseudo_scalar()
> ```

49	ELIMINATE_ABOVE	D

NAME

> eliminate_above :– eliminate upper rows in a matrix

SYNOPSIS

> ```
> matrix *eliminate_above(matrix *m, matrix *rhs);
> ```

ARGUMENTS

> 1: io ◇ matrix *m   ~ matrix
>
> 2: io ◇ matrix *rhs ~ columns of known data on right hand side, in a matrix

DESCRIPTION

> Use row elimination to reduce values in the last column of all rows above the last. Perform the same row operations on the right-hand side. Both matrices are modified.

RETURN VALUE

> The return value is the matrix on the right-hand side.

RECYCLING

> ~

ERROR RECOVERY

> It is an error if the matrices have a different number of rows. On error, the recovery strategy brexit(0) is invoked to initiate a user re-start. The routine has no mechanism to recover if a pivot with no finite inverse is encountered.

SEE ALSO

> ```
> eliminate_below(), rows_above(), solve_matrix()
> ```

50	ELIMINATE_BELOW	D

NAME

> eliminate_below :– eliminate lower rows in a matrix

SYNOPSIS

```
matrix *eliminate_below(matrix *m, matrix *rhs);
```

ARGUMENTS

1: io ⬦ matrix *m   ~ matrix

2: io ⬦ matrix *rhs ~ columns of known data on right hand side, in a matrix

DESCRIPTION

Use row elimination to reduce values in the first column of all rows below the first. Perform the same operations on the right-hand side. Both matrices are modified.

RETURN VALUE

The return value is the matrix on the right-hand side.

RECYCLING

~

ERROR RECOVERY

It is an error if the matrices have a different number of rows. On error, the recovery strategy `brexit(0)` is invoked to initiate a user re-start. The routine has no mechanism to recover if a pivot with no finite inverse is encountered.

SEE ALSO

```
eliminate_above(), rows_below(), solve_matrix()
```

27.4.2 F–J

51	FACTORSOF_QUATERNION	F

NAME

factorsof_quaternion :– produce vector factors of quaternion

SYNOPSIS

```
number factorsof_quaternion(number x, number factors[2],
int d);
```

ARGUMENTS

1: id ⬦ number x   ~ quaternion

2: o ⬦ number factors[2] ~ array for two factors

3: i ⬦ int d   ~ disposal flag

DESCRIPTION

Construct an array of numbers containing two vectors which multiplied together with central multiplication produce the quaternion specified. The factorisation is not unique because there are six unknown coefficients in the two vectors and only four known coefficients in the quaternion.

RETURN VALUE

The return value is a pointer to the array containing the two factors.

RECYCLING

The quaternion x is recycled according to the value of the disposal flag.

ERROR RECOVERY

It is an error if the signature contains 0 or positive 1. On error, the recovery strategy `brexit(6)` is invoked, terminating the execution.

SEE ALSO

```
reciprocalof_quaternion()
```

52	FILTER_NUMBER	F

NAME

 filter_number :– produce components passing through filter

SYNOPSIS

```
number filter_number(number x, int test(cliff_comp *c),
                     int d);
```

ARGUMENTS

 1: id ⋄ number x   ~ number
 2: i ⋄ int test(cliff_comp *c) ~ pointer to filter test function
 3: i ⋄ int d   ~ disposal flag

DESCRIPTION

 Construct a number containing all components which pass the test. The test function returns a value of 1 (true) for components which pass the test and 0 (false) for components which fail the test.

RETURN VALUE

 The return value is a number containing all components passing the test.

RECYCLING

 The number x is recycled according to the value of the disposal flag.

ERROR RECOVERY

 ~

SEE ALSO

```
clone_filter(), keep_filter(), trim_zero()
```

53	FIND_COMP	F

NAME

 find_comp :– find component in number

SYNOPSIS

```
cliff_comp *find_comp(number x, cliff_comp *a);
```

ARGUMENTS

 1: i ⋄ number x ~ number
 2: i ⋄ cliff_comp *a ~ pointer to structure containing component

DESCRIPTION

 Find the component in the number with the same unit as in the component specified.

RETURN VALUE

 The return value is a pointer to the component, or a null pointer if the component is not found.

RECYCLING

 ~

ERROR RECOVERY

 The pointer addresses volatile memory and may become stale if some other operation causes memory to expand. To avoid that, the data should be copied or modified immediately.

SEE ALSO

```
access_element(), get_coef(), merge_comp(), find_unit()
```

54	Find_Unit	F

NAME

> find_unit :– find component in number with matching unit

SYNOPSIS

```
cliff_comp *find_unit(number x, bitmap u);
```

ARGUMENTS

> 1: i ◇ number x ~ number
>
> 2: i ◇ bitmap u ~ integer bitmap for unit

DESCRIPTION

> Find the component in the number with the same unit as in the bitmap specified.

RETURN VALUE

> The return value is a pointer to the component, or a null pointer if the component is not found.

RECYCLING

> ~

ERROR RECOVERY

> The pointer addresses volatile memory and may become stale if some other operation causes memory to expand. To avoid that the data should be copied or modified immediately.

SEE ALSO

```
access_element(), find_comp(), get_coef(), merge_comp()
```

55	Format_Matrix	F

NAME

> format_matrix :– define printing layout for matrices

SYNOPSIS

```
int format_matrix(int k);
```

ARGUMENTS

> 1: i ◇ int k ~ type of layout

DESCRIPTION

> Define the layout for printing whole matrices. For type k=0, rows are printed one after the other, with elements from adjacent columns one per line down the page. For type k=1, columns are printed one after the other, with elements from adjacent rows one per line down the page.

RETURN VALUE

> The return value is the old value of the format.

RECYCLING

> ~

ERROR RECOVERY

> It is an error if the matrix-printing format is not set to a known value. On error, the recovery strategy brexit(0) to initiate a user re-start.

SEE ALSO

```
format_printing(), print_matrix()
```

56	FORMAT_PRINTING	F

NAME

format_printing :– declare numerical printing format

SYNOPSIS

```
int format_printing(char *fmt);
```

ARGUMENTS

1: i ◇ char *fmt ~ pointer to format string

DESCRIPTION

Set the default format for printing. The format string contains a single directive specifying the conversion of one double precision data value (such as %f or %e), optionally with declared precision (such as %.4f or %.5e). The format may be extended to indicate any rounding to be applied (# none, + up, 0 nearest, - down) and how the coefficient should be transformed (c complex, r real part, i imaginary part, m magnitude, p phase in radians). If the extended format is used, both rounding and transformation must be supplied. For example, the format string %.3e#m specifies the magnitude of number with three decimal places after the decimal point, with an exponent, and no rounding. When rounding is specified, it is to one of the adjacent integers.

RETURN VALUE

The return value is the maximum format string length permitted, not counting the null at the end.

RECYCLING

~

ERROR RECOVERY

Format strings with invalid syntax are not detected at compile time. It is an error to provide a format string greater than the maximum length permitted. On error, the recovery strategy brexit(0) is invoked to initiate a user re-start.

SEE ALSO

```
clone_format(), format_matrix()
```

57	FREE_CLIFFMEM	F

NAME

free_cliffmem :– free Clifford memory

SYNOPSIS

```
void free_cliffmem(int p);
```

ARGUMENTS

1: i ◇ int p ~ printing policy

DESCRIPTION

Free all memory allocated by create_cliffmem(), including any subsequent automatic memory expansion. A summary of memory usage is printed if the policy p=1. No summary is printed if p=0. The summary can be used to determine whether any leakage of memory has occurred within the Clifford memory structures during execution. It can also be used to determine the peak memory usage. If the peak value is requested at the time of initial memory allocation, no expansion of memory is necessary, and no memory allocated is left unused.

RETURN VALUE

~

RECYCLING

~

ERROR RECOVERY

It is an error to free the memory before it has been established. On error, the recovery strategy `brexit(2)` is invoked, terminating the execution.

SEE ALSO

`create_cliffmem(), recycle_allnumber(), recycle_number()`

| 58 | FRESH_ARRAY | F |

NAME

fresh_array :– fill array with numbers of zero value

SYNOPSIS

`number *fresh_array(int n, number *a);`

ARGUMENTS

1: i ◇ `int n`   ~ number of elements in array

2: o ◇ `number *a` ~ pointer to array in which to save numbers

DESCRIPTION

Fill an array with n numbers set to value zero. Initially, the array must be empty.

RETURN VALUE

The return value is a pointer to the array containing zeros.

RECYCLING

~

ERROR RECOVERY

It is an error to provide an array which is too small.

SEE ALSO

`clone_array(), copy_array(), fresh_number(), recycle_array()`

| 59 | FRESH_NUMBER | F |

NAME

fresh_number :– construct a number with zero value

SYNOPSIS

`number fresh_number(void);`

ARGUMENTS

~

DESCRIPTION

Construct a new number with zero value.

RETURN VALUE

The return value is the new number.

RECYCLING

~

ERROR RECOVERY

~

SEE ALSO

    primal_vector(), pseudo_number(), scalar_number(),
    simple_number(), zero()

60	FUNC_NUMBER	F

NAME

    func_number :– calculate function of number

SYNOPSIS

    number func_number(number x, cliff_comp
     *function(cliff_comp *a, cliff_comp *c));

ARGUMENTS

    1: ir ◊ number x   ~ number
    2: i ◊ cliff_comp *function   ~ pointer to function
          (cliff_comp *a, cliff_comp *c)

DESCRIPTION

    Apply the function to every component in the number.

RETURN VALUE

    The return value is the original number with the function applied.

RECYCLING

    The original number is recycled. It is overwritten *in situ* and returned in place of
    the new one.

ERROR RECOVERY

    It is an error to specify a function which changes the unit of a component. On error,
    the recovery strategy brexit(6) is invoked, terminating the execution.

SEE ALSO

    clonefuncof_number()

61	FUNDAMENTAL_SOLUTION3D	F

NAME

    fundamental_solution3D :– fundamental solution of k-Dirac operator in three
    dimensions

SYNOPSIS

    number fundamental_solution3D(cliff_comp *k, number r);

ARGUMENTS

    1: ir ◊ cliff_comp *k ~ wavenumber, pointer to structure containing
                              component
    2: i ◊ number r   ~ radial vector from origin

DESCRIPTION

    Calculate the fundamental solution of the (perturbed) k-Dirac operator in three
    dimensions.

RETURN VALUE

    The return value is a number containing the fundamental solution as a bivector.

RECYCLING

~

ERROR RECOVERY

It is an error to calculate the solution at the origin, where the value is unbounded.

SEE ALSO

`cauchy_kernel3D()`

| 62 | GET_COEF | G |

NAME

get_coef :– get coefficient

SYNOPSIS

`double *get_coef(number x, bitmap u, double c[2], int d);`

ARGUMENTS

1: id ◇ number x   ~ number
2: i ◇ bitmap u   ~ unit
3: o ◇ double c[2] ~ pointer to array in which to save complex coefficient
4: i ◇ int d   ~ disposal flag

DESCRIPTION

Get a copy of the coefficient for the component in the number which has the unit specified.

RETURN VALUE

The return value is a pointer to the value of the coefficient.

RECYCLING

The number x is recycled according to the value of the disposal flag d.

ERROR RECOVERY

~

SEE ALSO

`find_comp(), get_imag(), get_real()`

| 63 | GET_IMAG | G |

NAME

get_imag :– get imaginary part of coefficient

SYNOPSIS

`double get_imag(number x, bitmap u, int d);`

ARGUMENTS

1: id ◇ number x ~ number
2: i ◇ bitmap u ~ unit
3: i ◇ int d   ~ disposal flag

DESCRIPTION

Get a copy of the imaginary part of the coefficient for the component in the number which has the unit specified.

RETURN VALUE

The return value is the imaginary part of the coefficient.

RECYCLING

The number x is recycled according to the value of the disposal flag d.

ERROR RECOVERY

~

SEE ALSO

get_coef(), get_real(), imaginary_part()

64	GET_MEM_USAGE	G

NAME

get_mem_usage :– get memory usage

SYNOPSIS

void get_mem_usage(int n[2][3]);

ARGUMENTS

1: o ◇ int n[2][3] ~ pointer to pair of arrays with three elements each

DESCRIPTION

Provides a compact summary of memory usage, for both structural heap n[0][*] and data heap n[1][*], in the two-dimensional array supplied. Values provided are the following:

n[*][0] = memory currently in use,

n[*][1] = highest memory usage so far,

n[*][2] = total memory capacity currently allocated.

The values are always in non-decreasing order. The free memory can be calculated as by subtracting n[*][0] from n[*][2].

RETURN VALUE

~

RECYCLING

~

ERROR RECOVERY

~

SEE ALSO

free_cliffmem(), print_mem_usage(), print_num_info()

65	GET_REAL	G

NAME

get_real :– get real part of coefficient

SYNOPSIS

double get_real(number x, bitmap u, int d);

ARGUMENTS

1: id ◇ number x ~ number

2: i ◇ bitmap u ~ unit

3: i ◇ int d ~ disposal flag

DESCRIPTION

Get a copy of the real part of the coefficient for the component in the number which has the unit specified.

RETURN VALUE

The return value is the real part of the coefficient.

RECYCLING

The number x is recycled according to the value of the disposal flag d.

ERROR RECOVERY

~

SEE ALSO

```
get_coef(), get_imag(), real_part()
```

66	GRADE	G

NAME

grade :– grade of unit

SYNOPSIS

```
int grade(bitmap u);
```

ARGUMENTS

1: i ◇ bitmap u ~ unit

DESCRIPTION

Calculate the grade of the specified unit.

RETURN VALUE

The return value is the grade of the unit.

RECYCLING

~

ERROR RECOVERY

~

SEE ALSO

```
dimension()
```

67	GRADE_ONE	G

NAME

grade_one :– test for component in grade 1

SYNOPSIS

```
int grade_one(cliff_comp *c);
```

ARGUMENTS

1: i ◇ cliff_comp *c ~ pointer to structure containing component

DESCRIPTION

Test to determine whether the component is in grade 1.

RETURN VALUE

The return value is 0 (false) when the component is not in grade 1 and 1 (true) when it is in grade 1.

RECYCLING

~

ERROR RECOVERY

~

SEE ALSO

```
grade_two(), grade_zero()
```

68	GRADE_TWO	G

NAME

    grade_two :– test for component in grade 2

SYNOPSIS

    `int grade_two(cliff_comp *c);`

ARGUMENTS

    1: i ◇ `cliff_comp *c` ~ pointer to structure containing component

DESCRIPTION

    Test to determine whether the component is in grade 2.

RETURN VALUE

    The return value is 0 (false) when the component is not in grade 2 and 1 (true) when it is in grade 2.

RECYCLING

    ~

ERROR RECOVERY

    ~

SEE ALSO

    `grade_one(), grade_zero()`

69	GRADE_ZERO	G

NAME

    grade_zero :– test for component in grade 0

SYNOPSIS

    `int grade_zero(cliff_comp *c);`

ARGUMENTS

    1: i ◇ `cliff_comp *c` ~ pointer to structure containing component

DESCRIPTION

    Test to determine whether the component is in grade 0.

RETURN VALUE

    The return value is 0 (false) when the component is not in grade 0 and 1 (true) when it is in grade 0.

RECYCLING

    ~

ERROR RECOVERY

    ~

SEE ALSO

    `grade_one(), grade_two(), non_scalar()`

70	HALF	H

NAME

    half :– half the value of component

SYNOPSIS

    `cliff_comp *half(cliff_comp *a, cliff_comp *c);`

ARGUMENTS

    1: i ◇ cliff_comp *a ~ pointer to structure containing component

    2: o ◇ cliff_comp *c ~ pointer to structure in which to save half
                         the component

DESCRIPTION

    Calculate half the value of the component. Setting the pointers of both components to the same value overwrites the input with the output.

RETURN VALUE

    The return value is a pointer to the component containing the result.

RECYCLING

    ~

ERROR RECOVERY

    ~

SEE ALSO

    identity(), rscale(), twice()

| 71 | HERTZIAN_DIPOLE | H |

NAME

    hertzian_dipole :– calculate electromagnetic field from Hertzian dipole

SYNOPSIS

```
number hertzian_dipole(cliff_comp *Ih, cliff_comp *mu,
cliff_comp *k, number r);
```

ARGUMENTS

    1: i ◇ cliff_comp *Ih ~ dipole strength, pointer to structure containing
                        component

    2: i ◇ cliff_comp *mu ~ complex magnetic permeability, pointer to structure
                        containing component

    3: i ◇ cliff_comp *k ~ complex wavenumber, pointer to structure
                        containing component

    4: i ◇ number r ~ location of field

DESCRIPTION

    Calculate the electromagnetic field at the specified location r from a Hertzian dipole standing aligned vertically at the origin. The strength of the dipole Ih is the product of current I and length (height) of dipole h. The location where the field is calculated is specified in Cartesian coordinates in three dimensions.

RETURN VALUE

    The return value is the electromagnetic field.

RECYCLING

    ~

ERROR RECOVERY

    It is an error to calculate the field at the origin, where the value is unbounded.

SEE ALSO

    bessel_potential3D()

| 72 | HODGESTAROF_NUMBER | H |

NAME

hodgestarof_number :– produce Hogde's star of number

SYNOPSIS

`number hodgestarof_number(number x, int d);`

ARGUMENTS

1: id ◇ `number x` ~ number

2: i ◇ `int d` ~ disposal flag

DESCRIPTION

Construct a number containing the Hodge star of the number specified.

RETURN VALUE

The return value is a number containing the Hodge's star.

RECYCLING

The number x is recycled according to the value of the disposal flag d.

ERROR RECOVERY

~

SEE ALSO

`supplementof_number()`

| 73 | I_UNIT | I |

NAME

i_unit :– involution of unit

SYNOPSIS

`int i_unit(bitmap u);`

ARGUMENTS

1: i ◇ `bitmap u` ~ unit

DESCRIPTION

Calculate the effect of reversing the sign of all of the primal units in the unit specified.

RETURN VALUE

The return value is the sign induced by involution of the unit.

RECYCLING

~

ERROR RECOVERY

~

SEE ALSO

`c_unit(), r_unit()`

| 74 | IDENTITY | I |

NAME

identity :– clone of component

SYNOPSIS

`cliff_comp *identity(cliff_comp *a, cliff_comp *c);`

ARGUMENTS

    1: i ◇ `cliff_comp *a` ~ pointer to structure containing component

    2: o ◇ `cliff_comp *c` ~ pointer to structure in which to save value of component

DESCRIPTION

    Produces a clone of the component.

RETURN VALUE

    The return value is a pointer to the component containing the clone.

RECYCLING

    ~

ERROR RECOVERY

    ~

SEE ALSO

    `cloneof_number()`, `half()`, `inverse()`, `negation()`, `rscale()`, `twice()`

| 75 | IDENTITY_MATRIX | I |

NAME

    identity_matrix :– fill an empty matrix with the identity

SYNOPSIS

    `matrix *identity_matrix(matrix *m);`

ARGUMENTS

    1: o ◇ `matrix *m` ~ empty matrix

DESCRIPTION

    Fill the matrix with the scalar 1 on the leading diagonal, and 0 elsewhere.

RETURN VALUE

    The return value is the original matrix containing the identity.

RECYCLING

    ~

ERROR RECOVERY

    It is an error to initialise a sub-matrix. On error, the recovery strategy `brexit(0)` is invoked to initiate a user re-start. It is an error if the matrix is not empty initially. No check is made.

SEE ALSO

    `check_identity()`, `random_matrix()`

| 76 | IMAGINARY_PART | I |

NAME

    imaginary_part :– imaginary part of component

SYNOPSIS

    `cliff_comp *imaginary_part(cliff_comp *a, cliff_comp *c);`

ARGUMENTS

    1: i ◇ `cliff_comp *a` ~ pointer to structure containing complex component

    2: o ◇ `cliff_comp *c` ~ pointer to structure in which to save imaginary part

DESCRIPTION

Extract the imaginary part of a component.

RETURN VALUE

The return value is a pointer to the component containing the imaginary part. Setting the pointers of both components to the same value overwrites the input with the output.

RECYCLING

~

ERROR RECOVERY

~

SEE ALSO

imaginarypartof_number(), no_real(), real_part(), squareof_magnitude()

77	IMAGINARYPARTOF_NUMBER	I

NAME

imaginarypartof_number :– produce imaginary part of number

SYNOPSIS

number imaginarypartof_number(number x, int d);

ARGUMENTS

1: id ◇ number x ~ number
2: i ◇ int d ~ disposal flag

DESCRIPTION

Construct a number containing the imaginary part of the number specified.

RETURN VALUE

The return value is a number containing the imaginary part only.

RECYCLING

The number x is recycled according to the value of the disposal flag d. If d=1, the original number is overwritten *in situ* and returned in place of the new one.

ERROR RECOVERY

~

SEE ALSO

imaginary_part(), realpartof_number()

78	INNER	I

NAME

inner :– inner product of components

SYNOPSIS

cliff_comp *inner(cliff_comp *a, cliff_comp *b,
                  cliff_comp *c);

ARGUMENTS

1: i ◇ cliff_comp *a ~ multiplier, pointer to structure containing component
2: i ◇ cliff_comp *b ~ multiplicand, pointer to structure containing component
3: o ◇ cliff_comp *c ~ pointer to structure in which to save product

DESCRIPTION

Calculate the inner (interior) product of two components. Setting the pointers of either input component to the same value as the output component overwrites the input with the output.

RETURN VALUE

The return value is a pointer to the component containing the inner (interior) product.

RECYCLING

~

ERROR RECOVERY

~

SEE ALSO

central (), inner_mul (), outer ()

79	INNER_MUL	I

NAME

inner_mul :– calculate inner product of numbers

SYNOPSIS

number inner_mul (number x, number y, int d);

ARGUMENTS

1: id ◇ number x ~ multiplier
2: id ◇ number y ~ multiplicand
3: i ◇ int d   ~ disposal flag

DESCRIPTION

Multiply two numbers using the inner multiplication.

RETURN VALUE

The return value is the inner product.

RECYCLING

The numbers x and y are recycled according to the value of the disposal flag.

ERROR RECOVERY

~

SEE ALSO

central_mul (), inner (), outer_mul (), scalar_mul ()

80	INNERSPACE_NUMBER	I

NAME

innerspace_number :– produce inner projection of number

SYNOPSIS

number innerspace_number (number x, number n, int d);

ARGUMENTS

1: id ◇ number x ~ vector or other number
2: id ◇ number n ~ unit vector
3: i ◇ int d   ~ disposal flag

DESCRIPTION

Apply inner projection operator to a number. The projection operator eliminates all components of the number perpendicular to the unit vector supplied. Any resultant vector components are parallel to the unit vector. Together the `innerspace_number()` and `outerspace_number()` projection operators can be used to split vectors into two orthogonal components.

RETURN VALUE

The return value is the result of the projection.

RECYCLING

The numbers x and n are recycled according to the value of the disposal flag.

ERROR RECOVERY

It is an error to specify unit vectors which are not of unit length.

SEE ALSO

```
innerspace_vector(), normalise_vector(),
outerspace_number(), reflect_number(), rotate_number()
```

81	INNERSPACE_VECTOR	I

NAME

innerspace_vector :– produce inner projection of number

SYNOPSIS

```
number innerspace_vector(number x, number n, int d);
```

ARGUMENTS

1: id ◇ number x ~ vector or other number

2: id ◇ number n ~ unit vector

3: i ◇ int d   ~ disposal flag

DESCRIPTION

Apply inner projection operator to a number. For numbers in grade 1, the projection operator eliminates all components of the number perpendicular to the unit vector supplied. In this case, the resultant vector is parallel to the unit vector. For numbers in higher grades, the result is different for different signatures. Which signature is more appropriate depends on the application. Together the `innerspace_vector()` and `outerspace_vector()` projection operators can be used to split vectors into two orthogonal components.

RETURN VALUE

The return value is the result of the projection.

RECYCLING

The numbers x and n are recycled according to the value of the disposal flag.

ERROR RECOVERY

It is an error to specify unit vectors which are not of unit length.

SEE ALSO

```
innerspace_number(), normalise_vector(),
outerspace_vector(), reflect_vector(), rotate_number()
```

| 82 | INV_BICOMPLEX | I |

NAME

inv_bicomplex :– invert a bicomplex number checking for zero denominator

SYNOPSIS

```
number inv_bicomplex(number x);
```

ARGUMENTS

1: i ◇ number x ~ bicomplex number

DESCRIPTION

Take the inverse of a bicomplex number, generating an exception in the case where the inverse has no finite value.

RETURN VALUE

The return value is the inverse of the bicomplex number.

RECYCLING

~

ERROR RECOVERY

It is an error if the bicomplex number has no finite inverse. On error, the recovery strategy brexit(0) is invoked to initiate a user re-start.

SEE ALSO

```
declare_inverse(), inv_pauli(), inv_quaternion(),
inv_scalar(), inv_xemfield(), inverseof_bicomplex()
```

| 83 | INV_PAULI | I |

NAME

inv_pauli :– invert a Pauli matrix checking for zero denominator

SYNOPSIS

```
number inv_pauli(number x);
```

ARGUMENTS

1: i ◇ number x ~ Pauli matrix

DESCRIPTION

Take the inverse of a Pauli matrix, generating an exception in the case where the inverse has no finite value.

RETURN VALUE

The return value is the inverse of the Pauli matrix.

RECYCLING

~

ERROR RECOVERY

It is an error if the Pauli matrix has no finite inverse. On error, the recovery strategy brexit(0) is invoked to initiate a user re-start.

SEE ALSO

```
declare_inverse(), inv_bicomplex(), inv_quaternion(),
inv_xemfield(), inverseof_pauli()
```

| 84 | INV_QUATERNION | I |

NAME

inv_quaternion :– invert a quaternion checking for zero denominator

SYNOPSIS

```
number inv_quaternion(number x);
```

ARGUMENTS

1: i ◊ number x ~ quaternion

DESCRIPTION

Take the inverse of a quaternion, generating an exception in the case where the inverse has no finite value.

RETURN VALUE

The return value is inverse of the quaternion.

RECYCLING

~

ERROR RECOVERY

It is an error if the quaternion has no finite inverse. On error, the recovery strategy `brexit(0)` is invoked to initiate a user re-start.

SEE ALSO

```
declare_inverse(), inv_bicomplex(), inv_pauli(),
inv_scalar(), inv_xemfield(), inverseof_quaternion()
```

85	INV_SCALAR	I

NAME

inv_scalar :– invert a scalar checking for zero denominator

SYNOPSIS

```
inv_scalar(number x);
```

ARGUMENTS

1: i ◊ number x ~ scalar number

DESCRIPTION

Take the inverse of a scalar, generating an exception in the case where the inverse has no finite value.

RETURN VALUE

The return value is the inverse of the scalar.

RECYCLING

~

ERROR RECOVERY

It is an error if the scalar has no finite inverse. On error, the recovery strategy `brexit(0)` is invoked to initiate a user re-start.

SEE ALSO

```
declare_inverse(), inv_bicomplex(), inv_pauli(),
inv_quaternion(), inv_xemfield(), inverseof_component()
```

86	INV_UNIT	I

NAME

inv_unit :– inverse of unit

SYNOPSIS

```
int inv_unit(bitmap u);
```

ARGUMENTS

    1: i ◊ bitmap u ~ unit

DESCRIPTION

    Calculate the sign of the inverse of the specified unit.

RETURN VALUE

    The return value is the sign of the inverse (under central multiplication) of the unit.

RECYCLING

    ~

ERROR RECOVERY

    It is an error to take the inverse of a unit with zero signature. On error, the recovery strategy brexit(6) is invoked, terminating the execution.

SEE ALSO

    `inverse(), inverseof_component()`

87	INV_XEMFIELD	I

NAME

    inv_xemfield :– invert an extended electromagnetic field checking for zero denominator

SYNOPSIS

    `number  inv_xemfield(number x);`

ARGUMENTS

    1: i ◊ number x ~ extended electromagnetic field

DESCRIPTION

    Take the inverse of an extended electromagnetic field, generating an exception in the case where the inverse has no finite value.

RETURN VALUE

    The return value is the inverse of the extended electromagnetic field.

RECYCLING

    ~

ERROR RECOVERY

    It is an error if the extended electromagnetic field has no finite inverse. On error, the recovery strategy brexit(0) is invoked to initiate a user re-start.

SEE ALSO

    `declare_inverse(), inv_bicomplex(), inv_pauli(),`
    `inv_quaternion(), inverseof_xemfield()`

88	INVERSE	I

NAME

    inverse :– inverse of component

SYNOPSIS

    `cliff_comp *inverse(cliff_comp *a, cliff_comp *c);`

ARGUMENTS

    1: i ◊ cliff_comp *a ~ pointer to structure containing component

    2: o ◊ cliff_comp *c ~ pointer to structure in which to save inverse

DESCRIPTION

Calculate the inverse of a component for central (Clifford) multiplication. Setting the pointers of both components to the same value overwrites the input with the output.

RETURN VALUE

The return value is a pointer to the component containing the inverse.

RECYCLING

~

ERROR RECOVERY

It is an error to take the inverse of a component with value zero. On error, the recovery strategy `brexit(0)` is invoked to initiate a user re-start.

SEE ALSO

`identity()`, `inverseof_component()`, `negation()`

| 89 | INVERSE_TRANSFORM | I |

NAME

inverse_transform :– inverse fast Fourier transform

SYNOPSIS

`number *inverse_transform(int n, number *a, number *b);`

ARGUMENTS

1: i ◇ `int n`   ~ number of elements in array (power of 2)

2: ir ◇ `number *a` ~ pointer to array containing spectrum

3: o ◇ `number *b` ~ pointer to array in which to save waveform

DESCRIPTION

Calculates the fast inverse Fourier transform for an array for which the number of elements is a power of 2. The spectral values are the n complex frequency components f_k, in the order of frequency $k = 0, 0 < k < n/2, k = \pm n/2, -n/2 < k < 0$. The waveform is placed in the second array. Initially, the second array must be empty.

RETURN VALUE

The return value is a pointer to the array containing the waveform.

RECYCLING

The spectrum array is recycled.

ERROR RECOVERY

It is an error to provide an array which does not have the number of elements specified. It is an error if the size of the arrays is not a power or 2. On error, the recovery strategy `brexit(0)` is invoked to initiate a user re-start.

SEE ALSO

`transform()`

| 90 | INVERSEOF_BICOMPLEX | I |

NAME

inverseof_bicomplex :– produce inverse of bicomplex number

SYNOPSIS

`number inverseof_bicomplex(number x, int d);`

ARGUMENTS

        1: id ◇ `number` x ~ bicomplex number

        2: i ◇ `int d` ~ disposal flag

DESCRIPTION

        Construct a bicomplex number containing the inverse of the bicomplex number specified.

RETURN VALUE

        The return value is the inverse.

RECYCLING

        The bicomplex number x is recycled according to the value of the disposal flag.

ERROR RECOVERY

        It is an error if the bicomplex number is 0, or the signature contains 0 or positive 1. Depending on the error, either the recovery strategy `brexit(0)` is invoked to initiate a user re-start or the recovery strategy `brexit(6)` is invoked, terminating the execution.

SEE ALSO

        `inverse(), inv_bicomplex(), reciprocalof_bicomplex()`

91	INVERSEOF_COMPONENT	I

NAME

        inverseof_component :– produce inverse of number with single component

SYNOPSIS

        `number inverseof_component(number x, int d);`

ARGUMENTS

        1: id ◇ `number` x ~ simple number (single non-zero component)

        2: i ◇ `int d` ~ disposal flag

DESCRIPTION

        Calculate the inverse of a number consisting of a single non-zero component.

RETURN VALUE

        The return value is the inverse of the component.

RECYCLING

        The number x is recycled according to the value of the disposal flag.

ERROR RECOVERY

        It is an error to take the inverse of a number which has more than one component, or if the coefficient or the signature is zero. Depending on the error, either the recovery strategy `brexit(0)` is invoked to initiate a user re-start or the recovery strategy `brexit(6)` is invoked, terminating the execution.

SEE ALSO

        `inv_scalar(), inverse(), inverseof_complex(),`
        `inverseof_emfield(), inverseof_pauli(),`
        `inverseof_quaternion(), inverseof_vector(),`
        `reciprocalof_component()`

92	INVERSEOF_EMFIELD	I

NAME

        inverseof_emfield :– produce inverse of electromagnetic field

SYNOPSIS

```
number inverseof_emfield(number x, int d);
```

ARGUMENTS

    1: id ◇ `number` x ~ electromagnetic field

    2: i ◇ `int d`   ~ disposal flag

DESCRIPTION

Construct an electromagnetic field containing the inverse of the electromagnetic field specified. Not every electromagnetic field has and inverse. Some electromagnetic fields, such as plane waves, have non-zero magnitude but have no inverse.

RETURN VALUE

The return value is the inverse (under central multiplication) of the electromagnetic field.

RECYCLING

The electromagnetic field x is recycled according to the value of the disposal flag.

ERROR RECOVERY

It is an error if the electromagnetic field has no inverse, or the signature contains 0 or positive 1. Depending on the error, either the recovery strategy `brexit(0)` is invoked to initiate a user re-start or the recovery strategy `brexit(6)` is invoked, terminating the execution.

SEE ALSO

`inverse()`, `inverseof_xemfield()`, `reciprocalof_emfield()`

| 93 | INVERSEOF_PAULI | I |

NAME

inverseof_pauli :– produce inverse of Pauli matrix

SYNOPSIS

```
number inverseof_pauli(number x, int d);
```

ARGUMENTS

    1: id ◇ `number` x ~ Pauli matrix

    2: i ◇ `int d`   ~ disposal flag

DESCRIPTION

Construct a Pauli matrix containing the inverse of the Pauli matrix specified.

RETURN VALUE

The return value contains the inverse (under central multiplication) of the Pauli matrix.

RECYCLING

The Pauli matrix x is recycled according to the value of the disposal flag.

ERROR RECOVERY

It is an error if the Pauli matrix is zero, or the signature contains 0 or positive 1. Depending on the error, either the recovery strategy `brexit(0)` is invoked to initiate a user re-start or the recovery strategy `brexit(6)` is invoked, terminating the execution.

SEE ALSO

`inv_pauli()`, `inverse()`, `reciprocalof_pauli()`

94	INVERSEOF_QUATERNION	I

NAME

> inverseof_quaternion :– produce inverse of quaternion

SYNOPSIS

> ```
> number inverseof_quaternion(number x, int d);
> ```

ARGUMENTS

> 1: id ◇ number x ~ quaternion
> 2: i ◇ int d   ~ disposal flag

DESCRIPTION

> Construct a quaternion containing the inverse of the quaternion specified.

RETURN VALUE

> The return value contains the inverse (under central multiplication) of the quaternion.

RECYCLING

> The quaternion x is recycled according to the value of the disposal flag.

ERROR RECOVERY

> It is an error if the quaternion is zero or the signature contains 0 or positive 1. Depending on the error, either the recovery strategy brexit(0) is invoked to initiate a user re-start or the recovery strategy brexit(6) is invoked, terminating the execution.

SEE ALSO

> ```
> inv_quaternion(), inverse(), factorsof_quaternion(),
> reciprocalof_quaternion()
> ```

95	INVERSEOF_VECTOR	I

NAME

> inverseof_vector :– produce inverse of vector

SYNOPSIS

> ```
> number inverseof_vector(number x, int d);
> ```

ARGUMENTS

> 1: id ◇ number x ~ vector
> 2: i ◇ int d   ~ disposal flag.

DESCRIPTION

> Construct a vector containing the inverse of the vector specified.

RETURN VALUE

> The return value contains the inverse (under central multiplication) of the vector.

RECYCLING

> The vector x is recycled according to the value of the disposal flag.

ERROR RECOVERY

> It is an error if the vector has length zero, or the signature is mixed or contains zero. Depending on the error, either the recovery strategy brexit(0) is invoked to initiate a user re-start or the recovery strategy brexit(6) is invoked, terminating the execution.

SEE ALSO

> ```
> inverse(), reciprocalof_vector()
> ```

96	INVERSEOF_XEMFIELD	I

NAME

inverseof_xemfield :– produce inverse of extended electromagnetic field

SYNOPSIS

```
number inverseof_xemfield(number x, int d);
```

ARGUMENTS

1: id ◇ number x ~ extended electromagnetic field

2: i ◇ int d   ~ disposal flag

DESCRIPTION

Construct an extended electromagnetic field containing the inverse of the extended electromagnetic field specified.

RETURN VALUE

The return value is the inverse (under central multiplication) of the extended electromagnetic field.

RECYCLING

The extended electromagnetic field x is recycled according to the value of the disposal flag.

ERROR RECOVERY

It is an error if the extended electromagnetic field has no inverse, or the signature contains 0 or positive 1. Depending on the error, either the recovery strategy brexit(0) is invoked to initiate a user re-start or the recovery strategy brexit(6) is invoked, terminating the execution.

SEE ALSO

```
inv_xemfield(), inverse(), inverseof_emfield(),
reciprocalof_xemfield()
```

97	INVOLUTION	I

NAME

involution :– involution of component

SYNOPSIS

```
cliff_comp *involution(cliff_comp *a, cliff_comp *c);
```

ARGUMENTS

1: i ◇ cliff_comp *a ~ pointer to structure containing component

2: o ◇ cliff_comp *c ~ pointer to structure in which to save involution

DESCRIPTION

Calculate the involution of a component. Setting the pointers of both components to the same value overwrites the input with the output.

RETURN VALUE

The return value is a pointer to the component containing the involution.

RECYCLING

~

ERROR RECOVERY

~

SEE ALSO

```
clifford_conjugation(), involutionof_number(), reversion()
```

98	INVOLUTIONOF_NUMBER	I

NAME

involutionof_number :– produce involution of number

SYNOPSIS

```
number involutionof_number(number x, int d);
```

ARGUMENTS

1: id ◇ number x ~ number

2: i ◇ int d   ~ disposal flag

DESCRIPTION

Construct a number containing the involution of the number specified.

RETURN VALUE

The return value is a number containing the involution.

RECYCLING

The number x is recycled according to the value of the disposal flag d. If d=1, the original number is overwritten *in situ* and returned in place of the new one.

ERROR RECOVERY

~

SEE ALSO

```
cliffordconjugateof_number(), involution(),
reversionof_number()
```

99	IS_BIVECTOR	I

NAME

is_bivector :– test number to see if it is a bivector

SYNOPSIS

```
int is_bivector(number x);
```

ARGUMENTS

1: i ◇ number x ~ number

DESCRIPTION

Test to determine if the number is a bivector.

RETURN VALUE

The return value is 1 (true) when the number is a bivector and 0 (false) when the number is a non-bivector.

RECYCLING

~

ERROR RECOVERY

A bivector of zero area has no bivector components and is not recognised as a bivector.

SEE ALSO

```
filter_number(), is_scalar(), is_vector()
```

100	IS_MULTI	I

NAME

is_multi :– test number to see if it has multiple components

SYNOPSIS

      `int is_multi(number x);`

ARGUMENTS

      1: i ⋄ number x ~ number

DESCRIPTION

      Test to determine if the number has multiple components.

RETURN VALUE

      The return value is 1 (true) when the number has more than one component and 0 (false) when the number has a single component.

RECYCLING

      ~

ERROR RECOVERY

      ~

SEE ALSO

      `filter_number(), is_single()`

101	IS_NONZERO	I

NAME

      is_nonzero :– test number to see if it is non-zero

SYNOPSIS

      `int is_nonzero(number x);`

ARGUMENTS

      1: i ⋄ number x ~ number

DESCRIPTION

      Test to determine if the number has a non-zero value.

RETURN VALUE

      The return value is 1 (true) when the number is non-zero and 0 (false) when the number is 0.

RECYCLING

      ~

ERROR RECOVERY

      ~

SEE ALSO

      `is_zero(), filter_number()`

102	IS_SCALAR	I

NAME

      is_scalar :– test number to see if it is a scalar

SYNOPSIS

      `int is_scalar(number x);`

ARGUMENTS

      1: i ⋄ number x ~ number

DESCRIPTION

      Test to determine if the number is a scalar.

RETURN VALUE

The return value is 1 (true) when the number is a scalar and 0 (false) when the number is a non-scalar.

RECYCLING

~

ERROR RECOVERY

~

SEE ALSO

`filter_number(), is_bivector(), is_vector()`

103	IS_SINGLE	I

NAME

is_single :– test number to see if it has only a single component

SYNOPSIS

`int is_single(number x);`

ARGUMENTS

1: i ◇ number x ~ number

DESCRIPTION

Test to determine if the number has a single component.

RETURN VALUE

The return value is 1 (true) when the number has a single component and 0 (false) when the number has more than one component.

RECYCLING

~

ERROR RECOVERY

~

SEE ALSO

`filter_number(), is_multi()`

104	IS_VECTOR	I

NAME

is_vector :– test number to see if it is a vector

SYNOPSIS

`int is_vector(number x);`

ARGUMENTS

1: i ◇ number x ~ number

DESCRIPTION

Test to determine if the number is a vector.

RETURN VALUE

The return value is 1 (true) when the number is a vector and 0 (false) when the number is a non-vector.

RECYCLING

~

ERROR RECOVERY

A vector of zero length has no vector components and is not recognised as a vector.

SEE ALSO

      `filter_number(), is_bivector(), is_scalar()`

105	IS_ZERO	I

NAME

      is_zero :– test number to see if it is zero

SYNOPSIS

      `int is_zero(number x);`

ARGUMENTS

      1: i ◇ `number x` ~ number

DESCRIPTION

      Test to determine if the number is zero.

RETURN VALUE

      The return value is 1 (true) when the number is 0, and 0 (false) when the number is non-zero.

RECYCLING

      ~

ERROR RECOVERY

      ~

SEE ALSO

      `filter_number(), is_nonzero()`

27.4.3 K–O

106	KEEP_FILTER	K

NAME

      keep_filter :– keep components passing through filter

SYNOPSIS

      `number keep_filter(number x, int test(cliff_comp *c));`

ARGUMENTS

      1: ir ◇ `number x`   ~ number

      2: i ◇ `int test(cliff_comp *c)` ~ pointer to filter test function

DESCRIPTION

      Set to zero all components which fail the test. The test function returns a value of 1 (true) for components which pass the test, and 0 (false) for components which fail the test. The original number is overwritten *in situ* and returned as the result.

RETURN VALUE

      The return value is a number containing all components passing the test.

RECYCLING

      The number x is recycled.

ERROR RECOVERY

      ~

SEE ALSO

      `clone_filter(), filter_number()`

107	KEEPONE_GRADE	K

NAME

keepone_grade :– keep one grade of number

SYNOPSIS

```
number keepone_grade(number x, int n);
```

ARGUMENTS

1: ir ◇ number x ~ number

2: i ◇ int n ~ grade

DESCRIPTION

Set to zero, all grades except the one specified. The number is overwritten *in situ* and returned as the result.

RETURN VALUE

The return value is a number containing only the specified grade.

RECYCLING

The number x is recycled.

ERROR RECOVERY

~

SEE ALSO

```
cloneone_grade(), grade(), onegradeof_number()
```

108	LATITUDEOF_NUMBER	L

NAME

latitudeof_number :– produce latitude of number

SYNOPSIS

```
number latitudeof_number(number x, int d);
```

ARGUMENTS

1: id ◇ number x ~ number

2: i ◇ int d ~ disposal flag

DESCRIPTION

Construct a number containing the latitude of the number specified. The latitude is the square root of the sums of the squares of all real and imaginary parts of each component multiplied by the squares of their unit. The signatures of the units are honoured. The result is a scalar (in grade 0).

RETURN VALUE

The return value is the latitude.

RECYCLING

The number x is recycled according to the value of the disposal flag.

ERROR RECOVERY

~

SEE ALSO

```
magnitudeof_number(), squareof_latitude()
```

109	LEFT	L

NAME

left :– left interior product of components

SYNOPSIS

```
cliff_comp *left(cliff_comp *a, cliff_comp *b,
                      cliff_comp *c);
```

ARGUMENTS

1: i ⬦ cliff_comp *a ~ multiplier, pointer to structure containing
component

2: i ⬦ cliff_comp *b ~ multiplicand, pointer to structure containing
component

3: o ⬦ cliff_comp *c ~ pointer to structure in which to save product

DESCRIPTION

Calculate the left interior (inner) product of two components. Setting the pointers
of either input component to the same value as the output component overwrites
the input with the output.

RETURN VALUE

The return value is a pointer to the component containing the left interior (inner)
product.

RECYCLING

~

ERROR RECOVERY

~

SEE ALSO

```
left_mul(), right(),
```

110	LEFT_MUL	L

NAME

left_mul :– calculate left interior product of numbers

SYNOPSIS

```
number left_mul(number x, number y, int d);
```

ARGUMENTS

1: id ⬦ number x ~ multiplier

2: id ⬦ number y ~ multiplicand

3: i ⬦ int d   ~ disposal flag

DESCRIPTION

Multiply two numbers using the left interior (inner) multiplication.

RETURN VALUE

The return value is the left interior (inner) product.

RECYCLING

The numbers x and y are recycled according to the value of the disposal flag.

ERROR RECOVERY

~

SEE ALSO

```
left(), right_mul()
```

111	LOG_COEF	L

NAME

log_coef :– natural logarithm of coefficient

SYNOPSIS

```
cliff_comp *log_coef(cliff_comp *a, cliff_comp *c);
```

ARGUMENTS

1: i ◇ cliff_comp *a ~ pointer to structure containing component

2: o ◇ cliff_comp *c ~ pointer to structure in which to save logarithm

DESCRIPTION

Calculate the natural logarithm of the complex coefficient, ignoring the value of the unit. Setting the pointers of both components to the same value overwrites the input with the output.

RETURN VALUE

The return value is a pointer to the component containing the logarithm. The unit of the return value is the same as the unit of the input.

RECYCLING

~

ERROR RECOVERY

It is an error to take the logarithm of a component with value zero. On error, the recovery strategy brexit(0) is invoked to initiate a user re-start.

SEE ALSO

```
power_coef(), square_root_coef()
```

112	LOWER_ROWS	L

NAME

lower_rows :– construct sub-matrix missing top row

SYNOPSIS

```
matrix *lower_rows(matrix *m, matrix *s);
```

ARGUMENTS

1: i ◇ matrix *m ~ matrix

2: o ◇ matrix *s ~ sub-matrix, undeclared

DESCRIPTION

Construct a sub-matrix without the top row. Initially, the sub-matrix should be undeclared.

RETURN VALUE

The return value is the sub-matrix within the original matrix.

RECYCLING

~

ERROR RECOVERY

It is an error to use the same matrix structure for both matrices, or to construct a sub-matrix from less than two rows. On error, the recovery strategy brexit(0) is invoked to initiate a user re-start. It is an error if the sub-matrix has been declared already. No check is made.

SEE ALSO

```
eliminate_below(), lower_submatrix(), upper_rows()
```

113	LOWER_SUBMATRIX	L

NAME

lower_submatrix :– construct sub-matrix missing top row and left column

SYNOPSIS

```
matrix *lower_submatrix(matrix *m, matrix *s);
```

ARGUMENTS

1: i ◇ matrix *m ~ matrix

2: o ◇ matrix *s ~ sub-matrix, undeclared

DESCRIPTION

Construct a sub-matrix without the top row or the left-hand column. Initially, the sub-matrix should be undeclared.

RETURN VALUE

The return value is the sub-matrix within the original matrix.

RECYCLING

~

ERROR RECOVERY

It is an error to use the same matrix structure for both matrices, or to construct a sub-matrix from less than two rows and columns. On error, the recovery strategy `brexit(0)` is invoked to initiate a user re-start. It is an error if the sub-matrix has been declared already. No check is made.

SEE ALSO

```
eliminate_below(), lower_rows(), upper_submatrix()
```

114	MAGNITUDEOF_BICOMPLEX	M

NAME

magnitudeof_bicomplex :– produce magnitude of bicomplex number

SYNOPSIS

```
number magnitudeof_bicomplex(number x, int d);
```

ARGUMENTS

1: id ◇ number x ~ bicomplex number

2: i ◇ int d   ~ disposal flag

DESCRIPTION

Construct a scalar containing the magnitude of the bicomplex number specified. The magnitude is the square root of the sums of the squares of all real and imaginary parts of all components. The signatures of the units are ignored. The result is a scalar (in grade 0).

RETURN VALUE

The return value is the magnitude.

RECYCLING

The bicomplex number x is recycled according to the value of the disposal flag.

ERROR RECOVERY

It is an error if the signature contains 0 or positive 1. On error, the recovery strategy `brexit(6)` is invoked, terminating the execution.

SEE ALSO

```
magnitudeof_number(), squareof_magnitude()
```

115	MAGNITUDEOF_EMFIELD	M

NAME

magnitudeof_emfield :– produce magnitude of electromagnetic field

SYNOPSIS

```
number magnitudeof_emfield(number x, int d);
```

ARGUMENTS

1: id ◇ number x ~ electromagnetic field

2: i ◇ int d   ~ disposal flag

DESCRIPTION

Construct a scalar containing the magnitude of the electromagnetic field specified. The magnitude is the square root of the sums of the squares of all real and imaginary parts of all components. The signatures of the units are ignored. The result is a scalar (in grade 0). Some electromagnetic fields, such as plane waves, have non-zero magnitude but have no inverse.

RETURN VALUE

The return value is the magnitude of the electromagnetic field.

RECYCLING

The electromagnetic field x is recycled according to the value of the disposal flag.

ERROR RECOVERY

It is an error if the signature contains 0 or positive 1. On error, the recovery strategy `brexit(6)` is invoked, terminating the execution.

SEE ALSO

```
magnitudeof_number(), squareof_magnitudeof()
```

116	MAGNITUDEOF_NUMBER	M

NAME

magnitudeof_number :– produce magnitude of number

SYNOPSIS

```
number magnitudeof_number(number x, int d);
```

ARGUMENTS

1: id ◇ number x ~ number

2: i ◇ int d   ~ disposal flag

DESCRIPTION

Construct a number containing the magnitude of the number specified. The magnitude is the square root of the sums of the squares of all real and imaginary parts of all components. The signatures of the units are ignored. The result is a scalar (in grade 0).

RETURN VALUE

The return value is the magnitude.

RECYCLING

The number x is recycled according to the value of the disposal flag.

ERROR RECOVERY

~

SEE ALSO

```
latitudeof_number(), squareof_magnitude()
```

117	MAGNITUDEOF_PAULI	M

NAME

magnitudeof_pauli :– produce magnitude of Pauli matrix

SYNOPSIS

```
number magnitudeof_pauli(number x, int d);
```

ARGUMENTS

    1: id ◇ number x ~ Pauli matrix

    2: i ◇ int d   ~ disposal flag

DESCRIPTION

Construct a scalar containing the magnitude of the Pauli matrix specified. The magnitude is the square root of the sums of the squares of all real and imaginary parts of all components. The signatures of the units are ignored. The result is a scalar (in grade 0).

RETURN VALUE

The return value is the magnitude of the Pauli matrix.

RECYCLING

The Pauli matrix x is recycled according to the value of the disposal flag.

ERROR RECOVERY

It is an error if the signature contains 0 or positive 1. On error, the recovery strategy `brexit(6)` is invoked, terminating the execution.

SEE ALSO

```
magnitudeof_number(), squareof_magnitude()
```

118	MAGNITUDEOF_QUATERNION	M

NAME

magnitudeof_quaternion :– produce magnitude of quaternion

SYNOPSIS

```
number magnitudeof_quaternion(number x, int d);
```

ARGUMENTS

    1: id ◇ number x ~ quaternion

    2: i ◇ int d   ~ disposal flag

DESCRIPTION

Construct a scalar containing the magnitude of the quaternion specified. The magnitude is the square root of the sums of the squares of all real and imaginary parts of all components. The signatures of the units are ignored. The result is a scalar (in grade 0).

RETURN VALUE

The return value is the magnitude.

RECYCLING

The quaternion x is recycled according to the value of the disposal flag.

ERROR RECOVERY

It is an error if the signature contains 0 or positive 1. On error, the recovery strategy `brexit(6)` is invoked, terminating the execution.

SEE ALSO

```
magnitudeof_number(), squareof_magnitude()
```

119	MAGNITUDEOF_VECTOR	M

NAME

magnitudeof_vector :– produce magnitude of vector

SYNOPSIS

```
number magnitudeof_vector(number x, int d);
```

ARGUMENTS

1: id ◇ number x ~ vector

2: i ◇ int d   ~ disposal flag

DESCRIPTION

Construct a scalar containing the magnitude of the vector specified. The magnitude is the square root of the sums of the squares of all real and imaginary parts of all components. The signatures of the units are ignored. The result is a scalar (in grade 0). For vectors with real coefficients, a unit vector can be constructed by dividing the vector by its magnitude. For vectors with complex coefficients that does not work as intended.

RETURN VALUE

The return value is the magnitude of the vector.

RECYCLING

The vector x is recycled according to the value of the disposal flag.

ERROR RECOVERY

It is an error if the signature is mixed or contains zero. On error, the recovery strategy brexit (6) is invoked, terminating the execution.

SEE ALSO

```
magnitudeof_number(), normalise_vector(),
squareof_magnitude()
```

120	MAKE_SERIES	M

NAME

make_series :– construct coefficients for power series of common functions

SYNOPSIS

```
void make_series(int p, int n);
```

ARGUMENTS

1: i ◇ int p ~ maximum power used in series

2: i ◇ int n ~ index for nth root function

DESCRIPTION

Construct and store internally arrays of coefficients for the power series of functions COS, SIN, COSH, SINH, EXP, LOG, CM_INV, SQRT, and ROOT. The series are constructed as p+1 coefficients from power zero to power p inclusive. The value of the index n is the root calculated when the ROOT function is invoked. For n=3, the root is the cube root. The function CM_INV is the central multiplicative inverse or reciprocal for central multiplication. It is only necessary to invoke the routine make_series() once, but it may be invoked repeatedly if the maximum power p or the index n needs to be changed.

RETURN VALUE

~

RECYCLING

~

ERROR RECOVERY

It is an error to specify the power p less than 1.

SEE ALSO

        poly_number(), series()

| 121 | MEANABSOF_ARRAY | M |

NAME

        meanabsof_array :– mean of absolute value of array

SYNOPSIS

        number meanabsof_array(int n, number *a);

ARGUMENTS

        1: i ◊ int n   ~ number of elements in array
        2: i ◊ number *a ~ pointer to array of numbers

DESCRIPTION

        Calculates the mean value from the magnitudes of all the of elements in the array.

RETURN VALUE

        The return value is the mean value. The return value is a scalar.

RECYCLING

        ~

ERROR RECOVERY

        It is an error to provide an array which does not have the number of elements specified.

SEE ALSO

        peakof_array(), rmsof_array(), sumof_array()

| 122 | MERGE_COMP | M |

NAME

        merge_comp :– merge component into number

SYNOPSIS

        number merge_comp(number x, cliff_comp *y_comp,
                          cliff_comp *combine(cliff_comp *a,
                                              cliff_comp *b),
                          cliff_comp *insert(cliff_comp *a,
                                             cliff_comp *c));

ARGUMENTS

        1: ir ◊ number x   ~ number
        2: i ◊ cliff_comp *y_comp   ~ pointer to structure
                                                        containing component
        3: i ◊ cliff_comp *combine(cliff_comp *a,   ~ pointer to combine
               cliff_comp *b)   routine
        4: i ◊ cliff_comp *insert(cliff_comp *a,   ~ pointer to insert
               cliff_comp *c)   routine

DESCRIPTION

        Combine a component into a number. If a component with the same unit already exists, the routine combine() is used to effect the merging. If no such unit already exits, the routine insert() is used.

RETURN VALUE

The return value is the new number.

RECYCLING

The input number x is recycled.

ERROR RECOVERY

~

SEE ALSO

find_comp()

| 123 | MUL_COMP | M |

NAME

mul_comp :– multiply number with component

SYNOPSIS

```
number mul_comp(number x,
                cliff_comp *multiply(cliff_comp *a,
                cliff_comp *b, cliff_comp *c),
                cliff_comp *y_comp);
```

ARGUMENTS

1: i ⋄ number x   ~ multiplier

2: i ⋄ cliff_comp *multiply(cliff_comp *a, ~ pointer to multiply
                            cliff_comp *b,   function
                            cliff_comp *c)

3: i ⋄ cliff_comp *y_comp   ~ multiplicand, pointer
                                                   to structure containing
                                                   component

DESCRIPTION

Multiply number onto component using the multiply function specified.

RETURN VALUE

The return value is the product generated using the multiply function specified.

RECYCLING

~

ERROR RECOVERY

~

SEE ALSO

comp_mul(), mul_number(), postscale_array()

| 124 | MUL_NUMBER | M |

NAME

mul_number :– multiply two numbers

SYNOPSIS

```
number mul_number(number x,
                  cliff_comp *multiply(cliff_comp *a,
                  cliff_comp *b, cliff_comp *c),
                  number y, int d);
```

ARGUMENTS

    1: id ◇ number x   ~ multiplier

    2: i ◇ cliff_comp *multiply(cliff_comp *a, ~ pointer to multiply

                             cliff_comp *b,   function

                             cliff_comp *c)

    3: id ◇ number y   ~ multiplicand

    4: i ◇ int d   ~ disposal flag

DESCRIPTION

    Multiply two numbers using the multiply function specified.

RETURN VALUE

    The return value is the product generated using the multiply function specified.

RECYCLING

    The numbers x and y are recycled according to the value of the disposal flag.

ERROR RECOVERY

    ~

SEE ALSO

    comp_mul(), mul_comp()

| 125 | MULTIPLY_MATRIX | M |

NAME

    multiply_matrix :– calculate product of two matrices

SYNOPSIS

```
matrix *multiply_matrix(matrix *a, matrix *b, matrix *c,
int d);
```

ARGUMENTS

    1: id ◇ matrix *a ~ multiplier, matrix

    2: id ◇ matrix *b ~ multiplicand, matrix

    3: o ◇ matrix *c ~ product, empty matrix

    4: i ◇ int d   ~ disposal flag

DESCRIPTION

    Multiply the second matrix (multiplicand) from the left by the first matrix (multiplier) and save the result in the third matrix.

RETURN VALUE

    The return value is the matrix containing the product.

RECYCLING

    ~

ERROR RECOVERY

    It is an error if the matrices have incompatible sizes. On error, the recovery strategy brexit(0) is invoked to initiate a user re-start. It is an error if the third matrix is not empty initially. No check is made.

SEE ALSO

    add_matrix(), central_mul(), subtract_matrix()

| 126 | NEGATION | N |

NAME

    negation :– –1 times value of component

SYNOPSIS

```
cliff_comp *negation(cliff_comp *a, cliff_comp *c);
```

ARGUMENTS

    1: i ◇ `cliff_comp *a` ~ pointer to structure containing component

    2: o ◇ `cliff_comp *c` ~ pointer to structure in which to save negative of component

DESCRIPTION

Calculate the negative of a component. Setting the pointers of both components to the same value overwrites the input with the output.

RETURN VALUE

The return value is a pointer to the component containing the negative of the input.

RECYCLING

~

ERROR RECOVERY

~

SEE ALSO

```
identity(), inverse(), negativeof_number(), rscale()
```

127	NEGATIVE_SIGNATURE	N

NAME

negative_signature :– test for component with negative signature

SYNOPSIS

```
int negative_signature(cliff_comp *c);
```

ARGUMENTS

    1: i ◇ `cliff_comp *c` ~ pointer to structure containing component

DESCRIPTION

Test to determine whether the component has a negative signature. The test passes if the number of primal units with –/0/+ signatures is odd/none/any.

RETURN VALUE

The return value is 0 (false) when the test fails and 1 (true) when the test passes.

RECYCLING

~

ERROR RECOVERY

~

SEE ALSO

```
null_signature(), positive_signature()
```

128	NEGATIVEOF_NUMBER	N

NAME

negativeof_number :– produce negative of number

SYNOPSIS

```
number negativeof_number(number x, int d);
```

ARGUMENTS

    1: id ◇ `number x` ~ number

    2: i ◇ `int d` ~ disposal flag

DESCRIPTION

Construct a number containing the negation of the number specified.

RETURN VALUE

The return value is a number with all signs changed everywhere.

RECYCLING

The number x is recycled according to the value of the disposal flag d. If d=1, the original number is overwritten *in situ* and returned in place of the new one.

ERROR RECOVERY

~

SEE ALSO

```
cloneof_number(), negation()
```

129	No_IMAGINARY	N

NAME

no_imaginary :– test for component with zero imaginary part

SYNOPSIS

```
int no_imaginary(cliff_comp *c);
```

ARGUMENTS

1: i ◇ cliff_comp *c ~ pointer to structure containing component

DESCRIPTION

Test to determine whether the imaginary part of a component is zero. If the imaginary part is zero, then the number is on the real axis of the complex plane.

RETURN VALUE

The return value is 0 (false) when the imaginary part is non-zero and 1 (true) when imaginary part is 0.

RECYCLING

~

ERROR RECOVERY

~

SEE ALSO

```
no_real(), real_part()
```

130	No_REAL	N

NAME

no_real :– test for component with zero real part

SYNOPSIS

```
int no_real(cliff_comp *c);
```

ARGUMENTS

1: i ◇ cliff_comp *c ~ pointer to structure containing component

DESCRIPTION

Test to determine whether the real part of a component is zero. If the real part is zero, then the number is on the imaginary axis of the complex plane.

RETURN VALUE

The return value is 0 (false) when the real part is non-zero and 1 (true) when real part is 0.

R<small>ECYCLING</small>

~

E<small>RROR</small> R<small>ECOVERY</small>

~

S<small>EE</small> A<small>LSO</small>

`imaginary_part(), no_imaginary()`

131	N<small>ON_</small>S<small>CALAR</small>	N

N<small>AME</small>

non_scalar :– test for non-scalar component

S<small>YNOPSIS</small>

`int non_scalar(cliff_comp *c);`

A<small>RGUMENTS</small>

1: i ◇ `cliff_comp *c` ~ pointer to structure containing component

D<small>ESCRIPTION</small>

Test to determine whether the component is not a scalar.

R<small>ETURN</small> V<small>ALUE</small>

The return value is 0 (false) when the component is in grade 0, and 1 (true) when the component is in any other grade.

R<small>ECYCLING</small>

~

E<small>RROR</small> R<small>ECOVERY</small>

~

S<small>EE</small> A<small>LSO</small>

`grade_zero()`

132	N<small>ORMALISE_</small>N<small>UMBER</small>	N

N<small>AME</small>

normalise_number :– produce number normalised to unit magnitude

S<small>YNOPSIS</small>

`number normalise_number(number x, number y, int d);`

A<small>RGUMENTS</small>

1: id ◇ `number x` ~ number

2: id ◇ `number y` ~ reference number

3: i ◇ `int d` ~ disposal flag

D<small>ESCRIPTION</small>

Scale the number so that its magnitude is the same as the magnitude of the reference number. The magnitude is the sum of the square of all real and imaginary parts of all components. The signature of units is not taken into account.

R<small>ETURN</small> V<small>ALUE</small>

The return value is the normalised value of the number.

R<small>ECYCLING</small>

The numbers x and y are recycled according to the value of the disposal flag.

ERROR RECOVERY

It is an error if the number has zero value, or the signature contains zero. On error, the recovery strategy `brexit(6)` is invoked, terminating execution.

SEE ALSO

```
inverseof_component(), magnitudeof_number(),
normalise_vector()
```

133	NORMALISE_VECTOR	N

NAME

normalise_vector :– produce vector which squares to unity

SYNOPSIS

```
number normalise_vector(number x, int d);
```

ARGUMENTS

1: id ◇ `number x` ~ vector

2: i ◇ `int d` ~ disposal flag

DESCRIPTION

Construct a unit vector from the vector specified. The square of a unit vector is either ±1, depending on the signature. For vectors with real coefficients, a unit vector can be constructed by dividing the vector by its magnitude. For vectors with complex coefficients, this does not work as intended.

RETURN VALUE

The return value contains the unit vector.

RECYCLING

The vector x is recycled according to the value of the disposal flag.

ERROR RECOVERY

It is an error if the vector has length zero, or the signature is mixed or contains zero. On error, the recovery strategy `brexit(6)` is invoked, terminating the execution.

SEE ALSO

```
inverseof_component(), magnitudeof_vector(),
normalise_number()
```

134	NOT_GRADE_ONE	N

NAME

not_grade_one :– test for non-vector component

SYNOPSIS

```
int not_grade_one(cliff_comp *c);
```

ARGUMENTS

1: i ◇ `cliff_comp *c` ~ pointer to structure containing component

DESCRIPTION

Test to determine whether the component is not a vector.

RETURN VALUE

The return value is 0 (false) when the component is in grade 1, and 1 (true) when the component is in any other grade.

RECYCLING

~

ERROR RECOVERY

~

SEE ALSO

grade_one(), non_scalar(), not_grade_two()

| 135 | NOT_GRADE_TWO | N |

NAME

not_grade_two :– test for non-bivector component

SYNOPSIS

int not_grade_two(cliff_comp *c);

ARGUMENTS

1: i ◇ cliff_comp *c ~ pointer to structure containing component

DESCRIPTION

Test to determine whether the component is not a bivector.

RETURN VALUE

The return value is 0 (false) when the component is in grade 2, and 1 (true) when the component is in any other grade.

RECYCLING

~

ERROR RECOVERY

~

SEE ALSO

grade_two(), non_scalar(), not_grade_one()

| 136 | NULL_SIGNATURE | N |

NAME

null_signature :– test for component with null signature

SYNOPSIS

int null_signature(cliff_comp *c);

ARGUMENTS

1: i ◇ cliff_comp *c ~ pointer to structure containing component

DESCRIPTION

Test to determine whether the component has a null signature. The test passes if the number of primal units with $-/0/+$ signatures is any/some/any.

RETURN VALUE

The return value is 0 (false) when the test fails and 1 (true) when the test passes.

RECYCLING

~

ERROR RECOVERY

~

SEE ALSO

negative_signature(), positive_signature()

137	ONEGRADEOF_NUMBER	O

NAME

onegradeof_number :– produce one grade of number

SYNOPSIS

```
number onegradeof_number(number x, int n, int d);
```

ARGUMENTS

1: id ◇ number x ~ number
2: i ◇ int n   ~ grade
3: i ◇ int d   ~ disposal flag

DESCRIPTION

Construct a number containing the one grade specified.

RETURN VALUE

The return value is a number containing only the specified grade.

RECYCLING

The number x is recycled according to the value of the disposal flag.

ERROR RECOVERY

~

SEE ALSO

```
cloneone_grade(), filter_number(), grade(), keepone_grade()
```

138	OUTER	O

NAME

outer :– outer product of components

SYNOPSIS

```
cliff_comp *outer(cliff_comp *a, cliff_comp *b,
                  cliff_comp *c);
```

ARGUMENTS

1: i ◇ cliff_comp *a ~ multiplier, pointer to structure containing
component
2: i ◇ cliff_comp *b ~ multiplicand, pointer to structure containing
component
3: o ◇ cliff_comp *c ~ pointer to structure in which to save product

DESCRIPTION

Calculate the outer (exterior) product of two components. Setting the pointers of
either input component to the same value as the output component overwrites the
input with the output.

RETURN VALUE

The return value is a pointer to the component containing the outer (wedge, exterior) product.

RECYCLING

~

ERROR RECOVERY

~

SEE ALSO

```
central(), inner(), outer_mul(), regress()
```

139	OUTER_MUL	O

NAME

outer_mul :– calculate outer product of numbers

SYNOPSIS

`number outer_mul(number x, number y, int d);`

ARGUMENTS

1: id ◇ number x ~ multiplier
2: id ◇ number y ~ multiplicand
3: i ◇ int d   ~ disposal flag

DESCRIPTION

Multiply two numbers using the outer (exterior, wedge) multiplication.

RETURN VALUE

The return value is the outer product.

RECYCLING

The numbers x and y are recycled according to the value of the disposal flag.

ERROR RECOVERY

~

SEE ALSO

`central_mul(), inner_mul(), outer(), regressive_mul()`

140	OUTERSPACE_NUMBER	O

NAME

outerspace_number :– produce outer projection of number

SYNOPSIS

`number outerspace_number(number x, number n, int d);`

ARGUMENTS

1: id ◇ number x ~ vector or other number
2: id ◇ number n ~ unit vector, normal to surface if in three dimensions
3: i ◇ int d   ~ disposal flag

DESCRIPTION

Apply outer projection (rejection) operator to a number. The projection operator eliminates all geometric components of the number parallel to the unit vector supplied. In three dimensions, the unit vector plays the role of a vector normal to a surface in which the projection lies. Together the `outerspace_number()` and `innerspace_number()` projection operators can be used to split numbers into two orthogonal components.

RETURN VALUE

The return value is the result of the projection.

RECYCLING

The numbers x and n are recycled according to the value of the disposal flag.

ERROR RECOVERY

It is an error to specify a unit vector which is not of unit length.

SEE ALSO

`innerspace_number(), normalise_vector(),`
`outerspace_vector(), reflect_number(), rotate_number()`

141	OUTERSPACE_VECTOR	O

NAME

outerspace_vector :– produce outer projection of vector

SYNOPSIS

```
number outerspace_vector(number x, number n, int d);
```

ARGUMENTS

1: id ◇ number x ~ vector or other number

2: id ◇ number n ~ unit vector, normal to surface if in three dimensions

3: i ◇ int d   ~ disposal flag

DESCRIPTION

Apply outer projection (rejection) operator to a number. For numbers in grade 1, the projection operator eliminates all geometric components of the number parallel to the unit vector supplied. In three dimensions, the unit vector plays the role of a vector normal to a surface in which the projection lies. For numbers in higher grades, the result is different for different signatures. Which signature is more appropriate depends on the application. Together the `outerspace_vector()` and `innerspace_vector()` projection operators can be used to split vectors into two orthogonal components.

RETURN VALUE

The return value is the result of the projection.

RECYCLING

The numbers x and n are recycled according to the value of the disposal flag.

ERROR RECOVERY

It is an error to specify a unit vector which is not of unit length.

SEE ALSO

```
innerspace_vector(), normalise_vector(),
outerspace_number(), reflect_vector(), rotate_number()
```

27.4.4 P–T

142	PACK_03ARRAY	P

NAME

pack_03array :– pack complex array of three dimensions into spacetime

SYNOPSIS

```
double (*pack_03array(double r1, double i1, double r2,
                                 double i2,
                                 double r3, double i3,
                                 double (*a)[2]))[2];
```

ARGUMENTS

1: i ◇ double r1   ~ x component, real part

2: i ◇ double i1   ~ x component, imaginary part

3: i ◇ double r2   ~ y component, real part

4: i ◇ double i2   ~ y component, imaginary part

5: i ◇ double r3   ~ z component, real part

6: i ◇ double i3   ~ z component, imaginary part

7: o ◇ double (*a)[2] ~ pointer to array of four complex values

DESCRIPTION

Pack three complex values into array. The first element of the array is set to zero. The remaining three components are set to the x, y, and z values supplied. The array may be later used to construct a number as an electromagnetic field.

RETURN VALUE

The return value is a pointer to the array a.

RECYCLING

~

ERROR RECOVERY

It is an error to provide an array which is too small.

SEE ALSO

```
pack_03vector(), pack_3array(), pack_R3array()
```

143	PACK_03VECTOR	P

NAME

pack_03vector :– pack complex vector of three dimensions into spacetime

SYNOPSIS

```
number pack_03vector(double (*a)[2]);
```

ARGUMENTS

1: i ◇ double (*a)[2] ~ pointer to array of four complex values

DESCRIPTION

Construct a number as a three-dimensional vector with complex values in four-dimensional space–time.

RETURN VALUE

The return value is the number.

RECYCLING

~

ERROR RECOVERY

~

SEE ALSO

```
pack_03array(), pack_3vector(), pack_R03vector(),
unpack_03vector()
```

144	PACK_3ARRAY	P

NAME

pack_3array :– pack complex array of three dimensions

SYNOPSIS

```
double (*pack_3array(double r1, double i1, double r2,
                                  double i2,
                                  double r3, double i3,
                                  double (*a)[2]))[2];
```

ARGUMENTS

1: i ⬦ double	r1	~ x component, real part
2: i ⬦ double	i1	~ x component, imaginary part
3: i ⬦ double	r2	~ y component, real part
4: i ⬦ double	i2	~ y component, imaginary part
5: i ⬦ double	r3	~ z component, real part
6: i ⬦ double	i3	~ z component, imaginary part
7: o ⬦ double	(*a)[2]	~ pointer to array of three complex vales

DESCRIPTION

Pack three complex values into an array. The array may be later used to construct a number as a vector.

RETURN VALUE

The return value is a pointer to the array a, for the purpose of nesting this routine within the argument list of another routine.

RECYCLING

~

ERROR RECOVERY

It is an error to provide an array which is too small.

SEE ALSO

pack_03array(), pack_3vector(), pack_R3array()

145	PACK_3VECTOR	P

NAME

pack_3vector :– pack complex vector of three dimensions

SYNOPSIS

number pack_3vector(double (*a)[2]);

ARGUMENTS

1: i ⬦ double (*a)[2] ~ pointer to array of three complex values

DESCRIPTION

Construct a number as a three-dimensional vector with complex values.

RETURN VALUE

The return value is the number.

RECYCLING

~

ERROR RECOVERY

~

SEE ALSO

pack_03vector(), pack_3array(), pack_R3vector(), unpack_3vector()

146	PACK_BICOMPLEX	P

NAME

pack_bicomplex :– pack bicomplex number from two complex numbers

SYNOPSIS

```
number pack_bicomplex(double r1, double i1, double rh,
                      double ih);
```

ARGUMENTS

1: i ◊ double `r1` ~ scalar (1) component, real part

2: i ◊ double `i1` ~ scalar (1) component, imaginary part

3: i ◊ double `rh` ~ bicomplex (h) component, real part

4: i ◊ double `ih` ~ bicomplex (h) component, imaginary part

DESCRIPTION

Pack two regular complex numbers into a single bicomplex number.

RETURN VALUE

The return value is the bicomplex number.

RECYCLING

~

ERROR RECOVERY

It is an error if the signature contains 0 or positive 1. On error, the recovery strategy `brexit(6)` is invoked, terminating the execution.

SEE ALSO

`pack_Rbicomplex()`, `print_bicomplex()`

147	PACK_COMP	P

NAME

pack_comp :– pack component

SYNOPSIS

```
cliff_comp *pack_comp(cliff_comp *c, double x, double y,
                      bitmap u);
```

ARGUMENTS

1: o ◊ cliff_comp `*c` ~ pointer to structure containing component

2: i ◊ double `x` ~ real part

3: i ◊ double `y` ~ imaginary part

4: i ◊ bitmap `u` ~ unit

DESCRIPTION

Pack a complex coefficient and a unit together into a component.

RETURN VALUE

The return value is a pointer to the component.

RECYCLING

~

ERROR RECOVERY

~

SEE ALSO

`unpack_comp()`, `simple_number()`

148	PACK_COMPONENTS	P

NAME

pack_components :– pack number from coefficients and units

SYNOPSIS

```
number pack_components(int n, double (*y_complex)[2],
bitmap *y_unit);
```

ARGUMENTS

1: i ◇ `int n`   ~ number of components
2: i ◇ `double (*y_complex)[2]` ~ pointer to array of complex coefficients
3: i ◇ `bitmap *y_unit`   ~ pointer to array of units

DESCRIPTION

Packs an array of complex coefficients and an array of units into a number.

RETURN VALUE

The return value is the number.

RECYCLING

~

ERROR RECOVERY

It is an error to provide an array which does not have the number of elements specified. It is an error if the number of values specified is less than 1. On error, the recovery strategy `brexit(0)` is invoked to initiate a user re-start.

SEE ALSO

```
pack_number(), unpack_components()
```

149	PACK_EMFIELD	P

NAME

pack_emfield :– pack electromagnetic field

SYNOPSIS

```
number pack_emfield(double mu[2], double epsilon[2],
double (*H)[2], double (*E)[2]);
```

ARGUMENTS

1: i ◇ `double mu[2]`   ~ complex magnetic permeability
2: i ◇ `double epsilon[2]` ~ complex electric permittivity
3: i ◇ `double (*H)[2]`   ~ magnetic field, pointer to array of 3 complex values
4: i ◇ `double (*E)[2]`   ~ electric field, pointer to array of 3 complex values

DESCRIPTION

Construct an electromagnetic field from its material properties μ and ϵ and from the magnetic field and the electric field. The magnetic field and the electric field are both complex vectors of three dimensions. The material properties are both complex scalars of type `double`.

RETURN VALUE

The return value is the electromagnetic field.

RECYCLING

~

ERROR RECOVERY

~

SEE ALSO

```
pack_spacetime(), unpack_emfield()
```

150	PACK_NUMBER	P

NAME

pack_number :– pack number from components

SYNOPSIS

```
number pack_number(int n, cliff_comp *c);
```

ARGUMENTS

1: i ◊ int n ~ number of components

2: i ◊ cliff_comp *c ~ pointer to array of structures containing components

DESCRIPTION

Pack an array of components into a number.

RETURN VALUE

The return value is the number.

RECYCLING

~

ERROR RECOVERY

It is an error to provide an array which does not have the number of elements specified. It is an error if the number of values specified is less than 1. On error, the recovery strategy brexit(0) is invoked to initiate a user re-start.

SEE ALSO

```
pack_components(), unpack_number()
```

151	PACK_NVECTOR	P

NAME

pack_Nvector :– pack complex vector of N dimensions

SYNOPSIS

```
number pack_Nvector(int n, double (*a)[2]);
```

ARGUMENTS

1: i ◊ int n ~ number of values

2: i ◊ double (*a)[2] ~ pointer to array of n complex values

DESCRIPTION

Construct a number using the array of n complex values.

RETURN VALUE

The return value is the number.

RECYCLING

~

ERROR RECOVERY

It is an error to provide an array which does not have the number of elements specified. It is an error if the number of values specified is less than one or greater than the Clifford dimension. On error, the recovery strategy brexit(0) is invoked to initiate a user re-start.

SEE ALSO

```
pack_03vector(), pack_3vector(), pack_RNvector(),
unpack_number()
```

152	PACK_PAULI	P

NAME

pack_pauli :– pack Pauli matrix

SYNOPSIS

```
number pack_pauli(double r1, double i1, double rX,
                  double iX,
                  double rY, double iY, double rZ,
                  double iZ);
```

ARGUMENTS

1: i ◇ `double r1` ~ identity Pauli matrix, real part
2: i ◇ `double i1` ~ identity Pauli matrix, imaginary part
3: i ◇ `double rX` ~ unit Pauli matrix X, real part
4: i ◇ `double iX` ~ unit Pauli matrix X, imaginary part
5: i ◇ `double rY` ~ unit Pauli matrix Y, real part
6: i ◇ `double iY` ~ unit Pauli matrix Y, imaginary part
7: i ◇ `double rZ` ~ unit Pauli matrix Z, real part
8: i ◇ `double iZ` ~ unit Pauli matrix Z, imaginary part

DESCRIPTION

Pack complex values into a Pauli matrix.

RETURN VALUE

The return value is the Pauli matrix.

RECYCLING

~

ERROR RECOVERY

It is an error if the signature contains 0 or positive 1. On error, the recovery strategy `brexit(6)` is invoked, terminating the execution.

SEE ALSO

`print_pauli()`

153	PACK_QUATERNION	P

NAME

pack_quaternion :– pack quaternion

SYNOPSIS

```
number pack_quaternion(double r1, double i1, double rI,
                       double iI,
                       double rJ, double iJ, double rK,
                       double iK);
```

ARGUMENTS

1: i ◇ `double r1` ~ identity unit quaternion, real part
2: i ◇ `double i1` ~ identity unit quaternion, imaginary part
3: i ◇ `double rI` ~ unit quaternion I, real part
4: i ◇ `double iI` ~ unit quaternion I, imaginary part
5: i ◇ `double rJ` ~ unit quaternion J, real part
6: i ◇ `double iJ` ~ unit quaternion J, imaginary part
7: i ◇ `double rK` ~ unit quaternion K, real part
8: i ◇ `double iK` ~ unit quaternion K, imaginary part

DESCRIPTION

Pack complex values into a quaternion.

RETURN VALUE

The return value is the quaternion.

RECYCLING

~

ERROR RECOVERY

It is an error if the signature contains 0 or positive 1. On error, the recovery strategy `brexit(6)` is invoked, terminating the execution.

SEE ALSO

`print_quaternion()`

154	PACK_R03VECTOR	P

NAME

pack_R03vector :– pack real vector of three dimensions into spacetime

SYNOPSIS

`number pack_R03vector(double *a);`

ARGUMENTS

1: i ◇ `double *a` ~ pointer to array of three (double) real values

DESCRIPTION

Construct a number as a three-dimensional vector with real values in four-dimensional space–time. The lowest (temporal) dimension is set to zero. The upper three dimensions are set to the three spatial components of the vector.

RETURN VALUE

The return value is the number.

RECYCLING

~

ERROR RECOVERY

~

SEE ALSO

`pack_03vector(), pack_R3array(), pack_R3vector()`

155	PACK_R3ARRAY	P

NAME

pack_R3array :– pack real array of three dimensions

SYNOPSIS

`double *pack_R3array(double r1, double r2, double r3,`
                     `double *a);`

ARGUMENTS

1: i ◇ `double r1` ~ x component
2: i ◇ `double r2` ~ y component
3: i ◇ `double r3` ~ z component
4: o ◇ `double *a` ~ pointer to array of three (double) real values

DESCRIPTION

Pack three real values into array. The array may be later used to construct a number as a vector.

RETURN VALUE

The return value is a pointer to the array a.

RECYCLING

~

ERROR RECOVERY

It is an error to provide an array which is too small.

SEE ALSO

```
pack_03array(), pack_3array(), pack_R3vector()
```

156	PACK_R3VECTOR	P

NAME

pack_R3vector :– pack real vector of three dimensions

SYNOPSIS

```
number pack_R3vector(double *a);
```

ARGUMENTS

1: i ◇ double *a ~ pointer to array of three (double) real values

DESCRIPTION

Construct a number as a three-dimensional vector with real values.

RETURN VALUE

The return value is the number.

RECYCLING

~

ERROR RECOVERY

~

SEE ALSO

```
pack_3vector(), pack_R03vector(), pack_R3array()
```

157	PACK_RBICOMPLEX	P

NAME

pack_Rbicomplex :– pack bicomplex number from two real numbers

SYNOPSIS

```
number pack_Rbicomplex(double r1, double rh);
```

ARGUMENTS

1: i ◇ double r1 ~ scalar (1) component, real part
2: i ◇ double rh ~ bicomplex (h) component, real part

DESCRIPTION

Pack two real numbers into a single bicomplex number.

RETURN VALUE

The return value is the bicomplex number.

RECYCLING

~

ERROR RECOVERY

It is an error if the signature contains 0 or positive 1. On error, the recovery strategy `brexit(6)` is invoked, terminating the execution.

SEE ALSO

`pack_bicomplex()`, `print_bicomplex()`

| 158 | PACK_RNVECTOR | P |

NAME

pack_RNvector :– pack real vector of N dimensions

SYNOPSIS

`number pack_RNvector(int n, double *a);`

ARGUMENTS

1: i ◇ `int n` ~ number of values
2: i ◇ `double *a` ~ pointer to array of n real values

DESCRIPTION

Construct a number using the array of n real values.

RETURN VALUE

The return value is the number.

RECYCLING

~

ERROR RECOVERY

It is an error to provide an array which does not have the number of elements specified. It is an error if the number of values specified is less than one or greater than the Clifford dimension. On error, the recovery strategy `brexit(0)` is invoked to initiate a user re-start.

SEE ALSO

`pack_Nvector()`, `pack_R03vector()`, `pack_R3vector()`

| 159 | PACK_SIGNATURE | P |

NAME

pack_signature :– pack signature array

SYNOPSIS

`int *pack_signature(int n, int *lambda, int sig);`

ARGUMENTS

1: i ◇ `int n` ~ number of dimensions
2: o ◇ `int *lambda` ~ pointer to array in which to save signatures
3: i ◇ `int sig` ~ value of signature

DESCRIPTION

Initialise an array of signatures for primal units $e_0 \ldots e_{n-1}$ by copying the value of signature `sig` into array `lambda` of dimension n. The value of `sig` must be either -1, or $+1$. All primal units take the same value of signature. To make the signatures available for use the array must be linked into the mathematical environment using the sub-routine `declare_primal()`. The array is not copied so that changing any of its values changes the signatures for the corresponding primal units.

RETURN VALUE

The return value is a pointer to the array containing the signatures.

RECYCLING

~

ERROR RECOVERY

It is an error to provide an array which does not have the number of elements specified. It is an error for the dimension to be not greater than zero, or for the value of the signatures to be other than −1 or +1. On error, the recovery strategy brexit(0) is invoked to initiate a user re-start.

SEE ALSO

access_primal(), declare_primal()

160	PACK_SPACETIME	P

NAME

pack_spacetime :– pack spatial and temporal parts of electromagnetic field

SYNOPSIS

number pack_spacetime(double (*S)[2], double (*T)[2]);

ARGUMENTS

1: i ◇ double (*S)[2] ~ spatial part of electromagnetic field, pointer to array of 3 complex values

2: i ◇ double (*T)[2] ~ temporal part of electromagnetic field, pointer to array of 3 complex values

DESCRIPTION

Construct an electromagnetic field from its spatial and temporal parts. The spatial part is the magnetic field scaled by $\sqrt{\mu}$, and the temporal part is the electric field scaled by $\sqrt{\epsilon}$. Each of these is a complex vector of three dimensions.

RETURN VALUE

The return value is the electromagnetic field.

RECYCLING

~

ERROR RECOVERY

~

SEE ALSO

pack_emfield(), unpack_spacetime()

161	PACK_UNIT	P

NAME

pack_unit :– pack unit from array

SYNOPSIS

bitmap pack_unit(int n, int *primal);

ARGUMENTS

1: i ◇ int n   ~ size of array

2: i ◇ int *primal ~ pointer to array of numbers of the primal units

DESCRIPTION

Constructs the unit from the numbers identifying the primal units as contained in the specified array. The numbers for the primal units start from zero. If the dimension is set to zero, the primal unit for grade 0 (scalars) is constructed.

RETURN VALUE

The return value is the unsigned integer bitmap of the unit.

RECYCLING

~

ERROR RECOVERY

It is an error to provide an array which does not have the number of elements specified. It is an error if the number of values specified is less than zero or greater than the Clifford dimension. It is an error if the numbers provided are not between zero and one less than the Clifford dimension inclusive. On error, the recovery strategy `brexit(0)` is invoked to initiate a user re-start.

SEE ALSO

`spack_unit()`, `unpack_unit()`

162	PEAKOF_ARRAY	P

NAME

peakof_array :– peak value of array

SYNOPSIS

```
number peakof_array(int n, number *a);
```

ARGUMENTS

1: i ◇ `int n`   ~ number of elements in array
2: i ◇ `number *a` ~ pointer to array of numbers

DESCRIPTION

Finds the largest value from the magnitudes of all the of elements in the array.

RETURN VALUE

The return value is the peak value. The return value is a scalar.

RECYCLING

~

ERROR RECOVERY

It is an error to provide an array which does not have the number of elements specified.

SEE ALSO

`meanabsof_array()`, `rmsof_array()`, `sumof_array()`

163	POLAR	P

NAME

polar :– convert component from Cartesian to polar

SYNOPSIS

```
cliff_comp *polar(cliff_comp *a, cliff_comp *c);
```

ARGUMENTS

      1: i ◇ `cliff_comp *a` ~ pointer to structure containing Cartesian
                            component

      2: o ◇ `cliff_comp *c` ~ pointer to structure in which to save polar
                            component

DESCRIPTION

Convert the component from Cartesian to polar. The radial part is returned in the real part, and the angular part in the imaginary part. The angle is in radians. Setting the pointers of both components to the same value overwrites the input with the output.

RETURN VALUE

The return value is a pointer to the component containing the polar form.

RECYCLING

~

ERROR RECOVERY

Other operations are unaware of the conversion and treat the returned value as Cartesian.

SEE ALSO

`cartesian(), degrees(), radians(), squareof_magnitude()`

164	POLY_NUMBER	P

NAME

poly_number :– calculate polynomial (power) series

SYNOPSIS

```
number poly_number(int n, number *a, number x, number c,
int d);
```

ARGUMENTS

      1: i ◇ `int n`   ~ number of elements in array
      2: i ◇ `number *a` ~ pointer to array of coefficients for powers
      3: i ◇ `number x` ~ number
      4: i ◇ `number c` ~ convergence centre of power series
      5: i ◇ `int d`   ~ disposal flag

DESCRIPTION

Calculates the sum of the series of powers from 0 to `n-1` of the number `(x-c)`, pre-multiplied by the coefficients in the array `a[n]`. Usually, the number `x` is a variable, and the number `c` is a constant for any particular power series, often 0 or 1.

RETURN VALUE

The return value is the sum.

RECYCLING

The values in the array `a[n]` and the numbers `x` and `c` are recycled according to the value of the disposal flag.

ERROR RECOVERY

It is an error to provide an array which does not have the number of elements specified.

SEE ALSO

    make_series(), series()

| 165 | POSITIVE_SIGNATURE | P |

NAME

    positive_signature :– test for component with positive signature

SYNOPSIS

    int positive_signature(cliff_comp *c);

ARGUMENTS

    1: i ◇ cliff_comp *c ~ pointer to structure containing component

DESCRIPTION

    Test to determine whether the component has a positive signature. The test passes
    if the number of primal units with −/0/+ signatures is even/none/any.

RETURN VALUE

    The return value is 0 (false) when the test fails and 1 (true) when the test passes.

RECYCLING

    ~

ERROR RECOVERY

    ~

SEE ALSO

    negative_signature(), null_signature()

| 166 | POSTSCALE_ARRAY | P |

NAME

    postscale_array :– produce product of array by number

SYNOPSIS

    number *postscale_array(int n, number *a, number s,
                            number *b);

ARGUMENTS

    1: i ◇ int n   ~ number of elements in arrays
    2: i ◇ number *a ~ multiplier, pointer to array of numbers
    3: i ◇ number s ~ multiplicand, scale factor
    4: o ◇ number *b ~ product, pointer to array in which to save scaled values

DESCRIPTION

    Multiply (as multiplier) all elements in the array using central multiplication onto
    (as multiplicand) the same scale factor. The products are saved into the second
    array. Initially, the second array must be empty.

RETURN VALUE

    The return value is a pointer to the array containing the results.

RECYCLING

    ~

ERROR RECOVERY

    It is an error to provide an array which does not have the number of elements
    specified.

SEE ALSO

> mul_comp(), prescale_array()

| 167 | POWER_COEF | P |

NAME

> power_coef :– raise coefficient to power

SYNOPSIS

> cliff_comp *power_coef(double p, cliff_comp *b,
>   cliff_comp *c);

ARGUMENTS

> 1: i ◇ double p   ~ exponent
> 2: i ◇ cliff_comp *b ~ base, pointer to structure containing component
> 3: o ◇ cliff_comp *c ~ pointer to component in which to save result

DESCRIPTION

> Raise the base to the exponent specified, ignoring the value of the unit. Setting the pointers of both components to the same value overwrites the input with the output.

RETURN VALUE

> The return value is a pointer to the component containing the result. The unit of the return value is the same as the unit of the input.

RECYCLING

> ~

ERROR RECOVERY

> ~

SEE ALSO

> log_coef(), poly_number(), square_root_coef()

| 168 | PRESCALE_ARRAY | P |

NAME

> prescale_array :– produce product of number by array

SYNOPSIS

> number *prescale_array(number s, int n, number *a,
>   number *b);

ARGUMENTS

> 1: i ◇ number s ~ multiplier, scale factor
> 2: i ◇ int n   ~ number of elements in arrays
> 3: i ◇ number *a ~ multiplicand, pointer to array of numbers
> 4: o ◇ number *b ~ product, pointer to array in which to save scaled values

DESCRIPTION

> Multiply (as multiplier) the same scale factor using central multiplication onto (as multiplicand) all elements in the array. The products are saved into the second array. Initially, the second array must be empty.

RETURN VALUE

> The return value is a pointer to the array containing the results.

RECYCLING

~

ERROR RECOVERY

It is an error to provide an array which does not have the number of elements specified.

SEE ALSO

```
comp_mul(), postscale_array(), prescale_row()
```

169	PRESCALE_ROW	P

NAME

prescale_row :– left multiply a row in a matrix by a number

SYNOPSIS

```
matrix *prescale_row(number s, int i, matrix *m);
```

ARGUMENTS

1: i ◇ number s ~ multiplier, scale factor

2: i ◇ int i ~ multiplicand, row

3: io ◇ matrix *m ~ matrix

DESCRIPTION

Pre-scale (multiply from the left) the row (i) by the factor s. The row is overwritten with the result. If unknown, the total number of rows can be obtained using the routine rows(). The index i for the rows start from zero and go up to one less than the total. No checks are made to determine if the index is out of bounds.

RETURN VALUE

The return value is the original matrix with modified values.

RECYCLING

~

ERROR RECOVERY

It is an error to use indexes which are out of bounds.

SEE ALSO

```
addscaledto_row(), row_mul_col(), subscaledfrom_row(),
weightedsumof_array()
```

170	PRIMAL	P

NAME

primal :– primal unit

SYNOPSIS

```
bitmap primal(int n);
```

ARGUMENTS

1: i ◇ int n ~ number of unit

DESCRIPTION

Constructs the bitmap for the primal unit of the specified number (0–31).

RETURN VALUE

The return value is the unsigned integer bitmap of the primal unit.

RECYCLING

> ~

ERROR RECOVERY

> ~

SEE ALSO

> `primal_vector(), psuedo_scalar()`

| 171 | PRIMAL_VECTOR | P |

NAME

> primal_vector :– construct a number with one primal unit

SYNOPSIS

> `number primal_vector(double x, double y, int n);`

ARGUMENTS

> 1: i ◇ `double x` ~ real part
> 2: i ◇ `double y` ~ imaginary part
> 3: i ◇ `int n`   ~ number of unit

DESCRIPTION

> Construct a new number with the real and imaginary parts and with the primal unit as specified. The unit is in integer form, taking values from zero to the one less than the Clifford dimension.

RETURN VALUE

> The return value is the new number.

RECYCLING

> ~

ERROR RECOVERY

> It is an error to specify a unit number greater or equal to the Clifford dimension. On error, the recovery strategy `brexit(0)` is invoked to initiate a user re-start.

SEE ALSO

> `fresh_number(), primal(), pseudo_number(), scalar_number(), simple_number()`

| 172 | PRINT_ARRAY | P |

NAME

> print_array :– print array

SYNOPSIS

> `number *print_array(int n, number *a, int cmax);`

ARGUMENTS

> 1: i ◇ `int n`   ~ size of array
> 2: i ◇ `number *a` ~ pointer to array of numbers
> 3: i ◇ `int cmax` ~ maximum number of components per line

DESCRIPTION

> Print array of numbers using the default format. The third argument `cmax` specifies the maximum number of components to be printed on each line.

RETURN VALUE

> The return value is a pointer to the original array.

RECYCLING

~

ERROR RECOVERY

~

SEE ALSO

> print_magnitudeof_array(), print_matrix(),
> print_onecoefficientof_array()

| 173 | PRINT_BICOMPLEX | P |

NAME

> print_bicomplex :– print bicomplex number

SYNOPSIS

> number print_bicomplex(number x);

ARGUMENTS

> 1: i ◇ number x ~ bicomplex number

DESCRIPTION

> Prints scalar and pseudo-scalar for two dimensions as a bicomplex number using the default format. Other components are ignored.

RETURN VALUE

> The return value is the original number.

RECYCLING

~

ERROR RECOVERY

~

SEE ALSO

> print_number(), printf_bicomplex()

| 174 | PRINT_BOUNDS | P |

NAME

> print_bounds :– print bounds

SYNOPSIS

> number print_bounds(number x);

ARGUMENTS

> 1: i ◇ number x ~ number

DESCRIPTION

> Recursively print the values of the bounds at each node, both internal (structural) nodes and terminal (data) nodes. The bounds define the range of units accommodated beyond the node. Only of interest for debugging internal operations.

RETURN VALUE

> The return value is the original number.

RECYCLING

~

ERROR RECOVERY

> It is an error to provide a token for a number which has been never assigned or recycled but not reassigned. The routine can fail if the tree structure is corrupted. No error checking is performed.

SEE ALSO

        print_nodes(), print_tree()

| 175 | PRINT_COEF | P |

NAME

        print_coef :– print coefficient

SYNOPSIS

        int print_coef(double x, double y);

ARGUMENTS

        1: i ◇ double x ~ real part
        2: i ◇ double y ~ imaginary part

DESCRIPTION

        Prints the real and imaginary parts of a coefficient formatted as a complex number
        or as magnitude, phase, real, or imaginary part only. The coefficient is printed using
        the default format.

RETURN VALUE

        The return value is the number of characters printed.

RECYCLING

        ~

ERROR RECOVERY

        ~

SEE ALSO

        get_coef(), print_single_coefficient(), print_unit(),
        printf_coef()

| 176 | PRINT_COLUMN | P |

NAME

        print_column :– print a column of a matrix

SYNOPSIS

        matrix *print_column(int j, matrix *m, int cmax);

ARGUMENTS

        1: i ◇ int j   ~ column
        2: i ◇ matrix *m ~ matrix
        3: i ◇ int cmax ~ maximum number of components per line

DESCRIPTION

        Print the specified column from a matrix.

RETURN VALUE

        The return value is the matrix.

RECYCLING

        ~

ERROR RECOVERY

        It is an error to use indexes which are out of bounds.

SEE ALSO

        print_element(), print_matrix(), print_row()

177	Print_Comp	P

NAME

print_comp :– print component

SYNOPSIS

`int print_comp(cliff_comp *c);`

ARGUMENTS

1: i ◊ `cliff_comp *c` ~ pointer to structure containing component

DESCRIPTION

Print coefficient and unit. The coefficient is formatted as a complex number or as magnitude, phase, real, or imaginary part only. The component is printed using the default format.

RETURN VALUE

The return value is the number of characters printed.

RECYCLING

~

ERROR RECOVERY

~

SEE ALSO

`print_coef()`, `print_unit()`, `printf_comp()`, `unpack_comp()`

178	Print_Element	P

NAME

print_element :– print an element of a matrix

SYNOPSIS

`matrix *print_element(int i, int j, matrix *m, int cmax);`

ARGUMENTS

1: i ◊ `int i` ~ row
2: i ◊ `int j` ~ column
3: i ◊ `matrix *m` ~ matrix
4: i ◊ `int cmax` ~ maximum number of components per line

DESCRIPTION

Print the specified element from a matrix.

RETURN VALUE

The return value is the matrix.

RECYCLING

~

ERROR RECOVERY

It is an error to use indexes which are out of bounds.

SEE ALSO

`print_column()`, `print_matrix()`, `print_row()`,
`printf_element()`

179	Print_Emfield	P

NAME

print_emfield :– print electromagnetic field

SYNOPSIS

```
number print_emfield(number x);
```

ARGUMENTS

1: i ⋄ number x ~ electromagnetic field

DESCRIPTION

Prints grade 2 for four dimensions as an electromagnetic field using the default format. Other components are ignored.

RETURN VALUE

The return value is the original number.

RECYCLING

~

ERROR RECOVERY

~

SEE ALSO

```
print_number(), printf_emfield()
```

180	PRINT_FORK	P

NAME

print_fork :– print fork (structural node)

SYNOPSIS

```
void print_fork(number f);
```

ARGUMENTS

1: i ⋄ number f ~ token for fork

DESCRIPTION

Print the value of the token and the bounds for the internal (structural) node and the tokens for the two branches. Only of interest for debugging internal operations.

RETURN VALUE

~

RECYCLING

~

ERROR RECOVERY

An error message is printed if the operation is attempted for a leaf (data node).

SEE ALSO

```
print_leaf()
```

181	PRINT_GRADEDLISTOF_COEFFICIENTS	P

NAME

print_gradedlistof_coefficients :– print list of coefficients in grades

SYNOPSIS

```
number print_gradedlistof_coefficients(number x,
                                        int cmax);
```

ARGUMENTS

1: i ⋄ number x ~ number
2: i ⋄ int cmax ~ maximum number of coefficients per line

DESCRIPTION

Print the coefficients for a number as a list without units grouped into grades, using the default format. The order of the coefficients is the same as the order of the units printed by `print_gradedlistof_units()`. The second argument `cmax` specifies the maximum number of coefficients to be printed on each line of output.

RETURN VALUE

The return value is the original number.

RECYCLING

~

ERROR RECOVERY

~

SEE ALSO

`print_gradedlistof_units()`, `print_listof_coefficients()`, `printf_gradedlistof_coefficients()`, `unpack_components()`

182	PRINT_GRADEDLISTOF_UNITS	P

NAME

print_gradedlistof_units :– print list of units in grades

SYNOPSIS

`number print_gradedlistof_units(number x, int cmax);`

ARGUMENTS

1: i ◇ number x ~ number

2: i ◇ int cmax ~ maximum number of units per line

DESCRIPTION

Print the units for a number as a list without coefficients grouped into grades. The order of the units is the same as the order of the coefficients printed by `print_gradedlistof_coefficients()` and `printf_gradedlistof_coefficients()` The second argument `cmax` specifies the maximum number of units to be printed on each line of output.

RETURN VALUE

The return value is the original number.

RECYCLING

~

ERROR RECOVERY

~

SEE ALSO

`print_gradedlistof_coefficients()`, `print_listof_units()`, `unpack_components()`

183	PRINT_LEAF	P

NAME

print_leaf :– print leaf (data node)

SYNOPSIS

`void print_leaf(number l);`

ARGUMENTS

> 1: i ◇ number 1 ~ token for leaf

DESCRIPTION

> Print value of the token the leaf (data node) and the values of the coefficient and unit. Only of interest for debugging internal operations.

RETURN VALUE

> ~

RECYCLING

> ~

ERROR RECOVERY

> An error message is printed if the operation is attempted for a fork (structural node).

SEE ALSO

> ```
> print_fork()
> ```

184	PRINT_LISTOF_COEFFICIENTS	P

NAME

> print_listof_coefficients :– print list of coefficients

SYNOPSIS

> ```
> number print_listof_coefficients(number x,  int cmax);
> ```

ARGUMENTS

> 1: i ◇ number x ~ number
>
> 2: i ◇ int cmax ~ maximum number of coefficients per line

DESCRIPTION

> Print the coefficients for a number as a list without units using the default format. The order of the coefficients is the same as the order of the units printed by `print_listof_units()`. The second argument `cmax` specifies the maximum number of coefficients to be printed on each line of output.

RETURN VALUE

> The return value is the original number.

RECYCLING

> ~

ERROR RECOVERY

> ~

SEE ALSO

> ```
> print_gradedlistof_coefficients(), print_listof_units(),
> printf_listof_coefficients(), unpack_components()
> ```

185	PRINT_LISTOF_UNITS	P

NAME

> print_listof_units :– print list of units

SYNOPSIS

> ```
> number print_listof_units(number x,  int cmax);
> ```

ARGUMENTS

> 1: i ◇ `number x` ~ number
> 2: i ◇ `int cmax` ~ maximum number of units per line

DESCRIPTION

> Print the units for a number as a list without coefficients. The order of the units is the same as the order of the coefficients printed by `print_listof _coefficients()` and `printf_listof_coefficients()`.

RETURN VALUE

> The return value is the original number.

RECYCLING

> ~

ERROR RECOVERY

> ~

SEE ALSO

> `print_gradedlistof_units()`, `print_listof_coefficients()`,
> `print_unit()`, `unpack_components()`

| 186 | PRINT_MAGNITUDEOF_ARRAY | P |

NAME

> print_magnitudeof_array :– print magnitude of array

SYNOPSIS

> `number *print_magnitudeof_array(int n, number *a);`

ARGUMENTS

> 1: i ◇ `int n`   ~ size of array
> 2: i ◇ `number *a` ~ pointer to array of numbers

DESCRIPTION

> Print array of numbers using the default format. Prints only the magnitude for each element of the array.

RETURN VALUE

> The return value is a pointer to the original array.

RECYCLING

> ~

ERROR RECOVERY

> ~

SEE ALSO

> `print_array()`, `print_magnitudeof_matrix()`

| 187 | PRINT_MAGNITUDEOF_MATRIX | P |

NAME

> print_magnitudeof_matrix :– print a whole matrix giving the magnitudes of the numbers

SYNOPSIS

> `matrix *print_magnitudeof_matrix(matrix *m);`

ARGUMENTS

> 1: i ◇ `matrix *m` ~ matrix

DESCRIPTION

>Print the magnitude of all elements in a matrix using the layout set through `for-mat_matrix()`.

RETURN VALUE

>The return value is the matrix.

RECYCLING

>~

ERROR RECOVERY

>It is an error if the matrix printing format is not set to a known value. On error, the recovery strategy `brexit(6)` is invoked to terminate execution.

SEE ALSO

>`print_magnitudeof_array(), print_matrix(),`
>`print_onecoefficientof_matrix()`

| 188 | PRINT_MATRIX | P |

NAME

>print_matrix :– print a whole matrix

SYNOPSIS

>`matrix *print_matrix(matrix *m, int cmax);`

ARGUMENTS

>1: i ◇ `matrix *m` ~ matrix
>2: i ◇ `int cmax` ~ maximum number of components per line

DESCRIPTION

>Print all components of all elements in a matrix using the layout set through `for-mat_matrix()`. At most, `cmax` components are printed on any one line.

RETURN VALUE

>The return value is the matrix.

RECYCLING

>~

ERROR RECOVERY

>It is an error if the matrix printing format is not set to a known value. On error, the recovery strategy `brexit(6)` is invoked to terminate execution.

SEE ALSO

>`format_matrix(), print_column(), print_element(),`
>`print_row(),`
>`print_magnitudeof_matrix(),`
>`print_onecoefficientof_matrix()`

| 189 | PRINT_MEM_USAGE | P |

NAME

>print_mem_usage :– print memory usage

SYNOPSIS

>`void print_mem_usage(char *text);`

ARGUMENTS

1: i ◇ char *text ~ pointer to message string

DESCRIPTION

Print a compact summary of memory usage. Intended for use when debugging in search of memory leaks from failure to recycle numbers when no longer used. The message string may be used to identify the location from where the routine is invoked. If no message is required, the argument may be set to either (char *) NULL or zero.

RETURN VALUE

~

RECYCLING

~

ERROR RECOVERY

~

SEE ALSO

free_cliffmem(), get_mem_usage(), print_num_info()

190	PRINT_NODES	P

NAME

print_nodes :– print data nodes

SYNOPSIS

number print_nodes(number x);

ARGUMENTS

1: i ◇ number x ~ number

DESCRIPTION

Recursively print the numeric values of the tokens of the terminal (data) nodes in the tree. The tokens define the position of the nodes within the list maintained by the memory management system. Only of interest for debugging internal operations.

RETURN VALUE

The return value is the original number.

RECYCLING

~

ERROR RECOVERY

It is an error to provide a token for a number which has been never assigned or recycled but not reassigned. The routine can fail if the tree structure is corrupted. No error checking is performed.

SEE ALSO

print_bounds(), print_tree()

191	PRINT_NUM_INFO	P

NAME

print_num_info :– print information about number

SYNOPSIS

> `number print_num_info(number x);`

ARGUMENTS

> 1: i ◊ number x ~ number

DESCRIPTION

> Prints brief information about the storage allocated to a number. The information includes the value of the token, the number of components used, and the maximum number of components when all are non-zero.

RETURN VALUE

> The return value is the original number.

RECYCLING

> ~

ERROR RECOVERY

> It is an error to provide a token for a number which has been never assigned or recycled but not reassigned.

SEE ALSO

> `print_mem_usage()`

| 192 | PRINT_NUMBER | P |

NAME

> print_number :– print number

SYNOPSIS

> `number print_number(number x, int cmax);`

ARGUMENTS

> 1: i ◊ number x ~ number
>
> 2: i ◊ int cmax ~ maximum number of components per line

DESCRIPTION

> Print the number as a sequence of components using the default format. The second argument `cmax` specifies the maximum number of components to be printed on each line of output.

RETURN VALUE

> The return value is the original number.

RECYCLING

> ~

ERROR RECOVERY

> Any value of `cmax` less than 1 will result in each component printed on a separate line.

SEE ALSO

> `print_element()`, `print_listof_coefficients()`,
> `print_listof_units()`, `printf_number()`, `unpack_number()`

| 193 | PRINT_ONECOEFFICIENTOF_ARRAY | P |

NAME

> print_onecoefficientof_array :– print one coefficient of array

SYNOPSIS

```
number *print_onecoefficientof_array(int n, number *a,
                                      bitmap u);
```

ARGUMENTS

1: i ◇ int n   ~ size of array
2: i ◇ number *a ~ pointer to array of numbers
3: i ◇ bitmap u ~ unit of component

DESCRIPTION

Print array of numbers using the default format. Prints only the coefficient for one of the components for each element of the array. The third argument u specifies the unit of the component.

RETURN VALUE

The return value is a pointer to the original array.

RECYCLING

~

ERROR RECOVERY

~

SEE ALSO

```
print_array(), print_onecoefficientof_matrix(),
printf_onecoefficientof_array()
```

| 194 | PRINT_ONECOEFFICIENTOF_MATRIX | P |

NAME

print_onecoefficientof_matrix :– print a whole matrix giving the coefficients of one unit

SYNOPSIS

```
matrix *print_onecoefficientof_matrix(matrix *m,
                                      bitmap u);
```

ARGUMENTS

1: i ◇ matrix *m ~ matrix
2: i ◇ bitmap u ~ unit

DESCRIPTION

Print the coefficient with the given unit for all elements in a matrix using the layout set through `format_matrix()`.

RETURN VALUE

The return value is the matrix.

RECYCLING

~

ERROR RECOVERY

It is an error if the matrix printing format is not set to a known value. On error, the recovery strategy `brexit(6)` is invoked to terminate execution.

SEE ALSO

```
print_magnitudeof_matrix(), print_matrix(),
print_onecoefficientof_array(),
printf_onecoefficientof_matrix()
```

195	PRINT_PAULI	P

NAME

       print_pauli :– print Pauli matrix

SYNOPSIS

```
number print_pauli(number x);
```

ARGUMENTS

       1: i ◇ number x ~ Pauli matrix

DESCRIPTION

       Prints grade 0 and 2 for three dimensions as a Pauli matrix using the default format. Other components are ignored.

RETURN VALUE

       The return value is the original number.

RECYCLING

       ~

ERROR RECOVERY

       ~

SEE ALSO

```
print_number(), printf_pauli()
```

196	PRINT_QUATERNION	P

NAME

       print_quaternion :– print quaternion

SYNOPSIS

```
number print_quaternion(number x);
```

ARGUMENTS

       1: i ◇ number x ~ quaternion

DESCRIPTION

       Prints grade 0 and 2 for three dimensions as a quaternion using the default format. Other components are ignored.

RETURN VALUE

       The return value is the original number.

RECYCLING

       ~

ERROR RECOVERY

       ~

SEE ALSO

```
print_number(), printf_quaternion()
```

197	PRINT_ROW	P

NAME

       print_row :– print a row of a matrix

SYNOPSIS

```
matrix *print_row(int i, matrix *m, int cmax);
```

ARGUMENTS

      1: i ◇ int i   ~ row

      2: i ◇ matrix *m ~ matrix

      3: i ◇ int cmax ~ maximum number of components per line

DESCRIPTION

      Print the specified row from a matrix.

RETURN VALUE

      The return value is the matrix.

RECYCLING

      ~

ERROR RECOVERY

      It is an error to use indexes which are out of bounds.

SEE ALSO

      `print_column(), print_element(), print_matrix()`

198	PRINT_SIGNATURE	P

NAME

      print_signature :– print signature array

SYNOPSIS

      `void print_signature(void);`

ARGUMENTS

      ~

DESCRIPTION

      Print the signature for all of the primal units.

RETURN VALUE

      ~

RECYCLING

      ~

ERROR RECOVERY

      It is an error to print the signature before it has been assigned. On error, the recovery strategy `brexit(6)` is invoked, terminating the execution.

SEE ALSO

      `access_primal(), declare_primal()`

199	PRINT_SINGLE_COEFFICIENT	P

NAME

      print_single_coefficient :– print single coefficient

SYNOPSIS

      `number print_single_coefficient(number x, bitmap u);`

ARGUMENTS

      1: i ◇ number x ~ number

      2: i ◇ bitmap u ~ unit of component

DESCRIPTION

      Print only the coefficient for a single component from the number using the default format. The second argument u specifies the unit for the component.

RETURN VALUE

    The return value is the original number.

RECYCLING

    ~

ERROR RECOVERY

    ~

SEE ALSO

    `get_coef(), print_coef(), printf_single_coefficient()`

200	PRINT_TREE	P

NAME

    print_tree :– print tree for number

SYNOPSIS

    `number print_tree(number x);`

ARGUMENTS

    1: i ◇ `number x` ~ number

DESCRIPTION

    Recursively prints a number in its tree structure, showing structural information at internal (structural) nodes as well as numerical values at terminal (data) nodes.

RETURN VALUE

    The return value is the original number.

RECYCLING

    ~

ERROR RECOVERY

    It is an error to provide a token for a number which has been never assigned or recycled but not reassigned. The routine can fail if the tree structure is corrupted. If the depth of recursion when printing the tree exceeds the dimension, the routine invokes `brexit(6)` and terminates the program unconditionally.

SEE ALSO

    `print_bounds(), print_nodes()`

201	PRINT_UNIT	P

NAME

    print_unit :– print unit

SYNOPSIS

    `int print_unit(bitmap u);`

ARGUMENTS

    1: i ◇ `bitmap u` ~ the unit as an integer bit pattern

DESCRIPTION

    Prints the unit formatted as a string of characters.

RETURN VALUE

    The return value is the number of characters printed.

RECYCLING

    ~

Error Recovery

~

See Also

  print_coef(), print_listof_units(), unpack_unit()

| 202 | PRINT_VECTOR | P |

Name

  print_vector :– print vector

Synopsis

  `number print_vector(number x);`

Arguments

  1: i ◊ number x ~ vector

Description

  Prints grade 1 of a number as a vector using the default format. Other grades are ignored. The dimension of the vector is taken from the dimension established for the numerical context using declare_primal().

Return Value

  The return value is the original number.

Recycling

  ~

Error Recovery

  ~

See Also

  declare_primal(), print_number(), printf_3vector(),
  printf_Nvector(), printf_vector(), unpack_03vector(),
  unpack_3vector(), unpack_R3vector()

| 203 | PRINTF_3VECTOR | P |

Name

  printf_3vector :– print vector of dimension 3 with specified format

Synopsis

  `number printf_3vector(char *fmt, number x);`

Arguments

  1: i ◊ char *fmt ~ pointer to format string
  2: i ◊ number x ~ vector in three dimensions

Description

  Prints grade 1 of a number as a vector using the format specified in the first argument. Other grades are ignored. The first three dimensions of the vector are printed.

Return Value

  The return value is the original number.

Recycling

  ~

Error Recovery

  ~

S<small>EE</small> A<small>LSO</small>

```
printf_Nvector(), printf_number(), printf_vector(),
unpack_3vector()
```

204	P<small>RINTF</small>_B<small>ICOMPLEX</small>	P

N<small>AME</small>

printf_bicomplex :– print bicomplex number with specified format

S<small>YNOPSIS</small>

```
number printf_bicomplex(char *fmt, number x);
```

A<small>RGUMENTS</small>

1: i ◇ char *fmt ~ pointer to format string

2: i ◇ number x ~ bicomplex number

D<small>ESCRIPTION</small>

Prints scalar and pseudo-scalar for two dimensions as a bicomplex number using the format specified in the first argument. Other components are ignored.

R<small>ETURN</small> V<small>ALUE</small>

The return value is the original number.

R<small>ECYCLING</small>

~

E<small>RROR</small> R<small>ECOVERY</small>

~

S<small>EE</small> A<small>LSO</small>

```
print_bicomplex(), printf_number()
```

205	P<small>RINTF</small>_C<small>OEF</small>	P

N<small>AME</small>

printf_coef :– print coefficient with specified format

S<small>YNOPSIS</small>

```
int printf_coef(char *fmt, double x, double y);
```

A<small>RGUMENTS</small>

1: i ◇ char *fmt ~ pointer to format string

2: i ◇ double x ~ real part

3: i ◇ double y ~ imaginary part

D<small>ESCRIPTION</small>

Prints the real and imaginary parts of a coefficient formatted as a complex number or as magnitude, phase, real, or imaginary part only. The coefficient is printed using the format specified in the first argument.

R<small>ETURN</small> V<small>ALUE</small>

The return value is the number characters printed.

R<small>ECYCLING</small>

~

E<small>RROR</small> R<small>ECOVERY</small>

~

S<small>EE</small> A<small>LSO</small>

```
get_coef(), print_coef(), print_unit(),
printf_single_coefficient()
```

206	Pᴿɪɴᴛꜰ_Cᴏᴍᴘ	P

NAME

printf_comp :– print component with specified format

SYNOPSIS

```
int printf_comp(char *fmt, cliff_comp *c);
```

ARGUMENTS

1: i◇ char *fmt   ~ pointer to format string

2: i◇ cliff_comp *c ~ pointer to structure containing component

DESCRIPTION

Prints coefficient and unit. The coefficient is formatted as a complex number or as magnitude, phase, real, or imaginary part only. The component is printed using the format specified in the first argument.

RETURN VALUE

The return value is the number of characters printed.

RECYCLING

~

ERROR RECOVERY

~

SEE ALSO

```
print_comp(), print_unit(), printf_coef(), unpack_comp()
```

207	Pᴿɪɴᴛꜰ_Eʟᴇᴍᴇɴᴛ	P

NAME

printf_element :– print an element of a matrix with specified format

SYNOPSIS

```
matrix *printf_element(char *fmt, int i, int j,
                       matrix *m, int cmax);
```

ARGUMENTS

1: i◇ char *fmt ~ pointer to format string

2: i◇ int i   ~ row

3: i◇ int j   ~ column

4: i◇ matrix *m ~ matrix

5: i◇ int cmax ~ maximum number of components per line

DESCRIPTION

Print an element from a matrix using the print format specified.

RETURN VALUE

The return value is the matrix.

RECYCLING

~

ERROR RECOVERY

It is an error to use indexes which are out of bounds.

SEE ALSO

```
print_element()
```

208	PRINTF_EMFIELD	P

NAME

printf_emfield :– print electromagnetic field with specified format

SYNOPSIS

```
number printf_emfield(char *fmt, number x);
```

ARGUMENTS

1: i ◇ char *fmt ~ pointer to format string

2: i ◇ number x ~ electromagnetic field

DESCRIPTION

Prints grade 2 for four dimensions as an electromagnetic field using the format specified in the first argument. Other components are ignored.

RETURN VALUE

The return value is the original number.

RECYCLING

~

ERROR RECOVERY

~

SEE ALSO

```
print_emfield(), printf_number(), printfs_emfield()
```

209	PRINTF_GRADEDLISTOF_COEFFICIENTS	P

NAME

printf_gradedlistof_coefficients :– print list of coefficients with specified format in grades

SYNOPSIS

```
number printf_gradedlistof_coefficients(char *fmt,
number x, int cmax);
```

ARGUMENTS

1: i ◇ char *fmt ~ pointer to format string

2: i ◇ number x ~ number

3: i ◇ int cmax ~ maximum number of coefficients per line

DESCRIPTION

Print the coefficients for a number as a list without units grouped into grades, using the format specified in the first argument. The order of the coefficients is the same as the order of the units printed by `print_gradedlistof_units()`. The third argument cmax specifies the maximum number of coefficients to be printed on each line of output.

RETURN VALUE

The return value is the original number.

RECYCLING

~

ERROR RECOVERY

~

SEE ALSO

        print_gradedlistof_coefficients(),
        print_gradedlistof_units(), printf_listof_coefficients(),
        unpack_components()

| 210 | PRINTF_LISTOF_COEFFICIENTS | P |

NAME

        printf_listof_coefficients :– print list of coefficients with specified format

SYNOPSIS

        number printf_listof_coefficients(char *fmt,
                                          number x, int cmax);

ARGUMENTS

        1: i◇ char *fmt ~ pointer to format string
        2: i◇ number x ~ number
        3: i◇ int cmax ~ maximum number of coefficients per line

DESCRIPTION

        Print the coefficients for a number as a list without units using the format specified
        in the first argument. The order of the coefficients is the same as the order of the
        units printed by print_listof_units(). The third argument cmax specifies
        the maximum number of coefficients to be printed on each line of output.

RETURN VALUE

        The return value is the original number.

RECYCLING

        ~

ERROR RECOVERY

        ~

SEE ALSO

        print_listof_coefficients(), print_listof_units(),
        printf_gradedlistof_coefficients(), unpack_components()

| 211 | PRINTF_NUMBER | P |

NAME

        printf_number :– print number with specified format

SYNOPSIS

        number printf_number(char *fmt, number x, int cmax);

ARGUMENTS

        1: i◇ char *fmt ~ pointer to format string
        2: i◇ number x ~ number
        3: i◇ int cmax ~ maximum number of components per line

DESCRIPTION

        Print the number as a sequence of components using the format specified in the
        first argument. The third argument cmax specifies the maximum number of com-
        ponents to be printed on each line of output.

RETURN VALUE

        The return value is the original number.

RECYCLING

~

ERROR RECOVERY

Any value of cmax less than 1 will result in each component printed on a separate line.

SEE ALSO

```
print_listof_coefficients(), print_listof_units(),
print_number(), printf_element(), unpack_number()
```

| 212 | PRINTF_NVECTOR | P |

NAME

printf_Nvector :– print vector of dimension N with specified format

SYNOPSIS

```
number printf_Nvector(char *fmt, int n, number x);
```

ARGUMENTS

1: i ⬦ char *fmt ~ pointer to format string

2: i ⬦ int n   ~ dimension N

3: i ⬦ number x ~ vector in N dimensions

DESCRIPTION

Prints grade 1 of a number as a vector using the format specified in the first argument. Other grades are ignored. The first N dimensions of the vector are printed.

RETURN VALUE

The return value is the original number.

RECYCLING

~

ERROR RECOVERY

~

SEE ALSO

```
printf_3vector(), printf_number(), printf_vector(),
unpack_components()
```

| 213 | PRINTF_ONECOEFFICIENTOF_ARRAY | P |

NAME

printf_onecoefficientof_array :– print one coefficient of array with specified format

SYNOPSIS

```
number *printf_onecoefficientof_array(char *fmt,
                          int n, number *a, bitmap u);
```

ARGUMENTS

1: i ⬦ char *fmt ~ pointer to format string

2: i ⬦ int n   ~ size of array

3: i ⬦ number *a ~ pointer to array of numbers

4: i ⬦ bitmap u ~ unit of component

DESCRIPTION

Print array of numbers using the format specified in the first argument. Prints only the coefficient for one of the components for each element of the array. The fourth argument u specifies the unit of the component.

RETURN VALUE

The return value is a pointer to the original array.

RECYCLING

~

ERROR RECOVERY

~

SEE ALSO

```
print_array(), print_onecoefficientof_array(),
printf_onecoefficientof_matrix()
```

214	PRINTF_ONECOEFFICIENTOF_MATRIX	P

NAME

printf_onecoefficientof_matrix :– print a whole matrix giving the coefficients of one unit with specified format

SYNOPSIS

```
matrix *printf_onecoefficientof_matrix(char *fmt,
                                    matrix *m, bitmap u);
```

ARGUMENTS

1: i ◇ char *fmt ~ pointer to format string

2: i ◇ matrix *m ~ matrix

3: i ◇ bitmap u ~ unit

DESCRIPTION

Print the coefficient with the given unit for all elements in a matrix using the layout set through format_matrix() and using the print format specified.

RETURN VALUE

The return value is the matrix.

RECYCLING

~

ERROR RECOVERY

~

SEE ALSO

```
print_onecoefficientof_matrix(),
printf_onecoefficientof_array()
```

215	PRINTF_PAULI	P

NAME

printf_pauli :– print Pauli matrix with specified format

SYNOPSIS

```
number printf_pauli(char *fmt, number x);
```

ARGUMENTS

1: i ◇ char *fmt ~ pointer to format string

2: i ◇ number x ~ Pauli matrix

DESCRIPTION

Prints grade 0 and 2 for three dimensions as a Pauli matrix using the format specified in the first argument. Other components are ignored.

RETURN VALUE

The return value is the original number.

RECYCLING

~

ERROR RECOVERY

~

SEE ALSO

    print_pauli(), printf_number()

| 216 | PRINTF_QUATERNION | P |

NAME

printf_quaternion :– print quaternion with specified format

SYNOPSIS

    number printf_quaternion(char *fmt, number x);

ARGUMENTS

    1: i ◇ char *fmt ~ pointer to format string
    2: i ◇ number x ~ quaternion

DESCRIPTION

Prints grade 0 and 2 for three dimensions as a quaternion using the format specified in the first argument. Other components are ignored.

RETURN VALUE

The return value is the original number.

RECYCLING

~

ERROR RECOVERY

~

SEE ALSO

    print_quaternion(), printf_number()

| 217 | PRINTF_SINGLE_COEFFICIENT | P |

NAME

printf_single_coefficient :– print single coefficient with specified format

SYNOPSIS

    number printf_single_coefficient(char *fmt,
                                     number x, bitmap u);

ARGUMENTS

    1: i ◇ char *fmt ~ pointer to format string
    2: i ◇ number x ~ number
    3: i ◇ bitmap u ~ unit of component

DESCRIPTION

Print only the coefficient for a single component from the number using the format specified in the first argument. The third argument u specifies the unit for the component.

RETURN VALUE

The return value is the original number.

RECYCLING

~

ERROR RECOVERY

~

SEE ALSO

get_coef(), print_single_coefficient(), printf_coef()

218	PRINTF_VECTOR	P

NAME

printf_vector :– print vector with specified format

SYNOPSIS

number printf_vector(char *fmt, number x);

ARGUMENTS

1: i ◇ char *fmt ~ pointer to format string

2: i ◇ number x ~ vector

DESCRIPTION

Prints grade 1 of a number as a vector using the format specified in the first argument. Other grades are ignored. The dimension of the vector is taken from the dimension established for the numerical context using declare_primal().

RETURN VALUE

The return value is the original number.

RECYCLING

~

ERROR RECOVERY

~

SEE ALSO

declare_primal(), print_vector(), printf_3vector(), printf_Nvector(), printf_number(), unpack_03vector(), unpack_3vector(), unpack_R3vector()

219	PRINTFS_EMFIELD	P

NAME

printfs_emfield :– print electromagnetic field with specified format and separator

SYNOPSIS

number printfs_emfield(char *fmt, char *sep, number x);

ARGUMENTS

1: i ◇ char *fmt ~ pointer to format string

2: i ◇ char *sep ~ pointer to separator string

3: i ◇ number x ~ electromagnetic field

DESCRIPTION

Prints grade 2 for four dimensions as an electromagnetic field using the format specified in the first argument and the separator specified in the second argument. Other components are ignored. The separator is inserted between the magnetic and electric field components. A carriage return as a separator is useful when the two field components are too long to fit together on one line.

RETURN VALUE

The return value is the original number.

RECYCLING

~

ERROR RECOVERY

~

SEE ALSO

`printf_emfield(), printf_number()`

220	PSEUDO_NUMBER	P

NAME

pseudo_number :– construct a pseudoscalar

SYNOPSIS

`number pseudo_number(double x, double y);`

ARGUMENTS

1: i ◊ `double x` ~ real part

2: i ◊ `double y` ~ imaginary part

DESCRIPTION

Construct a new number with the real and imaginary parts as specified. The new number is a pseudo-scalar (in the highest grade).

RETURN VALUE

The return value is the new number.

RECYCLING

~

ERROR RECOVERY

~

SEE ALSO

`fresh_number(), primal_vector(), pseudo_scalar(), scalar_number(), simple_number()`

221	PSEUDO_SCALAR	P

NAME

pseudo_scalar :– pseudoscalar unit

SYNOPSIS

`bitmap pseudo_scalar(int n);`

ARGUMENTS

1: i ◊ `int n` ~ dimension

DESCRIPTION

Constructs the unit for the pseudo-scalar of the specified dimension.

RETURN VALUE

The return value is the unsigned integer bitmap unit of the pseudo-scalar.

RECYCLING

~

ERROR RECOVERY

~

SEE ALSO

dimension(), primal(), psuedo_number()

222	R_UNIT	R

NAME

r_unit :– reversion of unit

SYNOPSIS

int r_unit(bitmap u);

ARGUMENTS

1: i ◇ bitmap u ~ unit

DESCRIPTION

Calculate the effect of reversing the order of all of the primal units in the unit specified.

RETURN VALUE

The return value is the sign induced by reversion of the unit.

RECYCLING

~

ERROR RECOVERY

~

SEE ALSO

c_unit(), i_unit()

223	RADIANS	R

NAME

radians :– convert component in polar form from degrees to radians

SYNOPSIS

cliff_comp *radians(cliff_comp *a, cliff_comp *c);

ARGUMENTS

1: i ◇ cliff_comp *a ~ pointer to structure containing polar component in degrees

2: o ◇ cliff_comp *c ~ pointer to structure in which to save polar component in radians

DESCRIPTION

Convert the component from degrees to radians. The radial part is taken from the real part, and the angular part from the imaginary part. The angle is initially in degrees. Setting the pointers of both components to the same value overwrites the input with the output.

RETURN VALUE

The return value is a pointer to the component containing the polar form with the angle in radians.

RECYCLING

~

ERROR RECOVERY

~

SEE ALSO

degrees(), polar()

| 224 | RANDOM_BICOMPLEX | R |

NAME

random_bicomplex :– random bicomplex number

SYNOPSIS

```
number random_bicomplex(void);
```

ARGUMENTS

~

DESCRIPTION

Generate a random bicomplex number with real and imaginary parts of all coefficients in the range from −1 to +1 with a uniform distribution.

RETURN VALUE

The return value is the random bicomplex number.

RECYCLING

~

ERROR RECOVERY

~

SEE ALSO

```
random_complex(), random_grade(), random_number(),
random_quaternion(), random_scalar(), random_seed(),
random_vector()
```

| 225 | RANDOM_COMPLEX | R |

NAME

random_complex :– generate component containing random complex scalar

SYNOPSIS

```
number random_complex(cliff_comp *c);
```

ARGUMENTS

1: o ◇ cliff_comp *c ~ pointer to structure containing component

DESCRIPTION

Generate a random complex component with real and imaginary parts of the coefficient in the range from −1 to +1 with a uniform distribution. The unit of the component is set for a scalar (grade 0).

RETURN VALUE

The return value is the random complex component.

RECYCLING

~

ERROR RECOVERY

~

SEE ALSO

```
random_bicomplex()
```

| 226 | RANDOM_GRADE | R |

NAME

random_grade :– random grade

SYNOPSIS

```
number random_grade(int n);
```

ARGUMENTS

1: i ◇ int n ~ grade in which to generate number

DESCRIPTION

Generate a random number in a specified grade with real and imaginary parts of all coefficients in the range from −1 to +1 with a uniform distribution.

RETURN VALUE

The return value is the random number.

RECYCLING

~

ERROR RECOVERY

~

SEE ALSO

```
random_bicomplex(), random_number(), random_quaternion(),
random_scalar(), random_seed(), random_vector()
```

227	RANDOM_MATRIX	R

NAME

random_matrix :– fill an empty matrix with random values

SYNOPSIS

```
matrix *random_matrix(matrix *m, int n, int *grades);
```

ARGUMENTS

1: o ◇ matrix *m   ~ empty matrix
2: i ◇ int n   ~ number of grades
3: i ◇ int *grades ~ array of grades

DESCRIPTION

Fill the matrix with the random numbers with values of real and imaginary parts in the range from −1 to +1. Only the grades listed in the array are initialised. All other grades are left empty.

RETURN VALUE

The return value is the original matrix containing random numbers.

RECYCLING

~

ERROR RECOVERY

It is an error to initialise a sub-matrix. On error, the recovery strategy brexit(0) is invoked to initiate a user re-start. It is an error if the matrix is not empty initially. No check is made. It is an error to specify grades incompatible with the dimension of the number context.

SEE ALSO

```
identity_matrix(), random_number(), random_seed()
```

228	RANDOM_NUMBER	R

NAME

random_number :– random number

SYNOPSIS

```
number random_number(void);
```

ARGUMENTS

~

DESCRIPTION

Generate a random number with real and imaginary parts of all coefficients in the range from −1 to +1 with a uniform distribution.

RETURN VALUE

The return value is the random number.

RECYCLING

~

ERROR RECOVERY

~

SEE ALSO

```
random_bicomplex(), random_grade(), random_matrix(),
random_quaternion(), random_scalar(), random_seed(),
random_vector()
```

229	RANDOM_QUATERNION	R

NAME

random_quaternion :– random quaternion

SYNOPSIS

```
number random_quaternion(void);
```

ARGUMENTS

~

DESCRIPTION

Generate a random quaternion with real and imaginary parts of all coefficients in the range from −1 to +1 with a uniform distribution.

RETURN VALUE

The return value is the random quaternion.

RECYCLING

~

ERROR RECOVERY

~

SEE ALSO

```
random_bicomplex(), random_grade(), random_number(),
random_scalar(), random_seed(), random_vector()
```

230	RANDOM_SCALAR	R

NAME

random_scalar :– random scalar

SYNOPSIS

```
number random_scalar(void);
```

ARGUMENTS

~

DESCRIPTION

Generate a random scalar with real and imaginary parts of all coefficients in the range from −1 to +1 with a uniform distribution.

RETURN VALUE

The return value is the random scalar.

RECYCLING

~

ERROR RECOVERY

~

SEE ALSO

```
random_bicomplex(), random_grade(), random_number(),
random_quaternion(), random_seed(), random_vector()
```

231	RANDOM_SEED	R

NAME

random_seed :– seed for random numbers

SYNOPSIS

```
void random_seed(int k);
```

ARGUMENTS

1: i ◇ int k ~ seed for random numbers

DESCRIPTION

Initialise the seed for random numbers.

RETURN VALUE

~

RECYCLING

~

ERROR RECOVERY

~

SEE ALSO

```
random_bicomplex(), random_grade(), random_matrix(),
random_number(), random_quaternion(), random_scalar(),
random_vector()
```

232	RANDOM_VECTOR	R

NAME

random_vector :– random vector

SYNOPSIS

```
number random_vector(void);
```

ARGUMENTS

~

DESCRIPTION

Generate a random vector with real and imaginary parts of all coefficients in the range from −1 to +1 with a uniform distribution.

RETURN VALUE

The return value is the random vector.

RECYCLING

~

ERROR RECOVERY

~

SEE ALSO

```
random_bicomplex(), random_grade(), random_matrix(),
random_number(), random_quaternion(), random_scalar(),
random_seed()
```

233	REAL_PART	R

NAME

real_part :– real part of component

SYNOPSIS

```
cliff_comp *real_part(cliff_comp *a, cliff_comp *c);
```

ARGUMENTS

1: i ◇ `cliff_comp *a` ~ pointer to structure containing complex component

2: o ◇ `cliff_comp *c` ~ pointer to structure in which to save real part

DESCRIPTION

Extract the real part of a component.

RETURN VALUE

The return value is a pointer to the component containing the real part. Setting the pointers of both components to the same value overwrites the input with the output.

RECYCLING

~

ERROR RECOVERY

~

SEE ALSO

```
imaginary_part(), no_imaginary(), realpartof_number(),
squareof_magnitude()
```

234	REALPARTOF_NUMBER	R

NAME

realpartof_number :– produce real part of number

SYNOPSIS

```
number realpartof_number(number x, int d);
```

ARGUMENTS

1: id ◇ `number x` ~ number

2: i ◇ `int d` ~ disposal flag

DESCRIPTION

Construct a number containing the real part of the number specified.

RETURN VALUE

The return value is a number containing the real part only.

RECYCLING

The number x is recycled according to the value of the disposal flag d. If d=1, the original number is overwritten *in situ* and returned in place of the new one.

ERROR RECOVERY

~

SEE ALSO

imaginarypartof_number(), real_part()

| 235 | RECIPROCALOF_BICOMPLEX | R |

NAME

reciprocalof_bicomplex :– produce inverse of bicomplex number

SYNOPSIS

```
number *reciprocalof_bicomplex(number x, number ratio[2],
int d);
```

ARGUMENTS

1: id ◇ number x   ~ bicomplex number
2: o ◇ number ratio[2] ~ array for numerator and denominator
3: i ◇ int d   ~ disposal flag

DESCRIPTION

Construct an array of bicomplex numbers containing the numerator and denominator of the ratio for the inverse of the bicomplex number specified. The denominator is guaranteed to be a scalar. If there is no finite inverse, the denominator is zero.

RETURN VALUE

The return value is a pointer to the array containing numerator and denominator.

RECYCLING

The bicomplex number x is recycled according to the value of the disposal flag.

ERROR RECOVERY

It is an error if the signature contains 0 or positive 1. On error, the recovery strategy brexit(6) is invoked, terminating the execution.

SEE ALSO

inverseof_bicomplex()

| 236 | RECIPROCALOF_COMPONENT | R |

NAME

reciprocalof_component :– produce inverse of number with single component

SYNOPSIS

```
number reciprocalof_component(number x, number ratio[2],
int d);
```

ARGUMENTS

1: id ◇ number x   ~ simple number (single component)
2: o ◇ number ratio[2] ~ array for numerator and denominator
3: i ◇ int d   ~ disposal flag

DESCRIPTION

Construct an array of numbers containing the numerator and denominator of the ratio for the inverse of the single component specified. The denominator is guaranteed to be a scalar. If there is no finite inverse, the denominator is zero.

RETURN VALUE

The return value is a pointer to the array containing numerator and denominator.

RECYCLING

The number x is recycled according to the value of the disposal flag.

ERROR RECOVERY

It is an error to take the inverse of a number which has more than one component, or if the signature is zero. Depending on the error, either the recovery strategy `brexit(0)` is invoked to initiate a user re-start or the recovery strategy `brexit(6)` is invoked, terminating the execution.

SEE ALSO

```
reciprocalof_complex(), reciprocalof_emfield(),
reciprocalof_pauli(), reciprocalof_quaternion(),
reciprocalof_vector(), reciprocalof_xemfield()
```

| 237 | RECIPROCALOF_EMFIELD | R |

NAME

reciprocalof_emfield :– produce inverse of electromagnetic field

SYNOPSIS

```
number reciprocalof_emfield(number x, number ratio[2],
                            int d);
```

ARGUMENTS

1: id ◇ number x   ~ electromagnetic field
2: o ◇ number ratio[2] ~ array for numerator and denominator
3: i ◇ int d   ~ disposal flag

DESCRIPTION

Construct an array of numbers containing the numerator and denominator of the ratio for the inverse of the electromagnetic field specified. The denominator is guaranteed to be a scalar. If there is no finite inverse, the denominator is zero.

RETURN VALUE

The return value is a pointer to the array containing numerator and denominator.

RECYCLING

The electromagnetic field x is recycled according to the value of the disposal flag.

ERROR RECOVERY

It is an error if the signature contains 0 or positive 1. On error, the recovery strategy `brexit(6)` is invoked, terminating the execution.

SEE ALSO

```
inverseof_emfield(), reciprocalof_xemfield()
```

| 238 | RECIPROCALOF_PAULI | R |

NAME

reciprocalof_pauli :– produce inverse of Pauli matrix

SYNOPSIS

```
number reciprocalof_pauli(number x, number ratio[2],
                          int d);
```

ARGUMENTS

1: id ◇ number x   ~ Pauli matrix
2: o ◇ number ratio[2] ~ array for numerator and denominator
3: i ◇ int d   ~ disposal flag

DESCRIPTION

Construct an array of numbers containing the numerator and denominator of the ratio for the inverse of the Pauli matrix specified. The denominator is guaranteed to be a scalar. If there is no finite inverse, the denominator is zero.

RETURN VALUE

The return value is a pointer to the array containing numerator and denominator.

RECYCLING

The Pauli matrix x is recycled according to the value of the disposal flag.

ERROR RECOVERY

It is an error if the signature contains 0 or positive 1. On error, the recovery strategy brexit(6) is invoked, terminating the execution.

SEE ALSO

```
inverseof_pauli()
```

239	RECIPROCALOF_QUATERNION	R

NAME

reciprocalof_quaternion :– produce inverse of quaternion

SYNOPSIS

```
number reciprocalof_quaternion(number x, number ratio[2],
                               int d);
```

ARGUMENTS

1: id ◇ number x   ~ quaternion
2: o ◇ number ratio[2] ~ array for numerator and denominator
3: i ◇ int d   ~ disposal flag

DESCRIPTION

Construct an array of numbers containing the numerator and denominator of the ratio for the inverse of the quaternion specified. The denominator is guaranteed to be a scalar. If there is no finite inverse, the denominator is zero.

RETURN VALUE

The return value is a pointer to the array containing numerator and denominator.

RECYCLING

The quaternion x is recycled according to the value of the disposal flag.

ERROR RECOVERY

It is an error if the signature contains 0 or positive 1. On error, the recovery strategy brexit(6) is invoked, terminating the execution.

SEE ALSO

```
inverseof_quaternion(), factorsof_quaternion()
```

| 240 | RECIPROCALOF_VECTOR | R |

NAME

reciprocalof_vector :– produce inverse of vector

SYNOPSIS

```
number reciprocalof_vector(number x, number ratio[2],
                          int d);
```

ARGUMENTS

1: id ◇ number x   ~ vector
2: o ◇ number ratio[2] ~ array for numerator and denominator
3: i ◇ int d   ~ disposal flag.

DESCRIPTION

Construct an array of numbers containing the numerator and denominator of the ratio for the inverse of the vector specified. The denominator is guaranteed to be a scalar. If there is no finite inverse, the denominator is zero.

RETURN VALUE

The return value is a pointer to the array containing numerator and denominator.

RECYCLING

The vector x is recycled according to the value of the disposal flag.

ERROR RECOVERY

It is an error if the signature is mixed or contains zero. On error, the recovery strategy `brexit(6)` is invoked, terminating the execution.

SEE ALSO

```
inverseof_vector()
```

| 241 | RECIPROCALOF_XEMFIELD | R |

NAME

reciprocalof_xemfield :– produce inverse of extended electromagnetic field

SYNOPSIS

```
number reciprocalof_xemfield(number x, number ratio[2],
                             int d);
```

ARGUMENTS

1: id ◇ number x   ~ extended electromagnetic field
2: o ◇ number ratio[2] ~ array for numerator and denominator
3: i ◇ int d   ~ disposal flag.

DESCRIPTION

Construct an array of numbers containing the numerator and denominator of the ratio for the inverse of the extended electromagnetic field specified. The denominator is guaranteed to be a scalar. If there is no finite inverse, the denominator is zero.

RETURN VALUE

The return value contains the inverse (under central multiplication) of the vector.

RECYCLING

The extended electromagnetic field x is recycled according to the value of the disposal flag.

Error Recovery

It is an error if the signature contains 0 or positive 1. On error, the recovery strategy `brexit(6)` is invoked, terminating the execution.

See Also

`denom_xemfield()`, `inverseof_xemfield()`, `reciprocalof_emfield()`

242	RECOVERY_LEVEL	R

Name

recovery_level :– set recovery level

Synopsis

`void recovery_level(int n, int s, sigjmp_buf *buf);`

Arguments

1: i◇ `int n` ~ number of times to attempt recovery

2: i◇ `int s` ~ strategy for recovery

3: i◇ `sigjmp_buf *buf` ~ pointer to structure for storage of recovery jump information

Description

Set n as the number of attempts at recovery before terminating execution. Set s as the strategy to be implemented when attempting to recovery from an error.

$s=0$	Do not re-start, exit
$s \neq 0$	Attempt re-start
s odd	Prompt on re-start
$\lvert s \rvert = \{1,2\}$	Do not free memory
$\lvert s \rvert = \{3,4\}$	Recycle all numbers
$\lvert s \rvert = \{5,6\}$	Free all memory
$\lvert s \rvert = \{7,8\}$	Free all memory and print statistics

The prompt strategy allows the user to choose whether to continue with the attempt to re-start or to terminate execution. During recovery, execution is returned to the location supplied to the c-library sub-routine `sigsetjmp()`. The user should declare a static memory area `sigjmp_buf buf;` and then invoke `sigsetjmp()` with the two arguments `buf` and 1 (one) on the line where the program should be reactivated during an attempt at recovery. It is possible to invoke `sigsetjmp()` in multiple places in the program code (using always the same buffer `buf`). Recovery will be directed to the location of the most recent invocation. See the operating system documentation for `sigsetjmp()` by invoking the command `man sigsetjmp` at the operating system command prompt. See also the examples in Sections 15.3 and 15.4.

Return Value

~

Recycling

~

ERROR RECOVERY

It is an error to initiate a strategy which specifies recovery, then fail to establish the location for recovery using `sigsetjmp()` before an error occurs.

SEE ALSO

`brexit(), siglongjmp(), sigsetjmp()`

243	RECYCLE_ALLNUMBER	R

NAME

recycle_allnumber :– recycle all numbers

SYNOPSIS

`void recycle_allnumber(void);`

ARGUMENTS

~

DESCRIPTION

Recycle all memory in the Clifford mathematical environment. All numbers become lost. Numbers can be re-used after reassignment.

RETURN VALUE

~

RECYCLING

~

ERROR RECOVERY

~

SEE ALSO

`free_cliffmem(), recycle_array(), recycle_matrix(), recycle_number()`

244	RECYCLE_ARRAY	R

NAME

recycle_array :– recycle array of numbers

SYNOPSIS

`number *recycle_array(int n, number *a);`

ARGUMENTS

1: i ◇ `int n`   ~ number of elements in array
2: i ◇ `number *a` ~ pointer to array of numbers

DESCRIPTION

Recycle an array of n numbers.

RETURN VALUE

The return value is a pointer to the array.

RECYCLING

~

ERROR RECOVERY

It is an error to provide an array which does not have the number of elements specified.

SEE ALSO

    fresh_array(), recycle_matrix(), recycle_number()

| 245 | RECYCLE_MATRIX | R |

NAME

recycle_matrix :– recycle numbers in matrix and dispose of memory space allocated

SYNOPSIS

    matrix *recycle_matrix(matrix *m);

ARGUMENTS

1: io ⋄ matrix *m ~ matrix

DESCRIPTION

Recycle all of the numbers in a matrix, then free the memory allocated allocated to the matrix.

RETURN VALUE

The return value is the matrix, now empty and with no memory for data.

RECYCLING

The numbers stored in the matrix are all recycled, and the memory allocated for the elements in the matrix is freed.

ERROR RECOVERY

It is an error to recycle an empty matrix or a sub-matrix. On error, for a sub-matrix, the recovery strategy brexit(0) is invoked to initiate a user re-start.

SEE ALSO

    recycle_array(), recycle_number(), refresh_matrix()

| 246 | RECYCLE_NUMBER | R |

NAME

recycle_number :– recycle number

SYNOPSIS

    void recycle_number(number x);

ARGUMENTS

1: i ⋄ number x ~ number to be recycled

DESCRIPTION

Recycle the memory for the number x. After the operation, the value of the token contained in x no longer provides access to any valid Clifford number.

RETURN VALUE

~

RECYCLING

~

ERROR RECOVERY

It is possible to use the variable x to represent a new number only after its contents have been reassigned by an appropriate operation to create the value of a new token. Using the contents of the variable before reassignment is an error which may result in unintended behaviour.

SEE ALSO

```
free_cliffmem(), recycle_allnumber(), recycle_array(),
recycle_matrix()
```

247	REFLECT_NUMBER	R

NAME

reflect_number :– produce reflection of number

SYNOPSIS

```
number reflect_number(number x, number n, int d);
```

ARGUMENTS

1: id ◇ number x ~ vector or other number

2: id ◇ number n ~ unit vector, normal to surface if in three dimensions

3: i ◇ int d   ~ disposal flag

DESCRIPTION

Apply reflection operator to a number. The reflection operator behaves in a geometric sense reversing all geometric components of the number parallel to the unit vector supplied. In three dimensions, the unit vector plays the role of a vector normal to a surface which behaves as a mirror.

RETURN VALUE

The return value is the result of the reflection.

RECYCLING

The numbers x and n are recycled according to the value of the disposal flag.

ERROR RECOVERY

It is an error to specify a unit vector which is not of unit length.

SEE ALSO

```
innerspace_number(), normalise_vector(),
outerspace_number(), reflect_vector()rotate_number()
```

248	REFLECT_VECTOR	R

NAME

reflect_vector :– produce reflection of vector

SYNOPSIS

```
number reflect_vector(number x, number n, int d);
```

ARGUMENTS

1: id ◇ number x ~ vector or other number

2: id ◇ number n ~ unit vector, normal to surface if in three dimensions

3: i ◇ int d   ~ disposal flag

DESCRIPTION

Apply the reflection operator to a number. The reflection operator for a vector is applied, even if the number has components in other grades. For proper vectors, numbers in grade 1, the reflection operator behaves in a geometric sense reversing all geometric components of the number parallel to the unit vector supplied. In three dimensions, the unit vector plays the role of a vector normal to a surface which behaves as a mirror. For numbers in higher grades, the result is different for different signatures. In the case of electromagnetic fields, which occupy grade

2 of four dimensions with a negative signature, the reflection operator serves to reflect the field as if in a perfectly electrically conducting mirror with unit normal specified in the three spatial dimensions.

RETURN VALUE

The return value is the result of the reflection.

RECYCLING

The numbers x and n are recycled according to the value of the disposal flag.

ERROR RECOVERY

It is an error to specify a unit vector which is not of unit length.

SEE ALSO

```
innerspace_vector(), normalise_vector(),
outerspace_vector(), reflect_number(), rotate_number()
```

249	REFRESH_MATRIX	R

NAME

refresh_matrix :– recycle numbers in matrix but retain memory space allocated

SYNOPSIS

```
matrix *refresh_matrix(matrix *m);
```

ARGUMENTS

1: io ◊ matrix *m ~ matrix

DESCRIPTION

Recycle all of the numbers in a matrix but do not free the memory allocated to the matrix.

RETURN VALUE

The return value is the matrix, now empty but retaining memory for data.

RECYCLING

The numbers stored in the matrix are all recycled.

ERROR RECOVERY

It is an error to recycle an empty matrix or a sub-matrix. On error, for a sub-matrix, the recovery strategy brexit(0) is invoked to initiate a user re-start.

SEE ALSO

```
define_matrix(), recycle_matrix()
```

250	REGRESS	R

NAME

regress :– regressive product of components

SYNOPSIS

```
cliff_comp *regress(cliff_comp *a, cliff_comp *b,
                    cliff_comp *c);
```

ARGUMENTS

1: i ◊ cliff_comp *a ~ multiplier, pointer to structure containing component

2: i ◊ cliff_comp *b ~ multiplicand, pointer to structure containing component

3: o ◊ cliff_comp *c ~ pointer to structure in which to save product

DESCRIPTION

Calculate the regressive product of two components. Setting the pointers of either input component to the same value as the output component overwrites the input with the output.

RETURN VALUE

The return value is a pointer to the component containing the regressive product.

RECYCLING

~

ERROR RECOVERY

~

SEE ALSO

```
outer(), regressive_mul()
```

251	REGRESSIVE_MUL	R

NAME

regressive_mul :– calculate regressive product of numbers

SYNOPSIS

```
number regressive_mul(number x, number y, int d);
```

ARGUMENTS

1: id ◇ number x ~ multiplier
2: id ◇ number y ~ multiplicand
3: i ◇ int d   ~ disposal flag

DESCRIPTION

Multiply two numbers using the regressive multiplication.

RETURN VALUE

The return value is the regressive product.

RECYCLING

~

ERROR RECOVERY

The numbers x and y are recycled according to the value of the disposal flag.

SEE ALSO

```
outer_mul(), regress()
```

252	REVERSION	R

NAME

reversion :– reversion of component

SYNOPSIS

```
cliff_comp *reversion(cliff_comp *a, cliff_comp *c);
```

ARGUMENTS

1: i ◇ cliff_comp *a ~ pointer to structure containing component
2: o ◇ cliff_comp *c ~ pointer to structure in which to save reversion

DESCRIPTION

Calculate the reversion of a component. Setting the pointers of both components to the same value overwrites the input with the output.

RETURN VALUE

The return value is a pointer to the component containing the reversion.

RECYCLING

~

ERROR RECOVERY

~

SEE ALSO

`clifford_conjugation()`, `involution()`, `reversionof_number()`

253	REVERSIONOF_NUMBER	R

NAME

reversionof_number :– produce reversion of number

SYNOPSIS

`number reversionof_number(number x, int d);`

ARGUMENTS

1: id ◇ `number x` ~ number

2: i ◇ `int d` ~ disposal flag

DESCRIPTION

Construct a number containing the reversion of the number specified.

RETURN VALUE

The return value is a number containing the reversion.

RECYCLING

The number x is recycled according to the value of the disposal flag d. If d=1, the original number is overwritten *in situ* and returned in place of the new one.

ERROR RECOVERY

~

SEE ALSO

`cliffordconjugateof_number()`, `involutionof_number()`, `reversion()`

254	RIGHT	R

NAME

right :– right interior product of components

SYNOPSIS

`cliff_comp *right(cliff_comp *a, cliff_comp *b,`
`                  cliff_comp *c);`

ARGUMENTS

1: i ◇ `cliff_comp *a` ~ multiplier, pointer to structure containing component

2: i ◇ `cliff_comp *b` ~ multiplicand, pointer to structure containing component

3: o ◇ `cliff_comp *c` ~ pointer to structure in which to save product

DESCRIPTION

Calculate the right interior (inner) product of two components. Setting the pointers of either input component to the same value as the output component overwrites the input with the output. The return value contains the interior (inner) product.

RETURN VALUE

The return value is a pointer to the component containing the right interior (inner) product.

RECYCLING

~

ERROR RECOVERY

~

SEE ALSO

```
left(), right_mul()
```

| 255 | RIGHT_MUL | R |

NAME

right_mul :– calculate right interior product of numbers

SYNOPSIS

```
number right_mul(number x, number y, int d);
```

ARGUMENTS

1: id ◇ number x ~ multiplier
2: id ◇ number y ~ multiplicand
3: i ◇ int d   ~ disposal flag

DESCRIPTION

Multiply two numbers using the right interior (inner) multiplication.

RETURN VALUE

The return value is the right interior (inner) product.

RECYCLING

The numbers x and y are recycled according to the value of the disposal flag.

ERROR RECOVERY

~

SEE ALSO

```
left_mul(), right()
```

| 256 | RMSOF_ARRAY | R |

NAME

rmsof_array :– root mean square value of array

SYNOPSIS

```
number rmsof_array(int n, number *a);
```

ARGUMENTS

1: i ◇ int n   ~ number of elements in array
2: i ◇ number *a ~ pointer to array of numbers

DESCRIPTION

Calculates the root mean square (RMS) value from the magnitudes of all of the elements in the array.

RETURN VALUE

The return value is the RMS value. The return value is a scalar.

RECYCLING

~

ERROR RECOVERY

It is an error to provide an array which does not have the number of elements specified.

SEE ALSO

`meanabsof_array(), peakof_array(), sumof_array()`

257	ROTATE_NUMBER	R

NAME

rotate_number :– produce rotation of number

SYNOPSIS

```
number rotate_number(number x, number n1, number n2,
                     int d);
```

ARGUMENTS

1: id ◇ number x ~ vector or other number
2: id ◇ number n1 ~ start unit vector, normal to axis of rotation if in three dimensions
3: id ◇ number n2 ~ halfway unit vector, normal to axis of rotation if in three dimensions
4: i ◇ int d   ~ disposal flag

DESCRIPTION

Apply rotation operator to a number. The rotation operator behaves in a geometric sense rotating all geometric components of the number within the plane of the two unit vectors. Rotation is effected in the direction of the smallest angle from the first unit vector to the second, covering a distance of twice that angle. In three dimensions, the unit vectors play the role of normals to the axis of rotation.

RETURN VALUE

The return value is the result of the rotation.

RECYCLING

The numbers x, n1, and n2 are recycled according to the value of the disposal flag.

ERROR RECOVERY

It is an error to specify unit vectors which are not of unit length.

SEE ALSO

```
innerspace_number(), normalise_vector(),
outerspace_number(), reflect_number()
```

258	ROW_MUL_COL	R

NAME

row_mul_col :– multiply a row of a matrix by a column of another matrix

SYNOPSIS

```
number row_mul_col(int i, matrix *a, int j, matrix *b);
```

ARGUMENTS

1: i ◇ int i ~ multiplier, row of first matrix

2: i ◇ matrix *a ~ first matrix

3: i ◇ int j ~ multiplicand, column of second matrix

4: i ◇ matrix *b ~ second matrix

DESCRIPTION

Multiply the row (i) in the first matrix by the column (j) in the second matrix. If unknown, the total number of rows and columns can be obtained using the routines rows() and columns(). The indexes i, j for the rows and columns start from zero and go up to one less than the totals. No checks are made to determine if the indexes are out of bounds.

RETURN VALUE

The return value is the matrix product of single row and column.

RECYCLING

~

ERROR RECOVERY

It is an error if the row and column have different sizes. On error, the recovery strategy brexit(0) is invoked to initiate a user re-start. It is an error to use indexes which are out of bounds.

SEE ALSO

multiply_matrix(), prescale_row(), weightedsumof_array()

| 259 | Rows | R |

NAME

rows :– get number of rows for a matrix

SYNOPSIS

int rows(matrix *m);

ARGUMENTS

1: i ◇ matrix *m ~ matrix

DESCRIPTION

Get the number of rows for a matrix. The indexes for the rows in the matrix start at zero and go up to one less than this number.

RETURN VALUE

The return value is the number of rows in the matrix.

RECYCLING

~

ERROR RECOVERY

~

SEE ALSO

access_element(), columns(), stride()

| 260 | RSCALE | R |

NAME

rscale :– scale component by real value

SYNOPSIS

```
cliff_comp *rscale(double s, cliff_comp *b,
                   cliff_comp *c);
```

ARGUMENTS

1: i ⬦ `double s` ~ scale factor

2: i ⬦ `cliff_comp *b` ~ pointer to structure containing component

3: o ⬦ `cliff_comp *c` ~ pointer to structure in which to save scaled component

DESCRIPTION

Scale the component by a real value. Setting the pointers of both components to the same value overwrites the input with the output.

RETURN VALUE

The return value is a pointer to the component containing the scaled value.

RECYCLING

~

ERROR RECOVERY

~

SEE ALSO

`half()`, `identity()`, `negation()`, `scale_number()`, `twice()`, `zero()`

261	S_UNIT	S

NAME

s_unit :– signature of unit

SYNOPSIS

```
int s_unit(bitmap u);
```

ARGUMENTS

1: i ⬦ `bitmap u` ~ unit

DESCRIPTION

Calculate the signature of the specified unit.

RETURN VALUE

The return value is the signature of the unit, either ± 1, or 0.

RECYCLING

~

ERROR RECOVERY

~

SEE ALSO

`u_sign()`, `x_count()`

262	SCALAR_MUL	S

NAME

scalar_mul :– calculate scalar product of numbers

SYNOPSIS

```
double *scalar_mul(number x, number y, double s[2],
                   int d);
```

ARGUMENTS

      1: id ◇ number x   ~ multiplier

      2: id ◇ number y   ~ multiplicand

      3: o ◇ double s[2] ~ product, pointer to array in which to save complex result

      4: i ◇ int d   ~ disposal flag

DESCRIPTION

      Multiply two numbers using the inner multiplication.

RETURN VALUE

      The return value is a pointer to the inner product as a complex scalar with no unit attached.

RECYCLING

      The numbers x and y are recycled according to the value of the disposal flag.

ERROR RECOVERY

      ~

SEE ALSO

      `inner_mul()`

263	SCALAR_NUMBER	S

NAME

      scalar_number :– construct a scalar

SYNOPSIS

      `number scalar_number(double x, double y);`

ARGUMENTS

      1: i ◇ double x ~ real part

      2: i ◇ double y ~ imaginary part

DESCRIPTION

      Construct a new number with the real and imaginary parts as specified. The new number is a scalar (in grade 0).

RETURN VALUE

      The return value is the new number.

RECYCLING

      ~

ERROR RECOVERY

      ~

SEE ALSO

      `fresh_number()`, `pack_comp()`, `primal_vector()`,
      `pseudo_number()`, `simple_number()`

264	SCALE_NUMBER	S

NAME

      scale_number :– multiply number by scalar

SYNOPSIS

      `number scale_number(number x, number y, int d);`

ARGUMENTS

    1: id ◇ number x ~ first number

    2: id ◇ number y ~ second number

    3: i ◇ int d   ~ disposal flag

DESCRIPTION

    Scale one number by the other using central multiplication. At least one number must be a simple number (have only one component).

RETURN VALUE

    The return value is the central (Clifford) product of the two numbers.

RECYCLING

    The numbers x and y are recycled according to the value of the disposal flag.

ERROR RECOVERY

    It is an error if both numbers have more than one component. On error, the recovery strategy brexit(0) is invoked to initiate a user re-start.

SEE ALSO

    central_mul(), prescale_row(), rscale()

265	SERIES	S

NAME

    series :– evaluate power series for common functions

SYNOPSIS

    number series(label f, number x, int p);

ARGUMENTS

    1: i ◇ label f ~ function

    2: i ◇ number x ~ number

    3: i ◇ int p   ~ highest power to use in series

DESCRIPTION

    Calculates the power series for the functions COS, SIN, COSH, SINH, EXP, LOG, CM_INV, SQRT, and ROOT, using powers of x from zero to p. All of the series have a centre of convergence, near which the results are likely to be more accurate with fewer terms. The centre of convergence is zero for COS, SIN, COSH, SINH and EXP, and is one for LOG, CM_INV, SQRT, and ROOT. In principle, the functions COS, SIN, COSH, SINH and EXP converge for all real values of x, but when $|x| > 1$ intermediate terms may become large with loss of significance due to cancellation of almost equal values. The convergence for LOG, CM_INV, SQRT, and ROOT is usually limited to real values of x between 0 and 2, not including the limits. For these functions, an extended range of convergence is offered by the routine x_series().

RETURN VALUE

    The return value is the value of the function, in so far as the series is a good or otherwise representation of the function to the number of terms used for the particular value of the function's argument x.

RECYCLING

    ~

Error Recovery

It is an error if the highest power p requested is greater than the maximum power pre-calculated when earlier invoking `make_series()`. On error, the recovery strategy `brexit(0)` is invoked to initiate a user re-start. It is also an error if the function specified is not supported or if the series for the function was never initialised. On these errors, the recovery strategy `brexit(6)` is invoked, terminating the execution.

See Also

`make_series(), poly_number(), x_series(), xx_series()`

266	SigLongJmp	S

Name

siglongjmp :– performing a nonlocal goto

Synopsis

`void siglongjmp(sigjmp_buf buf, int val);`

Description

See the documentation within the operating system by invoking the command man `siglongjmp` at the operating system command prompt.

Return Value

~

Recycling

~

Error Recovery

~

See Also

`brexit(), recovery_level(), sigsetjmp()`

267	SigSetJmp	S

Name

setsigjmp :– performing a nonlocal goto

Synopsis

`int sigsetjmp(sigjmp_buf buf, int savesigs);`

Description

See the documentation within the operating system by invoking the command man `sigsetjmp` at the operating system command prompt.

Return Value

~

Recycling

~

Error Recovery

~

See Also

`brexit(), recovery_level(), siglongjmp()`

268	SIMPLE_NUMBER	S

NAME

simple_number :– construct a number with one component

SYNOPSIS

```
number simple_number(double x, double y, bitmap u);
```

ARGUMENTS

1: i ◇ double x ~ real part
2: i ◇ double y ~ imaginary part
3: i ◇ bitmap u ~ unit

DESCRIPTION

Construct a new number with the real and imaginary parts and with the unit as specified. The unit is in binary form. Use `spack_unit()` to convert to binary if specifying the unit in the form of a character string form.

RETURN VALUE

The return value is the new number.

RECYCLING

~

ERROR RECOVERY

~

SEE ALSO

```
fresh_number(), pack_comp(), primal_vector(),
pseudo_number(), scalar_number()
```

269	SOLVE_MATRIX	S

NAME

solve_matrix :– solve a linear system by LU decomposition

SYNOPSIS

```
matrix *solve_matrix(matrix *a, matrix *xb);
```

ARGUMENTS

1: io ◇ matrix *a ~ matrix A for linear system
2: io ◇ matrix *xb ~ columns of known data b on right hand side, in a matrix

DESCRIPTION

Solve the linear system $Ax = b$ with matrix A, unknowns x, and known data b using LU decomposition. Initially, the set of one or more right-hand side vectors stored in the matrix xb is initialised to the known data. After solution, the RHS matrix contains the values of the unknowns and the system matrix A (a) is reduced to the identity matrix, somewhere close to machine precision.

RETURN VALUE

The return value is the right-hand side matrix, now containing in columns the solution(s) x.

RECYCLING

~

ERROR RECOVERY

It is an error if the matrices have a different number of rows. On error, the recovery strategy brexit(0) is invoked to initiate a user re-start. The routine has no mechanism to recover if a pivot with no finite inverse is encountered.

SEE ALSO

    check_matrix(), eliminate_above(), eliminate_below(),
    solve_matrix_LU(), solve_matrix_UL()

270	SOLVE_MATRIX_LU	S

NAME

    solve_matrix_LU :– solve a linear system by LU decomposition

SYNOPSIS

    matrix *solve_matrix_LU(matrix *a, matrix *xb);

ARGUMENTS

    1: io ⋄ matrix *a ~ matrix A for linear system
    2: io ⋄ matrix *xb ~ columns of known data b on right hand side, in a matrix

DESCRIPTION

Solve the linear system $Ax = b$ with matrix A, unknowns x, and known data b using LU decomposition. Initially, the set of one or more right-hand side vectors stored in the matrix xb is initialised to the known data. After solution, the RHS matrix contains the values of the unknowns, and the system matrix A (a) is reduced to the identity matrix, somewhere close to machine precision.

RETURN VALUE

The return value is the right-hand side matrix, now containing in columns the solution(s) x.

RECYCLING

    ~

ERROR RECOVERY

It is an error if the matrices have a different number of rows. On error, the recovery strategy brexit(0) is invoked to initiate a user re-start. The routine has no mechanism to recover if a pivot with no finite inverse is encountered.

SEE ALSO

    check_identity(), eliminate_above(), eliminate_below(),
    solve_matrix(), solve_matrix_UL()

271	SOLVE_MATRIX_UL	S

NAME

    solve_matrix_UL :– solve a linear system by UL decomposition

SYNOPSIS

    matrix *solve_matrix_UL(matrix *a, matrix *xb);

ARGUMENTS

    1: io ⋄ matrix *a ~ matrix A for linear system
    2: io ⋄ matrix *xb ~ columns of known data b on right hand side, in a matrix

DESCRIPTION

Solve the linear system $Ax = b$ with matrix A, unknowns x, and known data b using UL decomposition. Initially, the set of one or more right-hand side vectors stored in the matrix xb is initialised to the known data. After solution, the RHS matrix contains the values of the unknowns, and the system matrix A (a) is reduced to the identity matrix, somewhere close to machine precision.

RETURN VALUE

The return value is the right-hand side matrix, now containing in columns the solution(s) x.

RECYCLING

~

ERROR RECOVERY

It is an error if the matrices have a different number of rows. On error, the recovery strategy `brexit(0)` is invoked to initiate a user re-start. The routine has no mechanism to recover if a pivot with no finite inverse is encountered.

SEE ALSO

`check_identity()`, `eliminate_above()`, `eliminate_below()`, `solve_matrix()`, `solve_matrix_LU()`

| 272 | SPACK_UNIT | S |

NAME

spack_unit :– pack unit from string of characters

SYNOPSIS

`bitmap spack_unit(char *evolved);`

ARGUMENTS

1: i ◊ `char *evolved` ~ pointer to unit in form of character string

DESCRIPTION

Constructs the unit from its character string.

RETURN VALUE

The return value is the unsigned integer bitmap of the primal units.

RECYCLING

~

ERROR RECOVERY

It is an error for the character string to contain invalid syntax. On error, the recovery strategy `brexit(0)` is invoked to initiate a user re-start.

SEE ALSO

`pack_unit()`, `sscan_unit()`, `unpack_unit()`

| 273 | SPATIAL_PART | S |

NAME

spatial_part :– test for spatial component

SYNOPSIS

`int spatial_part(cliff_comp *c);`

ARGUMENTS

1: i ◊ `cliff_comp *c` ~ pointer to structure containing component

DESCRIPTION

Test to determine whether the component is in the spatial part of an electromagnetic field (the magnetic field). The test passes if the component does not contain the lowest dimensional primal unit e_0.

RETURN VALUE

The return value is 0 (false) when the test fails and 1 (true) when the test passes.

RECYCLING

~

ERROR RECOVERY

~

SEE ALSO

`pack_spacetime()`, `temporal_part()`, `unpack_spacetime()`

274	SQUARE_ROOT	S

NAME

square_root :– square root of component

SYNOPSIS

`cliff_comp *square_root(cliff_comp *a, cliff_comp *c);`

ARGUMENTS

1: i ⋄ `cliff_comp *a` ~ pointer to structure containing component

2: o ⋄ `cliff_comp *c` ~ pointer to structure in which to save square root

DESCRIPTION

Calculate the square root of the component, honouring the value of the unit. Setting the pointers of both the components to the same value overwrites the input with the output.

RETURN VALUE

The return value is a pointer to the component containing the square root.

RECYCLING

~

ERROR RECOVERY

It is an error if the component is not a scalar. On error, the recovery strategy `brexit(0)` is invoked to initiate a user re-start.

SEE ALSO

`square_root_coef()`, `squareof_magnitude()`,
`squarerootof_scalar()`

275	SQUARE_ROOT_COEF	S

NAME

square_root_coef :– square root of coefficient

SYNOPSIS

```
cliff_comp *square_root_coef(cliff_comp *a,
                             cliff_comp *c);
```

ARGUMENTS

1: i ⋄ `cliff_comp *a` ~ pointer to structure containing component

2: o ⋄ `cliff_comp *c` ~ pointer to structure in which to save square root

DESCRIPTION

Calculate the square root of the complex coefficient, ignoring the value of the unit. Setting the pointers of both the components to the same value overwrites the input with the output.

RETURN VALUE

The return value is a pointer to the component containing the square root. The unit of the return value is the same as the unit of the input.

RECYCLING

~

ERROR RECOVERY

~

SEE ALSO

```
log_coef(), power_coef(), square_root(),
squareof_magnitude(), squarerootof_scalar()
```

276	SQUAREOF_LATITUDE	S

NAME

squareof_latitude :– square of latitude of component

SYNOPSIS

```
cliff_comp *squareof_latitude(cliff_comp *a, cliff_comp
*c);
```

ARGUMENTS

1: i ◊ cliff_comp *a ~ pointer to structure containing component

2: o ◊ cliff_comp *c ~ pointer to structure in which to save square of latitude

DESCRIPTION

Calculate the square of the magnitude of the component, honouring the value of the unit. Setting the pointers of both components to the same value overwrites the input with the output.

RETURN VALUE

The return value is a pointer to the component containing the result.

RECYCLING

~

ERROR RECOVERY

~

SEE ALSO

```
imaginary_part(), latitudeof_number(), real_part(),
square_root(), squareof_magnitude()
```

277	SQUAREOF_MAGNITUDE	S

NAME

squareof_magnitude :– square of magnitude of component

SYNOPSIS

```
cliff_comp *squareof_magnitude(cliff_comp *a, cliff_comp
*c);
```

ARGUMENTS

1: i ◊ cliff_comp *a ~ pointer to structure containing component

2: o ◊ cliff_comp *c ~ pointer to structure in which to save square of magnitude

DESCRIPTION

Calculate the square of the magnitude of the complex coefficient, ignoring the value of the unit. Setting the pointers of both components to the same value overwrites the input with the output.

RETURN VALUE

The return value is a pointer to the component containing the result. The unit of the return value is the same as the unit of the input.

RECYCLING

~

ERROR RECOVERY

~

SEE ALSO

```
imaginary_part(), magnitudeof_number(), real_part(),
square_root(), squareof_latitude()
```

278	SQUAREROOTOF_SCALAR	S

NAME

squarerootof_scalar :– produce square root of a scalar

SYNOPSIS

```
number squarerootof_scalar(number x, int d);
```

ARGUMENTS

1: id ◇ number x ~ scalar

2: i ◇ int d   ~ disposal flag

DESCRIPTION

Construct a number containing the square root of the scalar specified.

RETURN VALUE

The return value is the square root.

RECYCLING

The number x is recycled according to the value of the disposal flag.

ERROR RECOVERY

It is an error if the number is not a pure scalar (grade 0 only). On error, the recovery strategy `brexit(0)` is invoked to initiate a user re-start.

SEE ALSO

```
square_root()
```

279	SSCAN_UNIT	S

NAME

sscan_unit :– scan unit from string of characters

SYNOPSIS

```
char *sscan_unit(char *evolved);
```

ARGUMENTS

1: i ◇ char *evolved ~ pointer to unit in form of character string

DESCRIPTION

Scans and performs syntax check of the unit in the form of a character string.

RETURN VALUE

The return value is either an error message or a null pointer (zero) if the syntax is correct.

RECYCLING

~

ERROR RECOVERY

~

SEE ALSO

spack_unit()

280	STRIDE	S

NAME

stride :– get the length of the stride for a matrix

SYNOPSIS

int stride(matrix *m);

ARGUMENTS

1: i ◇ matrix *m ~ matrix

DESCRIPTION

Get the distance (as the number of elements) between the start of adjacent columns in a matrix. For a full (not sub) matrix, the value of the stride is the same as the number of rows. For a sub-matrix, the number of rows is less than the value of the stride.

RETURN VALUE

The return value is the distance between adjacent columns in the matrix.

RECYCLING

~

ERROR RECOVERY

~

SEE ALSO

access_element(), columns(), rows()

281	SUB_NUMBER	S

NAME

sub_number :– produce difference of two numbers

SYNOPSIS

number sub_number(number x, number y, int d);

ARGUMENTS

1: id ◇ number x ~ minuend

2: id ◇ number y ~ subtrahend

3: i ◇ int d   ~ disposal flag

DESCRIPTION

Calculate the difference between two numbers.

RETURN VALUE

The return value is the difference.

RECYCLING

The numbers x and y are recycled according to the value of the disposal flag.

ERROR RECOVERY

~

SEE ALSO

    add_number(), subfrom_number(), subtract_array(),
    subtract_matrix(), subtraction_from()

| 282 | SUBFROM_ARRAY | S |

NAME

subfrom_array :– subtract one array from another

SYNOPSIS

    number *subfrom_array(int n, number *a, number *b);

ARGUMENTS

1: i ◇ int n   ~ number of elements in array
2: io ◇ number *a ~ minuend, pointer to first array of numbers
3: i ◇ number *b ~ subtrahend, pointer to second array of numbers

DESCRIPTION

Subtract the second array away from the first array. The numbers in the first array are overwritten with the results.

RETURN VALUE

The return value is a pointer to the array containing the results.

RECYCLING

The numbers in the first array are recycled before they are overwritten.

ERROR RECOVERY

It is an error to provide an array which does not have the number of elements specified.

SEE ALSO

    addto_array(), subfrom_matrix(), subfrom_number(),
    subtract_array()

| 283 | SUBFROM_MATRIX | S |

NAME

subfrom_matrix :– subtract one matrix from another

SYNOPSIS

    matrix *subfrom_matrix(matrix *a, matrix *b);

ARGUMENTS

1: io ◇ matrix *a ~ minuend, matrix
2: i ◇ matrix *b ~ subtrahend, matrix

DESCRIPTION

Subtract the second matrix (subtrahend) from the first matrix (minuend) and save the result in the first matrix.

RETURN VALUE

The return value is the minuend matrix with modified values.

RECYCLING

~

ERROR RECOVERY

It is an error if the matrices have different sizes. On error, the recovery strategy `brexit(0)` is invoked to initiate a user re-start.

SEE ALSO

addto_matrix(), subfrom_array(), subfrom_number(),
subfrom_row(), subtract_matrix()

| 284 | SUBFROM_NUMBER | S |

NAME

subfrom_number :– subtract number from another

SYNOPSIS

number subfrom_number(number x, number y);

ARGUMENTS

1: ir ◇ number x ~ minuend
2: i ◇ number y ~ subtrahend

DESCRIPTION

Subtracts the second number away from the first number.

RETURN VALUE

The return value is the difference.

RECYCLING

The original number is recycled. It is overwritten *in situ* and returned in place of the new one.

ERROR RECOVERY

It is an error to subtract a number from itself. On error, the recovery strategy `brexit(6)` is invoked, terminating the execution.

SEE ALSO

addto_number(), sub_number(), subfrom_array(),
subfrom_matrix(), subfuncfrom_number(), subtraction_from()

| 285 | SUBFROM_ROW | S |

NAME

subfrom_row :– subtract one row of a matrix from another

SYNOPSIS

matrix *subfrom_row(int i, int j, matrix *m);

ARGUMENTS

1: i ◇ int i   ~ minuend, row to update by subtraction
2: i ◇ int j   ~ subtrahend, row to subtract
3: io ◇ matrix *m ~ matrix

DESCRIPTION

Subtract the second row (j, subtrahend) from the first row (i, minuend). The first of the two rows is overwritten with the result. If unknown, the total number of rows can be obtained using the routine rows(). The indexes i, j for the rows start from zero and go up to one less than the total. No checks are made to determine if the indexes are out of bounds.

RETURN VALUE

The return value is the original matrix with modified values.

RECYCLING

It is an error to use indexes which are out of bounds.

ERROR RECOVERY

~

SEE ALSO

```
addto_row(), rows(), subfrom_array(), subfrom_matrix(),
subfrom_number()
```

| 286 | SUBFUNCFROM_NUMBER | S |

NAME

subfuncfrom_number :– subtract function of number from another

SYNOPSIS

```
number subfuncfrom_number(number x, number y,
                    cliff_comp *function(cliff_comp
                            *a, cliff_comp *c));
```

ARGUMENTS

1: ir ◇ number x   ~ minuend
2: i ◇ number y   ~ number, argument of function
3: i ◇ cliff_comp *function   ~ pointer to function
   (cliff_comp *a, cliff_comp *c)

DESCRIPTION

Apply the function to every component in the second number and subtract the result away from the first number.

RETURN VALUE

The return value contains the result.

RECYCLING

The minuend is recycled. It is overwritten *in situ* and returned as the result.

ERROR RECOVERY

It is an error to subtract the function of a number from itself. On error, the recovery strategy brexit(6) is invoked, terminating the execution.

SEE ALSO

```
addfuncto_number(), subfrom_number()
```

| 287 | SUBSCALEDFROM_ROW | S |

NAME

subscaledfrom_row :– scale one row of matrix and subtract from another

SYNOPSIS

```
matrix *subscaledfrom_row(int i, number s, int j,
                        matrix *m);
```

ARGUMENTS
> 1: i ◇ int i   ~ minuend, row to update by subtraction
> 2: i ◇ number s ~ scale factor
> 3: i ◇ int j   ~ row to scale and subtract
> 4: io ◇ matrix *m ~ matrix

DESCRIPTION
> Pre-scale (multiply from the left) the second row (j) and subtract from the first row (i). The first of the two rows is overwritten with the result. The second row is not modified. If unknown, the total number of rows can be obtained using the routine rows(). The indexes i, j for the rows start from zero and go up to one less than the total. No checks are made to determine if the indexes are out of bounds.

RETURN VALUE
> The return value is the original matrix with modified contents.

RECYCLING
> ~

ERROR RECOVERY
> It is an error to use indexes which are out of bounds.

SEE ALSO
> addscaledto_row(), rows(), subfrom_row()

288	SUBTRACT_ARRAY	S

NAME
> subtract_array :– difference between two arrays

SYNOPSIS
> ```
> number *subtract_array(int n, number *a, number *b,
>                        number *c);
> ```

ARGUMENTS
> 1: i ◇ int n   ~ number of elements in arrays
> 2: i ◇ number *a ~ minuend, pointer to first array of numbers
> 3: i ◇ number *b ~ subtrahend, pointer to second array of numbers
> 4: o ◇ number *c ~ difference, pointer to array in which to save result

DESCRIPTION
> Calculate the difference between two arrays. The second array is subtracted from the first array, and the result is placed in the third array. Initially, the third array must be empty.

RETURN VALUE
> The return value is a pointer to the array containing the results.

RECYCLING
> ~

ERROR RECOVERY
> It is an error to provide an array which does not have the number of elements specified. It is an error if the third array is not empty initially.

SEE ALSO
> add_array(), sub_number(), subfrom_array(),
> subtract_matrix()

| 289 | SUBTRACT_MATRIX | S |

NAME

>   subtract_matrix :– calculate difference between two matrices

SYNOPSIS

```
matrix *subtract_matrix(matrix *a, matrix *b,
                        matrix *c, int d);
```

ARGUMENTS

>   1: id ◇ matrix *a ~ minuend, matrix
>   2: id ◇ matrix *b ~ subtrahend, matrix
>   3: o ◇ matrix *c ~ difference, empty matrix
>   4: i ◇ int d   ~ disposal flag

DESCRIPTION

>   Subtract the second matrix (subtrahend) from the first matrix (minuend) and save the result in the third matrix.

RETURN VALUE

>   The return value is the matrix containing the difference.

RECYCLING

>   ~

ERROR RECOVERY

>   It is an error if the matrices have different sizes. On error, the recovery strategy brexit(0) is invoked to initiate a user re-start. It is an error if the third matrix is not empty initially. No check is made.

SEE ALSO

```
add_matrix(), subfrom_matrix(), subtract_array(),
subtract_number()
```

| 290 | SUBTRACTION | S |

NAME

>   subtraction :– difference of two components

SYNOPSIS

```
cliff_comp *subtraction(cliff_comp *a, cliff_comp *b,
        cliff_comp *c);
```

ARGUMENTS

>   1: i ◇ cliff_comp *a ~ minuend, pointer to structure containing component
>   2: i ◇ cliff_comp *b ~ subtrahend, pointer to structure containing component
>   3: o ◇ cliff_comp *c ~ pointer to component in which to save difference

DESCRIPTION

>   Calculate the difference between two components. Setting the pointers of either input component to the same value as the output component overwrites the input with the output.

RETURN VALUE

>   The return value is a pointer to the component containing the difference.

RECYCLING

>   ~

ERROR RECOVERY

It is an error if the two input components do not have the same unit. On error, the recovery strategy `brexit(6)` is invoked, terminating the execution.

SEE ALSO

`addition(), sub_number(), subtraction_from()`

| 291 | SUBTRACTION_FROM | S |

NAME

subtraction_from :– subtract component from another

SYNOPSIS

```
cliff_comp *subtraction_from(cliff_comp *a, cliff_comp
*b);
```

ARGUMENTS

1: io ◇ `cliff_comp *a` ~ minuend, pointer to structure containing component

2: i ◇ `cliff_comp *b` ~ subtrahend, pointer to structure containing component

DESCRIPTION

Subtract the second component away from the first component. Setting the pointers of both components to the same value overwrites the input with the output.

RETURN VALUE

The return value is a pointer to the first component, which contains the difference.

RECYCLING

~

ERROR RECOVERY

It is an error if the two components do not have the same unit. On error, the recovery strategy `brexit(6)` is invoked, terminating the execution.

SEE ALSO

`addition_to(), subfrom_number(), subtraction()`

| 292 | SUMOF_ARRAY | S |

NAME

sumof_array :– sum of numbers in array

SYNOPSIS

```
number sumof_array(int n, number *a);
```

ARGUMENTS

1: i ◇ `int n` ~ number of elements in array

2: i ◇ `number *a` ~ pointer to array of numbers

DESCRIPTION

Calculate the sum of all the elements in the array.

RETURN VALUE

The return value is the sum.

RECYCLING

~

ERROR RECOVERY

It is an error to provide an array which does not have the number of elements specified.

SEE ALSO

```
meanabsof_array(), peakof_array(), rmsof_array(),
weightedsumof_array()
```

293	SUPPLEMENT	S

NAME

supplement :– supplement of component

SYNOPSIS

```
cliff_comp *supplement(cliff_comp *a, cliff_comp *c);
```

ARGUMENTS

1: i ◇ cliff_comp *a ~ pointer to structure containing component
2: o ◇ cliff_comp *c ~ pointer to structure in which to save supplement

DESCRIPTION

Calculate the supplement of a component. Setting the pointers of both components to the same value overwrites the input with the output.

RETURN VALUE

The return value is a pointer to the component containing the supplement.

RECYCLING

~

ERROR RECOVERY

~

SEE ALSO

```
inverse(), negation(), supplementof_number()
```

294	SUPPLEMENTOF_NUMBER	S

NAME

supplementof_number :– produce supplement of number

SYNOPSIS

```
number supplementof_number(number x, int d);
```

ARGUMENTS

1: id ◇ number x ~ number
2: i ◇ int d ~ disposal flag

DESCRIPTION

Construct a number containing the supplement of the number specified.

RETURN VALUE

The return value is a number containing Grassmann's supplement.

RECYCLING

The number x is recycled according to the value of the disposal flag d.

ERROR RECOVERY

~

SEE ALSO

```
hodgestarof_number(), supplement()
```

295	TEMPORAL_PART	T

NAME

temporal_part :– test for temporal component

SYNOPSIS

```
int temporal_part(cliff_comp *c);
```

ARGUMENTS

1: i ◇ cliff_comp *c ~ pointer to structure containing component

DESCRIPTION

Test to determine whether the component is in the temporal part of an electromagnetic field (the electric field). The test passes if the component contains the lowest dimensional primal unit e_0.

RETURN VALUE

The return value is 0 (false) when the test fails and 1 (true) when the test passes.

RECYCLING

~

ERROR RECOVERY

~

SEE ALSO

```
pack_spacetime(), spatial_part(), unpack_spacetime()
```

296	TRANSFORM	T

NAME

transform :– fast Fourier transform

SYNOPSIS

```
number *transform(int n, number *a, number *b);
```

ARGUMENTS

1: i ◇ int n   ~ number of elements in array (power of 2)

2: ir ◇ number *a ~ pointer to array containing waveform

3: o ◇ number *b ~ pointer to array in which to save spectrum

DESCRIPTION

Calculates the fast Fourier transform for an array for which the number of elements is a power of 2. The spectrum is placed in the second array. Initially, the second array must be empty.

RETURN VALUE

The return value is a pointer to the array containing the spectrum. The spectral values are the n complex frequency components f_k, in the order of frequency $k = 0$, $0 < k < n/2, k = \pm n/2, -n/2 < k < 0$.

RECYCLING

The waveform array is recycled.

ERROR RECOVERY

It is an error to provide an array which does not have the number of elements specified. It is an error if the size of the arrays is not a power or 2. On error, the recovery strategy brexit(0) is invoked to initiate a user re-start.

SEE ALSO

```
inverse_transform()
```

297	Trim_Zero	T

NAME

trim_zero :– produce number with values near zero trimmed

SYNOPSIS

```
number trim_zero(number x, number z, int d);
```

ARGUMENTS

1: id ◇ number x ~ number

2: id ◇ number z ~ reference scalar

3: i ◇ int d   ~ disposal flag

DESCRIPTION

Set to zero components which have magnitude less than the magnitude of the reference scalar. The magnitude is the square root of the sum of the squares of the real and imaginary parts. The signature of the components are not taken into account.

RETURN VALUE

The return value is the trimmed number.

RECYCLING

The numbers x and z are recycled according to the value of the disposal flag.

ERROR RECOVERY

~

SEE ALSO

```
filter_number(), is_nonzero(), is_zero()
```

298	Twice	T

NAME

twice :– twice the value of component

SYNOPSIS

```
cliff_comp *twice(cliff_comp *a, cliff_comp *c);
```

ARGUMENTS

1: i ◇ cliff_comp *a ~ pointer to structure containing component

2: o ◇ cliff_comp *c ~ pointer to structure in which to save twice the component

DESCRIPTION

Calculates twice the value of the component. Setting the pointers of both components to the same value overwrites the input with the output.

RETURN VALUE

The return value is a pointer to the component containing the result.

RECYCLING

~

ERROR RECOVERY

~

SEE ALSO

```
half(), identity(), rscale()
```

27.4.5 U–Z

299	U_SIGN	U

NAME

u_sign :– sign for product of units

SYNOPSIS

```
int u_sign(bitmap u, bitmap v);
```

ARGUMENTS

1: i ◇ bitmap u ~ first unit

2: i ◇ bitmap v ~ second unit

DESCRIPTION

Calculate the sign produced when two units are multiplied and primal units are restored to ascending order.

RETURN VALUE

The return value is the sign of the unit. If the product is non-zero the result is either ±1. Otherwise, the result is zero.

RECYCLING

~

ERROR RECOVERY

~

SEE ALSO

```
s_unit(), x_count()
```

300	UNPACK_03VECTOR	U

NAME

unpack_03vector :– unpack complex vector of three dimensions from spacetime

SYNOPSIS

```
double (*unpack_03vector(number x, double (*a)[2]))[2];
```

ARGUMENTS

1: i ◇ number x   ~ number

2: o ◇ double (*a)[2] ~ pointer to array in which to save three complex values

DESCRIPTION

Unpack a complex three-dimensional vector from four-dimensional space–time into an array of three complex values.

RETURN VALUE

The return value is a pointer to the array.

RECYCLING

~

ERROR RECOVERY

It is an error to provide an array which is too small.

SEE ALSO

```
pack_03vector(), unpack_3vector()
```

301	UNPACK_3VECTOR	U

NAME

unpack_3vector :– unpack complex vector of three dimensions

SYNOPSIS

```
double (*unpack_3vector(number x, double (*a)[2]))[2];
```

ARGUMENTS

1: i ◇ number x ～ number

2: o ◇ double (*a)[2] ～ pointer to array in which to save three complex values

DESCRIPTION

Unpack a complex three-dimensional vector into an array of three complex values.

RETURN VALUE

The return value is a pointer to the array.

RECYCLING

～

ERROR RECOVERY

It is an error to provide an array which is too small.

SEE ALSO

```
pack_3vector(), unpack_03vector(), unpack_R3vector()
```

302	UNPACK_COMP	U

NAME

unpack_comp :– unpack component

SYNOPSIS

```
void unpack_comp(cliff_comp *c, double *x, double *y,
                 bitmap *u);
```

ARGUMENTS

1: i ◇ cliff_comp *c ～ pointer to structure containing component

2: o ◇ double *x ～ pointer to real part

3: o ◇ double *y ～ pointer to imaginary part

4: o ◇ bitmap *u ～ pointer to unit

DESCRIPTION

Unpack a component into real and imaginary parts and unit.

RETURN VALUE

～

RECYCLING

～

ERROR RECOVERY

～

SEE ALSO

```
pack_comp(), print_comp()
```

303	UNPACK_COMPONENTS	U

NAME

unpack_components :– unpack number as coefficients and units

SYNOPSIS

```
int unpack_components(number x, double (*y_complex)[2],
bitmap *y_unit);
```

ARGUMENTS

1: i ◇ number x   ~ number
2: o ◇ double (*y_complex)[2] ~ pointer to array in which to save
                                          complex coefficients
3: o ◇ bitmap *y_unit   ~ pointer to array in which to save units

DESCRIPTION

Unpack a number into an array of complex coefficients and an array of units.

RETURN VALUE

The return value is the number of components unpacked.

RECYCLING

~

ERROR RECOVERY

It is an error to supply an array which is not large enough to contain all of the components. The sub-routine `count_components()` may be used to determine the minimum size of the two arrays before unpacking.

SEE ALSO

```
count_components(), print_listof_coefficients(),
print_listof_units(), unpack_number()
```

| 304 | UNPACK_EMFIELD | U |

NAME

unpack_emfield :– unpack electromagnetic field

SYNOPSIS

```
void unpack_emfield(number x, double mu[2],
                            double epsilon[2],
                            double (*H)[2],
                            double (*E)[2]);
```

ARGUMENTS

1: i ◇ number x   ~ electromagnetic field
2: i ◇ double mu[2]   ~ complex magnetic permeability
3: i ◇ double epsilon[2] ~ complex electric permittivity
4: o ◇ double (*H)[2]   ~ magnetic field, pointer to array in which to
                                    save 3 complex values
5: o ◇ double (*E)[2]   ~ electric field, pointer to array in which to save
                                    3 complex values

DESCRIPTION

Unpack the magnetic field and the electric field from the electromagnetic field specified and from the known values of the material properties μ and ϵ. The magnetic field and the electric field are both complex vectors of three dimensions. The material properties are both complex scalars of type `double`.

RETURN VALUE

The return values are the magnetic and electric fields.

RECYCLING

~

ERROR RECOVERY

It is an error to provide arrays which are too small.

SEE ALSO

`pack_emfield(), print_emfield(), unpack_spacetime()`

305	UNPACK_NUMBER	U

NAME

unpack_number :– unpack number as components

SYNOPSIS

`int unpack_number(number x, cliff_comp *c);`

ARGUMENTS

1: i ◇ number x   ~ number

2: o ◇ cliff_comp *c ~ pointer to array of structures in which to save
                                          components

DESCRIPTION

Unpack number into array of components.

RETURN VALUE

The return value is the number of components unpacked.

RECYCLING

~

ERROR RECOVERY

It is an error to supply an array which is not large enough to contain all of the components. The sub-routine `count_components()` may be used to determine the minimum size of the array before unpacking.

SEE ALSO

`count_components(), print_number(), unpack_components()`

306	UNPACK_R3VECTOR	U

NAME

unpack_R3vector :– unpack real vector of three dimensions

SYNOPSIS

`double *unpack_R3vector(number x, double *a);`

ARGUMENTS

1: i ◇ number x ~ number

2: o ◇ double *a ~ pointer to array in which to save three real values

DESCRIPTION

Unpack a real three-dimensional vector into an array of three real values.

RETURN VALUE

The return value is a pointer to the array.

RECYCLING

~

ERROR RECOVERY

It is an error to provide an array which is too small.

SEE ALSO

pack_R3vector(), unpack_3vector()

307	UNPACK_SPACETIME	U

NAME

unpack_spacetime :– unpack spatial and temporal parts from electromagnetic field

SYNOPSIS

void unpack_spacetime(number x, double (*S)[2], double (*T)[2]);

ARGUMENTS

1: i ◇ number x   ~ electromagnetic field
2: o ◇ double (*S)[2] ~ spatial part of electromagnetic field, pointer to array of 3 complex values
3: o ◇ double (*T)[2] ~ temporal part of electromagnetic field, pointer to array of 3 complex values

DESCRIPTION

Unpack the spatial and temporal parts from the electromagnetic field specified. The spatial part is the magnetic field scaled by $\sqrt{\mu}$, and the temporal part is the electric field scaled by $\sqrt{\epsilon}$. Each of these is a complex vector of three dimensions.

RETURN VALUE

~

RECYCLING

~

ERROR RECOVERY

It is an error to provide arrays which are too small.

SEE ALSO

pack_spacetime(), unpack_emfield()

308	UNPACK_UNIT	U

NAME

unpack_unit :– unpack unit into array

SYNOPSIS

int unpack_unit(bitmap u, int *primal);

ARGUMENTS

1: i ◇ bitmap u   ~ unit
2: o ◇ int *primal ~ pointer to array of numbers in which to save the primal units

DESCRIPTION

Fills the array with the numbers identifying the primal units as contained in the specified unit. If the size of the array is insufficient, values are written beyond the end of the array. An array of size equal to the Clifford dimension is always big enough.

RETURN VALUE

The return value is the dimension of the unit.

RECYCLING

~

ERROR RECOVERY

If the size of the array is insufficient, values are written beyond the end of the array.

SEE ALSO

`pack_unit(), print_unit()`

309	UPPER_ROWS		U

NAME

upper_rows :– construct sub-matrix missing bottom row

SYNOPSIS

`matrix *upper_rows(matrix *m, matrix *s);`

ARGUMENTS

1: i ◇ `matrix *m` ~ matrix

2: o ◇ `matrix *s` ~ sub-matrix, undeclared

DESCRIPTION

Construct a sub-matrix without the bottom row. Initially, the sub-matrix should be undeclared.

RETURN VALUE

The return value is the sub-matrix within the original matrix.

RECYCLING

~

ERROR RECOVERY

It is an error to use the same matrix structure for both matrices, or to construct a sub-matrix from less than two rows. On error, the recovery strategy `brexit(0)` is invoked to initiate a user re-start. It is an error if the sub-matrix has been declared already. No check is made.

SEE ALSO

`eliminate_above(), lower_rows(), upper_submatrix()`

310	UPPER_SUBMATRIX		U

NAME

upper_submatrix :– construct sub-matrix missing bottom row and right column

SYNOPSIS

`matrix *upper_submatrix(matrix *m, matrix *s);`

ARGUMENTS

1: i ◇ `matrix *m` ~ matrix

2: o ◇ `matrix *s` ~ sub-matrix, undeclared

DESCRIPTION

Construct a sub-matrix without the bottom row or the right-hand column. Initially, the sub-matrix should be undeclared.

RETURN VALUE

The return value is the sub-matrix within the original matrix.

RECYCLING

~

ERROR RECOVERY

It is an error to use the same matrix structure for both matrices or to construct a sub-matrix from less than two rows and columns. On error, the recovery strategy `brexit(0)` is invoked to initiate a user re-start. It is an error if the sub-matrix has been declared already. No check is made.

SEE ALSO

`eliminate_above(), lower_submatrix(), upper_rows()`

| 311 | WEIGHTEDSUMOF_ARRAY | W |

NAME

weightedsumof_array :– weighted sum of numbers in array

SYNOPSIS

`number weightedsumof_array(int n, number *a, cliff_comp *w);`

ARGUMENTS

1: i ◇ `int n` ~ number of elements in arrays
2: i ◇ `number *a` ~ pointer to array of numbers
3: i ◇ `cliff_comp *w` ~ weights, pointer to array of structures containing components

DESCRIPTION

Calculates the sum of all the of elements in the array post-multiplied by the array of weights. The weights are each single components.

RETURN VALUE

The return value is the weighted sum.

RECYCLING

~

ERROR RECOVERY

It is an error to provide an array which does not have the number of elements specified.

SEE ALSO

`row_mul_col(), sumof_array()`

| 312 | X_COUNT | X |

NAME

x_count :– primal exchange count for product of units

SYNOPSIS

`int x_count(bitmap u, bitmap v);`

ARGUMENTS

1: i ◇ `bitmap u` ~ first unit
2: i ◇ `bitmap v` ~ second unit

DESCRIPTION

Calculate the number of exchanges of pairs of primal units required to restore a product of two units to ascending order.

RETURN VALUE

The return value is the number of exchanges.

RECYCLING

~

ERROR RECOVERY

~

SEE ALSO

```
s_unit(), u_sign()
```

313	X_SERIES	X

NAME

x_series :– evaluate power series with extended range of convergence

SYNOPSIS

```
number x_series(label f, number x, int p);
```

ARGUMENTS

1: i ◇ label f ~ function

2: i ◇ number x ~ number

3: i ◇ int p   ~ highest power to use in series

DESCRIPTION

Calculates the power series for the functions LOG, CM_INV, SQRT, and ROOT, using powers of x from zero to p, with an extended range of convergence. The range does not extend to zero for the scalar part (grade 0). For the function CM_INV, a doubly extended range of convergence is offered by the routine xx_series().

RETURN VALUE

The return value is the value of the function in so far as the series is a good or otherwise representation of the function to the number of terms used for the particular value of the function's argument x.

RECYCLING

~

ERROR RECOVERY

It is an error if the highest power p requested is greater than the maximum power pre-calculated when earlier invoking make_series(). On error, the recovery strategy brexit(0) is invoked to initiate a user re-start. It is also an error if the function specified is not supported or if the series for the function was never initialised. On these errors, the recovery strategy brexit(6) is invoked, terminating the execution.

SEE ALSO

```
make_series(), series(), xx_series()
```

314	XX_SERIES	X

NAME

xx_series :– evaluate power series with doubly extended range of convergence

SYNOPSIS

```
number xx_series(label f, number x, int p);
```

ARGUMENTS

    1: i ◇ label f ~ function

    2: i ◇ number x ~ number

    3: i ◇ int p   ~ highest power to use in series

DESCRIPTION

Calculates the power series for the function CM_INV, using powers of x from zero to p, with a doubly extended range of convergence. The range extends to non-zero numbers which have zero value for the scalar part (grade 0).

RETURN VALUE

The return value is the value of the function in so far as the series is a good or otherwise representation of the function to the number of terms used for the particular value of the function's argument x.

RECYCLING

~

ERROR RECOVERY

It is an error if the highest power p requested is greater than the maximum power pre-calculated when earlier invoking make_series(). On error, the recovery strategy brexit(0) is invoked to initiate a user re-start. It is also an error if the function specified is not supported or if the series for the function was never initialised. On these errors, the recovery strategy brexit(6) is invoked, terminating the execution.

SEE ALSO

make_series(), series(), x_series()

315	ZERO	Z

NAME

zero :– set component to zero

SYNOPSIS

```
cliff_comp *zero(cliff_comp *c);
```

ARGUMENTS

    1: o ◇ cliff_comp *c ~ pointer to structure containing component

DESCRIPTION

Sets a component to the scalar value zero.

RETURN VALUE

The return value is a pointer to the component.

RECYCLING

~

ERROR RECOVERY

~

SEE ALSO

fresh_number()

27.5   Quick Reference

Reference

King, K. N. (2008), *C Programming: A Modern Approach*, 2nd edn, W. W. Norton, New York, US-NY.

A

Key to Example Code and Results

The examples run by the program code 'demo', 'application', and 'radiate' provided in the Clifford numerical suite appear in the book as the examples and figures in the chapters as listed in Tables A.1 to A.3.

The code `applicationx()` is invoked by issuing at the keyboard commands in the form 'application x ...', where x is an argument which takes a value from 0 to 12, and the dots represent additional arguments as required by the individual applications. Running an application without additional arguments when they are required leads to a message describing the arguments which should be provided. Applications 2 and 3 are the same as application 1, but with different data values.

The code `surface_appx()` is invoked by issuing commands at the keyboard in the form 'radiate x ...', where x is an argument which takes a value from 0 to 6 and the dots represent additional arguments as required by the individual applications.

The code `demo x()` is invoked by issuing commands at the keyboard in the form 'demo x', where x takes a value from 0 to 34. For the demo code no additional arguments are required, but some of the demos prompt for values to be typed in at the keyboard. Demos 30 to 33 are the same as demo 29, but with different data values.

Numerical Calculations in Clifford Algebra: A Practical Guide for Engineers and Scientists,
First Edition. Andrew Seagar.
© 2023 John Wiley & Sons Ltd. Published 2023 by John Wiley & Sons Ltd.

Table A.1 General application code and results.

code	chapter	examples	figures	comments
application0()	11 arrays of Clifford numbers	11.3(a,b)	11(1,2)	calculation with arrays
application1()	12 power series	12.2,12.3(a,b)		series functions
application2()	12 power series	12.4(a,b)		series functions
application3()	12 power series	12.5(a,b)		series functions
application4()	17 verification	17.1(a,b)	17(1-4)	check identities
application5()	18 lines not parallel	18.1(a-d),18.2		intersection of lines
application6()	19 perspective projection	19.1(a,b),19.2(a-c)	19(2)	perspective drawing
application7()	20 linear systems	20.1(a-c)		solution of 4x4 linear system
application8()	21 fast Fourier transform	21.1(a,b),21.2(a,b)		fast Fourier transform
application9()	22 Hertzian dipole	22.1(a-c)	22(2-3)	Hertzian dipole field on x-axis
application10()	22 Hertzian dipole	22(4,5)		Hertzian dipole field in xz-plane
application11()	23 FDTD	23.1(a-d)	23(3-9)	finite difference time domain
application12()	25 EM scattering	25.1	25(1-6)	matrix inversion

Table A.2 Surface and Cauchy application code and results.

code	chapter	examples	figures	comments
surface_app0()				plot sample points on tiles and edges of tiles
surface_app1()				plot edges of a solid, with refined tiles if requested
surface_app2()				check consistency of basis and interpolation
surface_app3()				check accuracy of dipole field interpolated
surface_app4()				check consistency of basis and integration
surface_app5()	24 Cauchy extension		24(3-7)	Cauchy extension without taming
surface_app6()	24 Cauchy extension	24.1(a-d)	24(6,7)	Cauchy extension with taming

Table A.3 Demonstration code and results.

code	chapter	examples	figures	comments
demo0 ()	1 introduction	1.3(a,b)		minimal
demo1 ()	2 input	2.1(a,b)		input - specific numbers
demo2 ()	2 input	2.2(a,b)		input - general numbers
demo3 ()	2 input	2.3,2.4,2.5(a,b)		keyboard entry of Clifford numbers with explicit checking
demo4 ()	3 output	3.1,3.2(a,b)		output - tree format
demo5 ()	3 output	3.3(a,b)		output - tree format
demo6 ()	3 output	3.4(a,b)		output - default formats
demo7 ()	3 output	3.5(a,b)		output - defined formats
demo8 ()	3 output	3.6(a,b)		output - rounding
demo9 ()	3 output	3.7(a,b)		output - parts
demo10 ()	3 output	3.8(a,b)		output - selected coefficients
demo11 ()	3 output	3.9(a,b)		output - raw formats
demo12 ()	3 output	3.10(a,b)		output - recovered values
demo13 ()	4 unary ops	4.1(a,b)		unary arithmetic operators
demo14 ()	5 binary ops	5.1(a,b)		binary arithmetic operators
demo15 ()	6 vectors, geometry	6.1(a,b)		vectors
demo16 ()	7 quaternions	7.1(a,b)		quaternions
demo17 ()	8 Pauli matrices	8.1(a,b)		Pauli matrices
demo18 ()	9 bicomplex numbers	9.1(a,b)		bicomplex numbers
demo19 ()	10 EM fields	10.1(a,b)		electromagnetic fields
demo20 ()	10 EM fields	10.2(a,b)		extended EM field
demo21 ()	12 power series	12.1(a,b)	12(1,2)	user's own series function
demo22 ()	13 matrices of Clifford numbers	13.2,13.3(a,b)		matrix inverse
demo23 ()	14 memory	14.1(a,b)		memory
demo24 ()	14 memory	14.2(a,b)		memory expansion
demo25 ()	14 memory	14.3(a,b)		explicit and implicit memory recycling
demo26 ()	14 memory	14.4(a,b)		implicit and nested memory recycling
demo27 ()	15 errors	15.1(a,b)		error without recovery
demo28 ()	15 errors	15.2(a,b)		error with recovery
demo29 ()	15 errors	15.3(a-c)		keyboard entry of Clifford numbers with error recovery

(Continued)

Table A.3 (Continued)

code	chapter	examples	figures	comments
demo30()	15 errors	15.3(d)		keyboard entry of Clifford numbers with error recovery
demo31()	15 errors	15.3(e)		keyboard entry of Clifford numbers with error recovery
demo32()	15 errors	15.3(f)		keyboard entry of Clifford numbers with error recovery
demo33()	15 errors	15.3(g)		keyboard entry of Clifford numbers with error recovery
demo34()	see demo16()			factorise quaternion

Index

Numerical Calculations in Clifford Algebra: A Practical Guide for Engineers and Scientists,
First Edition. Andrew Seagar.
© 2023 John Wiley & Sons Ltd. Published 2023 by John Wiley & Sons Ltd.